TOOLS FOR DROUGHT MITIGATION IN MEDITERRANEAN REGIONS

Water Science and Technology Library

VOLUME 44

The titles published in this series are listed at the end of this volume.

TOOLS FOR DROUGHT MITIGATION IN MEDITERRANEAN REGIONS

edited by

GIUSEPPE ROSSI

Department of Civil and Environmental Engineering,
University of Catania, Italy

ANTONINO CANCELLIERE

Department of Civil and Environmental Engineering,
University of Catania, Italy

LUIS S. PEREIRA

Agricultural Engineering Research Center,
Institute of Agronomy,
Technical University of Lisbon, Portugal

THEIB OWEIS

International Center for Agricultural Research in Dry Areas, Aleppo, Syria

MUHAMMAD SHATANAWI

Faculty of Agriculture, University of Jordan, Amman, Jordan

and

ABDELAZIZ ZAIRI

National Research Institute for Agricultural Engineering,
Water and Forestry, Tunis, Tunisia

Springer-Science+Business Media, B.V.

A C.I.P. Catalogue record for this book is available from the Library of Congress.

ISBN 978-94-010-3965-9 ISBN 978-94-010-0129-8 (eBook)
DOI 10.1007/978-94-010-0129-8

Printed on acid-free paper

Table of contents

Preface

Significant vulnerability of water systems to drought is a common issue of water resources management in Mediterranean regions. This is due both to the increasing occurrence and severity of drought events and to the growing demand for municipal, tourist, and agricultural uses. The INCO-DC project entitled "A Decision Support System for Mitigation of Drought Impacts in the Mediterranean Regions" (DSS-DROUGHT) addresses this issue, contributing to an improved management of water-supply systems for irrigation, which represents the most consumptive sector of water resources uses in the Mediterranean region.

In order to develop a comprehensive approach toward improved operation of irrigation systems under drought conditions, the project was developed around the following five main, strictly interconnected tasks:

– Identification of drought characteristics at a site and over a region
– Modelling irrigation management under conditions of water scarcity
– Modelling operation of water supply systems under drought conditions
– Integration of the developed methodologies within a Decision Support System software package
– Definition of requisites for Drought Watch Systems

The research resulted in an advancement of knowledge through in-depth analysis of innovative methodologies, the development of tools to help decision-makers in coping with droughts through the implementation of the developed procedures in software packages, and the application of these tools to the case-studies identified by partners in their countries.

The project was performed by four partners, namely:

- The Department of Civil and Environmental Engineering (formerly The Institute of Hydraulics, Hydrology and Water Management), University of Catania, Italy;
- The Institute of Agronomy, Technical University of Lisbon, Portugal;
- The National Research Institute for Agricultural Engineering, Water and Forestry, Tunis, Tunisia;
- The Water and Environment Research and Study Center, Jordan University, Jordan.

In addition, the International Center for Agricultural Research in Dry Areas (ICARDA), Aleppo (Syria) cooperated by preparing a case-study and participating in the dissemination of the results.

The successful application of methods and software packages for drought identification, for defining irrigation scheduling strategies, and for managing water supply systems in the different case-studies selected in Portugal, Italy, Tunisia, the Jordan Valley, and Syria showed the feasibility of an integrated approach to drought-mitigation, able to face the drought threat in European Mediterranean countries as well as in North Africa and the Near East countries. The developed tools represent a contribution to the implementation of an improved strategy for an integrated and sustainable water resources management in the countries of the two shores of the Mediterranean sea.

Results of the three-year activities carried out within the DSS-Drought Project were shared with specialists in water fields from the Mediterranean region at a workshop held in Aleppo, May 27-31, 2001, which was partially funded by the European Commission through the INCO-DC project and through a specific Accompanying Measures project and by the FAO Regional Office for the Near East. The goals of the workshop were to disseminate the research results and exchange experiences in coping with droughts with experts invited from several Mediterranean and Near East countries.

This book includes the lectures presented at the workshop by the participating project partners, and by the two invited speakers who presented their experiences in the analysis and mitigation of droughts in another EC-funded project and in investigations carried out in the USA.

The book is organized as follows: after an introduction, where issues related to drought mitigation approach and irrigation management under drought conditions are discussed, the results related to drought identification and monitoring in the Mediterranean region are presented. Then, irrigation management strategies under drought conditions are illustrated through the application to several case studies. Finally, results related to modelling of water supply systems under drought conditions are discussed.

Particular gratitude is expressed to everybody involved in the project and contributing to the workshop. Special thanks are due to Dr. Giuseppe Giuliano for the help in editing this book and to Mrs. Sandra Lazzarini for the technical assistance. The financial support of the European Commission through contract IC18-CT97-0169 and Accompanying Measures contract ICA3-2001-50006 is gratefully acknowledged.

June 2002

The editors

Part 1

Introduction

AN INTEGRATED APPROACH TO DROUGHT MITIGATION IN MEDITERRANEAN REGIONS

G. ROSSI
Department of Civil and Environmental Engineering, University of Catania, Italy

Abstract: Severe recent droughts that occurred in Mediterranean countries have shown that a proactive approach based on drought management plans to identify and implement adequate measures for coping with droughts is necessary. The EC INCO-DC project DSS-DROUGHT, developed by five organisations from Italy, Portugal, Tunisia, Jordan and Syria aimed at developing an appropriate Decision Support System able to cope with drought and in particular to improve the management of irrigation systems under drought conditions. The paper presents the main features of the integrated approach to drought mitigation developed within the project through different tasks, namely hydrological drought identification and monitoring, irrigation scheduling strategies, and water supply operation rules. The related software modules of the Decision Support System package resulted very useful tools for analysing droughts and developing appropriate mitigation measures. Also the requisites of a Drought Watch System defined in the project set the criteria for implementing early drought warning in Mediterranean regions.

Key words: Drought analysis, drought mitigation, Mediterranean Region.

1. Introduction

An appropriate water resources management which ensures a sustainable use of a resource limited in amount and affected by a growing pollution represents a major problem to be faced by all Mediterranean countries. Water problems have been recognized as key issues for the Mediterranean countries since the Barcelona Declaration for the Euro-Mediterranean Partnership on November 1995, in order to garantee a zone of peace, stability, economic prosperity where democracy and human rights are respected.

It is well know that the countries of the two shores of Mediterranean exhibit relevant differences with respect to climatic features and water resources availability both total and per capita amount, as well as to institutional and economic characteristics. In particular the countries of the South shore (North Africa and Near East) and the islands already suffer from water scarcity effects, aggravated by the dramatic impacts of long and severe recent droughts on economic, social and environmental systems. Thus in these countries, the fostering of the development of water infrastructures as well as the implementation of appropriate operation rules for drought mitigation are considered prioritary issues.

G. Rossi et al. (eds.), Tools for Drought Mitigation in Mediterranean Regions, 3-18.

Mediterranean regions of European countries, generally face a better situation because of a more favourable average renewable water availability and for the presence of water works able to transfer water in time and space. Nonetheless, the vulnerability of several large-scale water supply systems to drought is still high, thus requiring in many cases the updating of the original design provisions in order to cope with increasing risk of drought also through improved operation rules.

Despite the differences, common features about drought occurrence, preparedness and mitigation can be identified in all Mediterranean countries. Some of these common features are the following:

- frequent meteorological anomalies causing severe hydrological deficits occurred in the last decades in several regions of the two Mediterranean shores and the situation is expected to worsen as a consequences of the feared climatic changes;
- a high vulnerability to drought exists in several Mediterranean countries where the current water withdrawals represent a relevant percentage of average renewable water resources and they are expected to raise due to the urban and tourist growth as well as to the development or at least the preservation of irrigated land motivated by the food self-sufficiency willingness of many countries;
- the generalised response to drought hazard consists in a reactive approach using emergency actions instead a pro-active approach based on prevision and prevention tools;
- a limited exchange of experiences gained during the severe recent droughts yet occurs among the majority of the drought – prone countries.

A project, funded by the European Commission within the INCO-DC Program, with the acronym DSS-DROUGHT, has been developed with the aim of facing some of these issues and in particular the reduction of the vulnerability to drought of surface water supply systems devoted to irrigation use. The project has been carried out by four partners, namely:

- Department of Civil and Environmental Engineering (formerly Institute of Hydraulics, Hydrology and Water Management), University of Catania, Italy;
- Institute of Agronomy, Technical University of Lisbon, Portugal;
- National Research Institute for Agricultural Engineering, Water and Forestry, Tunis, Tunisia;
- Water and Environment Research and Study Center, Jordan University, Jordan.

In addition, the International Center for Agricultural Research in Dry Areas (ICARDA), Aleppo (Syria) cooperated to the Project by preparing a case study and participating to the dissemination of the results.

The DSS-DROUGHT project aimed at covering the multifaceted aspects of a successful decision making for drought management by an integrated approach including knowledge and monitoring of drought, assessment of the impacts and definition of mitigation measures.

The project has developed and tested in some Mediterranean regions a Decision Support System (DSS) specifically oriented to the management of water systems supplying water to irrigated districts under drought conditions and has defined the requisites and institutional framework for developing a Drought Watch System. In what follows a brief description of the objectives, methodologies and main results of the different tasks of the project are reported.

2. Integrated approach to drought mitigation

2.1. DROUGHT DEFINITIONS

Many definitions of drought are adopted in various fields, with reference to the components of the hydrological cycle considered in the analysis and to the different impacts on water users and ecosystems (Yevjevich et al., 1983, Easterling, 1988, Rossi et al., 1992, Wilhite et al., 2000).

In particular from a meteorological/climatic point of view, drought can be defined as a temporary reduction in water or moisture availability significantly below the normal or expected amount for a specified period, with an emphasis on the transitory character of the water deficiency with reference to the particular climatic zone.

From a hydrological point of view, drought is defined as a period of abnormally dry weather sufficiently prolonged for the lack of precipitation to cause serious water deficits in water bodies.

From an agricultural point of view drought is defined as depletion of soil moisture to a level that affect significantly crop and/or pasture yield.

According to a water-resource oriented definition which takes into account the water requirements related to biological, economical and social characteristics of a region, drought refers to a casual (random) condition of *severe reduction* of water supply availability (compared to normal value), extending along a *significant period* of time over a *large region*. It has been pointed out that the criteria to define *severe reduction*, *significant period* and *large region* are affected by subjectivity, as they stem from the demand level as well as from the perception of negative impacts of the water deficits (Rossi, 2000).

In any case, a general consensus exists in distinguishing drought from aridity, since *aridity* refers to a permanent climatic condition with very low annual or seasonal precipitation amount or a limited ratio of precipitation to potential evapotranspiration. Meteorological drought and aridity are frequently associated, because the arid and/or semi-arid regions usually present a high variability of precipitation. It is necessary also distinguish drought from *desertification,* which indicates a long-term and somehow irreversible process of decrease or destruction of biological soil potential caused by several factors (mainly soil degradation by human activities) and with drought occurrences acting as acceleration factors of the process.

2.2. DIFFICULTIES OF DROUGHT MITIGATION

Drought is a natural hazard but is also a man-affected phenomenon. It is recognised that drought is perceived as a disaster or as an adverse event only when it affects the life, the economic interests or the social well being of a human community: a lack of rainfall on an uninhabited ocean is not considered drought! Besides, the impacts of a drought of a certain severity can differ considerably according to the level of demand with respect to the water availability. Finally the drought impacts can be substantially modified by the strategies adopted for coping with drought risk.

It is necessary to ask why drought mitigation actions are difficult to be implemented. A preliminary answer should consider at least the following points:

1. Drought is a complex phenomenon and a lack of understanding of different natural and man-affected aspects exists at scientific level.
2. Drought characteristics (duration, severity) are difficult to forecast, and both time and space variability of drought is not generally well monitored.
3. Impacts of drought on different sectors of economy, society and environment are difficult to quantify.
4. Conflicts among different groups (municipalities, farmers, ecologists) complicates policy decision on drought management.
5. Water management under drought conditions is limited by legal constraints on water allocation as well as by inadequate decision processes.
6. Effective drought mitigation measures require horizontal co-ordination (between water supply organisations) and vertical co-ordination (between local and national levels).
7. Effective drought mitigation measures require that a plan for coping with drought effects be developed in advance and not during emergency.

From the above points, it is evident that difficulty for coping with drought problems derives mainly from:

- Inadequate knowledge of the phenomenon and lack of an effective monitoring;
- Complexity of the assessment of drought impacts;
- Delays in improving preparedness to drought risk management through innovations in legal, institutional, organisational aspects of water resources management.

Despite the difficulties, it is largely recognised that the possibility of an effective mitigation of adverse effects of drought is relevant and generally greater than the one of other natural disasters since:

- The man-affected aspects are in certain case prevailing;
- The most severe impacts are not a sudden response to a drought event, thus allowing their minimisation through an appropriate decision-making process.

2.3. FEATURES OF AN INTEGRATED APPROACH

The main features of an integrated approach to drought mitigation include:

- Identification of droughts on relevant hydrometeorological variables and water resources conditions with the aim of computing the water deficit risk and monitoring drought evolution;
- assessment of economic, environmental, societal impacts of drought based on the analysis of long time-series;
- definition and choice of available measures for coping with drought, to be distinguished as long-term measures on water supply and water demand oriented to reduce the system's vulnerability to drought and short-term measures oriented to face drought emergency, both studied in advance through long-term and contingency drought management plans (Rossi, 2000). In particular these mitigation measures include a mix of measures oriented to increase supply, reduce demands or minimise drought impacts.

Within such an integrated approach, the presence of a decision support system (DSS) can play a very important role to help water managers into their decisions.

3. DSS-DROUGHT project for drought mitigation in Mediterranean regions

3.1. OVERVIEW

According to the main features of an integrated approach to drought mitigation the main objective of the DSS-DROUGHT project has been to develop a DSS oriented to help decision makers in coping with drought, through an improved understanding of the drought phenomenon, an assessment of drought impact and an appropriate choice of the actions to be taken. Since in arid and semi-arid Mediterranean drought-prone regions the prevailing use of water resources is for irrigation purpose and most of the irrigation districts are supplied by surface water resources stored in reservoirs, the project focused on the management of water supply irrigation reservoirs under drought conditions.

In particular, the core of DSS includes three connected models:

a) to identify the regional drought events and determine their characteristics (duration, cumulative deficit, areal coverage of deficit);
b) to determine the most effective irrigation scheduling under water scarcity conditions, using an appropriate climate-soil-crop simulation model;
c) to define operating rules for storage facilities of the water supply system under drought conditions.

A second objective has been the investigation of the pre-requisites and institutional framework for developing an appropriate scheme of a Drought Watch System in Mediterranean countries.

The research activities have been developed according the following tasks:

- Identification of drought characteristics, including the selection of the methodology for the characterisation of severe droughts affecting large-scale water supply systems and the analysis of the variability of drought characteristics in Mediterranean areas on the basis of the selected case studies.

- Modelling irrigation management under drought conditions, aiming to select, develop and improve models to be used for determining appropriate irrigation scheduling at different levels of information availability according to different management objectives.
- Modelling of water system operation and evaluation of drought mitigation actions, oriented to define the operation rules of the water supply system, necessary for minimising negative drought impacts in different conditions as identified in the selected case- studies.
- Development of the Decision Support System software package, conceived as a tool to be easily used by decision makers in selecting the most suitable options in water resources management for irrigation, on the basis of the results of the different interconnected modules.
- Definition of the requisites for Drought Watch Systems, aiming to allow an early drought warning and a timely implementation of drought mitigation measures, on the basis of adequate hydrometeorological and water resource monitoring networks.

The result of the research led, for each of these tasks, to an advancement of knowledge through an in-depth analysis of innovative methodologies. Also a comparison of the applications of the software packages to several Case Studies was carried out, as reported in Table I. The location of the selected Case Studies is indicated in Figure 1. In what follows the objectives, methodologies and main results of each of the aforementioned tasks are briefly outlined. A detailed presentation of the investigations carried out by the partners within each task can be found in the other chapters of this book.

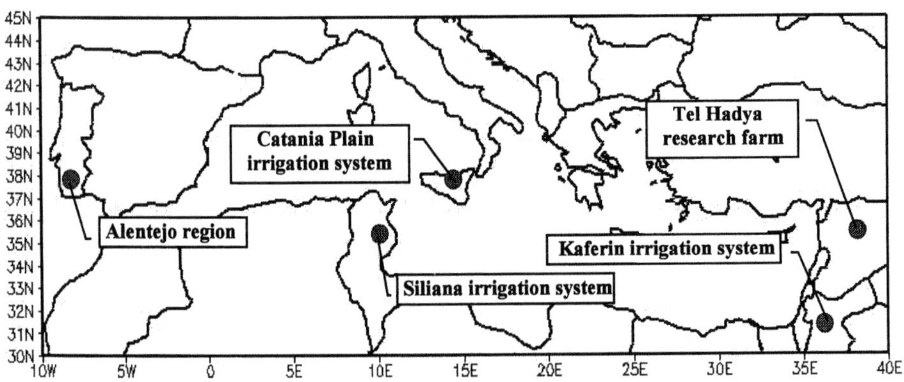

Fig. 1. Location of case studies

Table I. Application of the DSS models to the Case -Studies

Partner and Case-Study	**Drought identification by REDIM**				**Irrigation management by ISAREG**		**Operation rules of irrigation water system**
	Investigated region		**Rain gages**	**Record period**	**Model validation through field data**	**Application for determining deficit irrigation strategies**	
	Name	Area (km^2)	(No)				
University of Lisbon Alentejo region	Alentejo sub-reg. C Alentejo region	4,275 27.294	7 58	1932-95 1932-95		Vigia district for wheat, maize, sunflower and sugar beet crops. Results used for economic evaluation	
University of Catania Catania Plain irrigation system	Pozzillo reservoir catchment Catania Plain irrigation district Sicily region	577 470 25,461	4 4 171	1921-96 1921-96 1921-96	At Fontanazza farm for citrus orchards irrigated by microject system (1999-2000 data)	Catania Plain District for citrus orchards with or without suppletive private resources	Pozzillo reservoir operation rules obtained by RESOPTI (DP) and RESOPER (NN) modules of the DSS
INGREF,Tunis Siliana irrigation system	Siliana catchment	1,040	3	1967-97	At Hendi Zitoun experimental farm for winter wheat 1992-93: sprinkler irrigation; 1997-98: surface irrigation	Siliana district for winter wheat and horticultural crops (potato and tomato). Results used for economic evaluation	Lakhmes reservoir optimisation by Stochastic DP
University of Jordan Kaferin irrigation system	Kaferin wadi catchment Hishan wadi catchment	161 60	3 3	1973-96 1973-96		Jordan Valley for citrus, wheat and horticultural crops	
ICARDA, Aleppo Tel Hadya research farm					At Tel Hadya research farm for winter wheat	Tel Hadya farm for winter wheat	

3.2. IDENTIFICATION OF REGIONAL DROUGHTS

The first part of the Project regarded the selection of the methodology to be adopted for identifying severe droughts affecting large-scale water supply systems and the comparison of the drought characteristics in Mediterranean regions through the analysis of the selected case-studies.

Critical drought conditions occur when there is a severe water deficit for a long duration over large areas (Rossi et al., 1992). Therefore, drought analysis requires the choice of the methods for the identification of the local (i.e. at-site) drought and the procedures for describing the areal coverage of the water deficit computed in the selected stations within the investigated area.

The activities carried out in this task can be summarised as follows:
- review of the methods for the identification of at-site droughts;
- review of the methods for the identification of regional droughts;
- choice of the method for the computation of the drought characteristics based on the concept of run (in particular duration, cumulative deficit, areal coverage);
- development of the software package REDIM for preliminary analysis of time-series stationarity and computation of drought characteristics;
- selection of the areas (within each case-study selected by the Partners) for analysing drought characteristics through REDIM software;
- comparison of drought characteristics at the selected Mediterranean areas;
- assessment of the capability of the developed procedure to describe severe drought events.

A first result of the investigation was the choice of the run method as the appropriate methodology for local and regional drought identification and characterization on hydrometeorological series of different time scale (civil year, water year, trimester, etc.) . The applications demonstrated the convenience of such a method for computing successfully the drought characteristics of interest at-site (duration, cumulative deficit, intensity). Once the onset and termination of a drought have been identified over a hydrological series at a finite number of sites within the region, the deficits computed in each time interval are used as a basis to describe the regional drought characteristics, by selecting an appropriate areal threshold and by computing areal coverage of deficit and areal weighted deficit over the considered region (Rossi *et al.*, 1992). Such regional extension of run analysis resulted a good method for identification of droughts over a region at different time scales. A comparison carried out on Alentejo region between regional run analysis results and the Standardized Precipitation Index, also applied in a regional fashion, indicated that both methods produce similar results, despite the differences in the underlying theories (Paulo *et al.*, 2002).

The run method was implemented within the software REDIM, which includes three main parts: i) testing for stationarity and randomness in hydrological time series; ii) identification and characterization of local droughts; iii) identification and characterization of regional droughts. The specific features of the product and in particular the format of numerical and graphical output make it an user-friendly tool for an easy understanding of the drought threat (Rossi and Cancelliere, 2002).

Application of REDIM software in all case studies by the different partners of the project, allowed an interesting comparison of the most severe droughts identified in the various Mediterranean regions (Cancelliere and Rossi, 2002). In particular the preliminary stationarity analysis for precipitation series showed that most of the annual and trimester series at considered stations and all areal precipitation series over the regions presents a reduction over time, particularly relevant for the wet trimesters. The klinear trend, analysed through t-Student test applied to water year series indicated the presence of generally negative trends, with a few exceptions. In particular, the series of Alentejo region, Siliana catchment and Kaferin catchment have always negative trends and all other series presented at least one or more significant negative trends. Also the comparison of regional drought characteristics identified on the annual series, showed that in all regions long and severe droughts occurred. In particular the maximum duration was observed in Catania plain district (7 years) and Pozzillo catchment (6 years), while the greatest dimensionless cumulated deficits, with respect to mean areal precipitation, was observed in Hisban catchment, followed by Pozzillo, Alentejo and Catania Plain.

3.3. DEFINITION OF IRRIGATION SCHEDULING UNDER LIMITED SUPPLY

The advantages of an appropriate irrigation scheduling to increase crop yield and income, to save water, and to improve conditions for sustainable irrigated agriculture are generally recognised. Under limited water supply conditions, such as during severe droughts, the role of such irrigation scheduling is particularly crucial to optimise the use of available water so that the dramatic consequence of a severe water shortage are avoided or at least minimised. Indeed, when the available water resources are not sufficient for the adequate supply for all crops in the district during the entire growing cycle, irrigation scheduling can still guarantee acceptable levels of crop yield (though below the maximum) by either reducing the irrigated areas or by reducing and distributing in time the irrigation volumes appropriately (deficit irrigation).

The definition of an appropriate irrigation scheduling under limited supply can be performed by one of the several climate-soil crop simulation models developed for design and/or operation of irrigation projects (Pereira et al., 1995). These models differ in terms of input data requirements (on climate, soil, and crops), water balance features (e.g. soil water depletion process) and output specification such as number, date and depth of watering for different crop-soil units.

The activities carried out in this second task of the Project included:

- review of different climate-soil-crop models including ISAREG, already developed by the Portuguese partner;
- updating of the ISAREG model for simulation under different options of irrigation management and of related programs (KCISA for computation of the crop coefficient pattern; EVAPOT for computation of evapotranspiration) with the development of a new Windows version of the software (Pereira *et al.*, 2002);

- selection of experimental farms within the case studies and validation of the ISAREG model through field experiments for different crops;
- application of ISAREG model to analyse various types of deficit irrigation for several crops in Mediterranean regions presenting different climatic conditions and different restrictions in water supply;
- comparison of the results and conclusions on the performance of the ISAREG model for determining irrigation scheduling under water scarcity conditions.

A first relevant result of the task's activities was the validation of the ISAREG model for various crops in different experimental conditions (winter wheat in Tunisia and in Syria, citrus orchards in Italy). In agreement with previous investigation results in Portugal, the best fitting between simulated and observed soil water contents was obtained for winter wheat. The other field experiments confirmed that the model is able to simulate the soil water content along the entire crop season with acceptable errors. However the validation for citrus orchards in Catania Plain district confirmed that a correct estimate of crop coefficient pattern is a key factor (Alba *et al.*, 2002).

ISAREG model was applied to all case-studies in order to analyse and compare several irrigation strategies under drought condition. Several alternatives were evaluated through the combined use of indicators relative to reduction in irrigation demand, the consequent yield reduction and in Tunisian and Portuguese cases also in the farmer income reduction.

In particular the analysis in the Portuguese case (Vigia district) concerned the supplemental irrigation of winter wheat and deficit irrigation of maize and sunflower under center-pivot sprinkler irrigation by comparing different deficit irrigation strategies (Rodriguez *et al.*, 2002). Supplemental irrigation of wheat was also analysed for the Syrian case-study (Tel Hadya farm) by considering surface and set sprinkler irrigation (Oweis *et al.*, 2002). The analysis for the Tunisian case (Siliana project) concerned the irrigation of winter wheat and potato and tomato crops. In the Portoguese and Tunisian cases, a careful economic analysis was carried out for the investigated crops, computing gross margin for hectare and gross margin for unit volume of water applied.

Results from the investigated case studies indicated differences in responses to deficit irrigation of the different analysed crops under specific climatic conditions. Application of ISAREG model allowed to define the best strategies to face water scarcity through deficit irrigation, although in some cases (e.g. maize in Portuguese case study) it appears preferable to fully irrigate only a reduced cropped area.

The water price analysis has shown that the raise in water price has a similar effect to that resulting from adopting water availability constraints. The analysis makes it evident that field horticultural crops remunerate better the water used than cereal crops, thus indicating the need of a careful water price policy.

The application of the ISAREG model was carried out in the Italian case study for the simulation of deficit irrigation strategies for citrus orchards by considering two scheduling typologies: flexible schedules, that can be adopted by farms with storage facilities and rigid schedules, fixed by the district managers. Flexible schedules appear to have the potential to better face water deficits, although, when very severe restrictions are imposed the difference between the two scheduling options tend to vanish.

In the Jordan case study the analysis referred to different crops (wheat, potato, tomato and citrus), considering different irrigation scenarios. An analytical hierarchy process model was applied as a multicriteria technique for selecting the best crop according to different objectives (save water, economic return, etc.). The highest priority crop with respect to the goal of sustainable irrigation management resulted potato, which however is sensitive to salinity problems (Shatanawi *et al.*, 2002).

3.4. DEFINITION OF RESERVOIR OPERATION RULES UNDER DROUGHT CONDITIONS

Consolidated approaches to reservoir operation analysis are based either on optimisation or simulation models which in some cases can be jointly used (Wurbs, 1995). This part of the research had the objective of defining the operation rules of a water system devoted to agricultural use under drought conditions. In particular the optimal releases, obtained by an optimisation technique with a given objective function, represent the basis for deriving, through regression technique or other algorithms, operation rules to be provided to water supply system operators. The analysis considered a water system including a reservoir regulating surface waters and an irrigation district.

The main activities carried out were the following:

- analysis of reservoir operating policies and selection of performance indices for evaluating system behaviour;
- development of mathematical models for reservoir analysis (simulation and optimisation through Dynamic Programming, taking into account real management criteria);
- development of models for deriving operation rules of an irrigation supply reservoir by multiple regression and neural network approaches;
- development of a Stochastic Dynamic Programming model for the operation of an irrigation reservoir, considering two different operation objectives;
- validation of the models through the application to the selected case-studies.

Some of the developed models have been implemented within user-friendly software packages, namely:

- RESOPTI, computing the optimal reservoir releases, using Dynamic Programming, assuming as objective function the sum of square deficits;
- RESOPER computing the operation rules, based on the releases obtained through Dynamic Programming, by using either a multivariate regression approach or Neural Networks;
- RESSIMUL, simulating the reservoir operation, providing a set of reservoir performance indices specifically oriented to investigate the behaviour of the water system under drought conditions.

The results of the activities allowed to develop an appropriate methodology that combines the soil water balance model and Dynamic Programming optimisation for determining optimal releases. Such optimal releases are used to derive, by multiple linear regressions and/or neural network approaches, operation rules which are validated by means of a simulation model (Rossi *et al.*, 1999; Cancelliere *et al*, 2002a). The set of developed software packages represent useful tools to improve the operation of a water supply system for irrigation use and to effectively reduce the most adverse effects of water deficiency, by redistributing the deficits over a longer period, thus reducing the crop yield losses.

It was confirmed that considering an optimisation period which includes a severe drought leads to more conservative rules, which allow to face more effectively periods of water scarcity. Further, the need for taking explicitly into account the management constraints (e.g. fully satisfy the demand when stored water is above a fixed threshold), within the optimisation algorithm was also demonstrated, in order to avoid too conservative restrictions (Cancelliere *et al.*, 2002b).

An application of Stochastic Dynamic Programming for the management of a Tunisian reservoir showed the feasibility of the approach, which allows to take into account two different objectives, namely the minimization of water deficits during the summer irrigation season and the conservation of water in the reservoir to meet water demands in the next season (Lebdi *et al.*, 2002).

3.5. DECISION SUPPORT SYSTEM PACKAGE

Although a clear and unambiguous definition of "Decision Support System" is still lacking, two essential characteristics are generally recognised (Loucks and Da Costa, 1991; Reitsma *et al.*, 1996):

- a DSS is an interactive tool including computer-based information and modelling systems;
- a DSS has the purpose of aiding decision-makers contributing to understand the problem, to explore alternative courses of actions, to predict their impacts, to facilitate sensitivity analysis, etc.

The software packages related to the models developed in the various tasks of the project have been integrated within a comprehensive Decision Support System aiming at aiding decision makers in managing water systems devoted to irrigation use under drought conditions. A scheme of the DSS is depicted in Figure 2, from which it can be inferred that the six main computational modules exchange data through a common user interface. Besides the previously described modules, the DSS includes also the multicriteria assessment model NAIDE able to compare different drought mitigation alternatives (Munda, 1995).

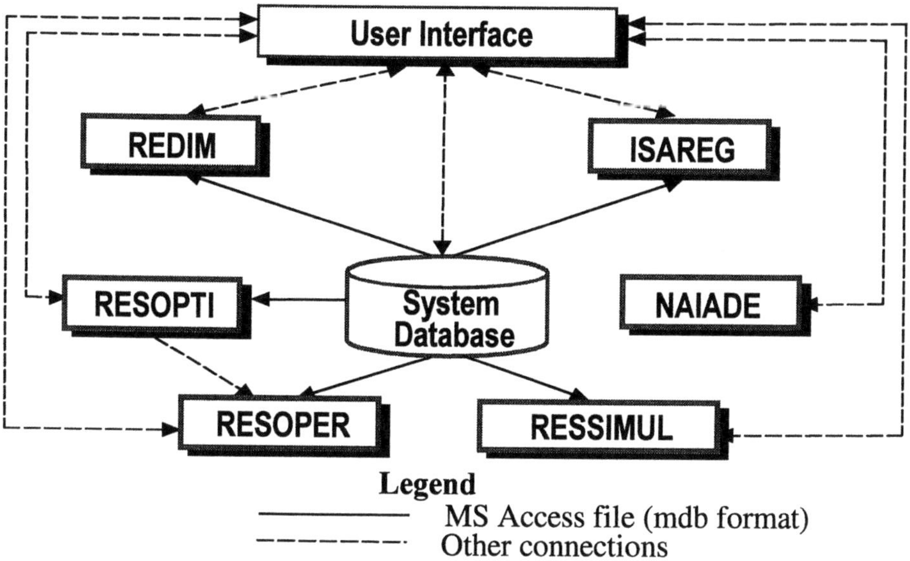

Fig. 2. General scheme of the DSS-DROUGHT package

3.6. REQUISITES OF A DROUGHT WATCH SYSTEM

The last part of the Project has been dedicated to the definition of the requisites of a Drought Watch System, which should serve as a tool for responsible organisations in a timely recognition of drought conditions and in the implementation of drought mitigation measures.

The activities carried out included:

- review of existing Drought Watch Systems;
- review of drought indicators able to provide an early warning of severe water shortages;
- identification of hydrometeorological and water resources information necessary for computing drought indicators;
- survey of present institutional and technical features of the organisations devoted to drought data acquisition and transfer to public officials responsible for drought management in partners countries;
- definition of the criteria for the development of a Drought Watch System at national level with international exchanges with other Drought Watch Systems in other countries.

The results of the survey showed large differences among the different countries (Portugal, Italy, Tunisia and Jordan) in terms of hydrometeorological network, systems for acquisition, storage and retrieval data (Rossi, 2002).

A proposal was developed for a national Drought Watch System including criteria for standard data acquisition and transfer, indices computing, and drought information dissemination. Figure 3 shows the structure of the proposed System where the flows of data to collect from national and international databases and flow of information to transfer to drought related organisations are reported.

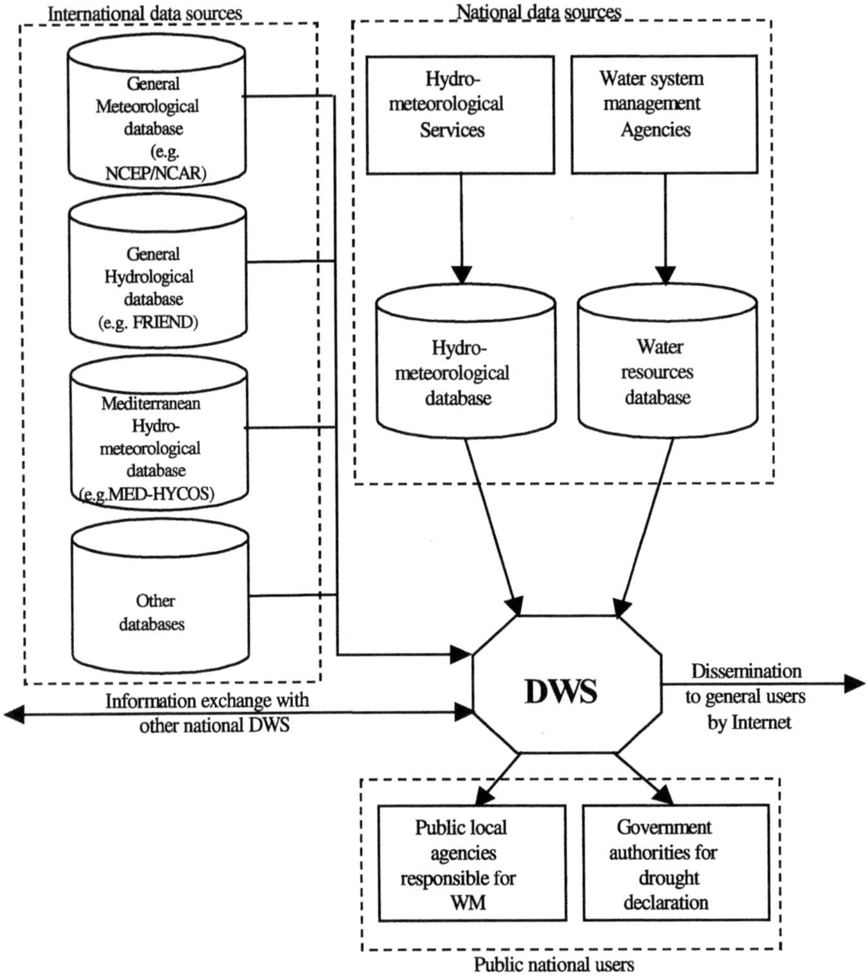

Fig. 3 Structure of a Drought Watch System for Mediterranean countries

4. Conclusions

The DSS-DROUGHT project funded by European Commission within the INCO-DC Program has obtained the following main results:

- the development of appropriate procedures to identify droughts affecting large-scale water supply systems, to define irrigation scheduling and to improve operation rules of water supply reservoirs under drought conditions;
- the development of the DSS software package and the analysis of the applicability of the proposed procedures to some Case Studies selected in the Mediterranean semi-arid countries involved in the Project (Portugal, Tunisia, Italy, Jordan and Syria);
- the definition of criteria for developing a Drought Watch System in the Mediterranean semi-arid regions.

The main outcome of the project is a comprehensive approach, developed with the contribution of significant experiences of different countries, to the problem of coping with drought in semiarid Mediterranean regions which present a high vulnerability to water deficit especially in agricultural sector. The use of the developed computer tools can significantly contribute to the preparedness to the drought events of the irrigation managers in the Mediterranean regions.

Acknowledgments

The research has been carried out within the European Community INCO-DC project "A DSS for drought mitigation in Mediterranean regions" (Research contract IC18-CT97-0169), with partial financial support from Italian National Research Council, Group for Prevention from Hydrogeological Disasters.

References

Alba, I., Rodrigues, P.N and Pereira, L.S.(2002) Irrigation scheduling simulation for citrus in Sicily to cope with water scarcity (This issue).

Cancelliere, A. and Rossi, G. (2002) Droughts in Sicily and comparison of identified droughts in Mediterranean regions (This issue).

Cancelliere, A., Ancarani, A., Giuliano, G. and Rossi, G. (2002a) A neural networks approach for deriving irrigation reservoir operating rules. *Water Resour. Manage.* 16: 71-88.

Cancelliere, A., Giuliano, G., Ancarani, A. and Rossi, G. (2002b) Derivation of operation rules for an irrigation water supply system by multiple linear regressions and neural networks. (This issue).

Easterling, W.E. (1988) Coping with drought hazard: recent progress and research priorities. In: F. Siccardi and R. L. Bras (Eds.) *Natural Disasters in European Mediterranean Countries*, NSF and NRC, Perugia (Italy), pp. 231-270.

Lebdi, F., Bergaoui, M. and Bouslimi, M.A. (2002) Optimisation of reservoir management in the semi-arid countries: Case of Lakhmess, Tunisia (This issue).

Loucks, D.P. and Da Costa, J.R. (1991) *Decision Support System. Water Resources Planning.* Nato ASI Series, Springer-Verlag, Berlin.

Munda, G. (1995) *Multicriteria evaluation in a fuzzy environment. Theory and applications in ecological economics,* Physica-Verlag, Heidelberg.

Oweis, T., Rodrigues, P.N. and Pereira, L.S. (2002) Simulation of supplemental irrigation strategies for wheat in Near East to cope with water scarcity. (This issue).

Paulo, A.A., Pereira, L.S. and Matias, P.G. (2002) Analysis of local and regional droughts in southern Portugal using the theory of runs and the standardised precipitation index. (This issue).

Pereira, L.S., Van Den Droek, B.J., Kabat, P. and Allen, R.G. (Eds.) (1995) *Crop-Water-Simulation Models in Practice*, Wageningen Pers, Wageningen.

Pereira, L.S., Teodoro, P.R., Rodrigues, P.N. and Teixeira, J.L. (2002) Irrigation scheduling simulation: the model ISAREG. (This issue).

Reitsma, R.F., Zagona, E.A., Chapra, S.C. and Strzepek, K.M. (1996) Decision Support Systems for Water Resources Management. In: L.W. Mays (Ed.) *Water Resources Handbook*, McGraw-Hill, New York, pp. 33, 1-35.

Rodrigues, P.N, Machado, T., Pereira, L.S., Teixeira, J.L, El Amami, H. and Zairi, A. (2002) Feasibility of deficit irrigation with center-pivot to cope with limited water supply in Alentejo, Portugal. (This issue).

Rossi, G. (2000) Drought mitigation measures: a comprehensive framework. In: J.V. Vogt and F. Somma (Eds.) *Drought and Drought Mitigation in Europe*, Kluwer Academic Publishers, Dordrecht, pp. 233-246.

Rossi, G. (2002) Requisites for a drought watch system. (This issue).

Rossi, G. and Cancelliere, A. (2002) At-site and regional drought identification by REDIM model (This issue).

Rossi, G., Benedini, M., Tsakiris, G. and Giakoumakis, S. (1992) On Regional drought estimation and analysis. *Water Resour. Manage.* 6: 249-277.

Rossi, G., Ancarani, A. and Cancelliere, A. (1999) A comparison between neural network approach and multiple regression procedures for deriving reservoir operation policies. In: V.P. Singh, Il Won Seo and J.W. Sonu (Eds.) *Water, Environment, Ecology, Socio-Economics, and Health Engineering.* Water Resources Publication, Colorado.

Shatanawi, M., Al-Zu'bi, Y. and Al-Jayoussi, O. (2002) Irrigation management dynamics in the Jordan valley under drought conditions . (This issue).

Yevjevich, V., Da Cunha L. and Vlachos, E. (Eds.) (1983) *Coping with Droughts.* Water Resources Publications, Littleton, Colorado.

Wilhite, D. A. (Ed.) (2000) *Drought. A global assessment for drought.* Westview Press, Boulder.

Wurbs, R.A. (1995) *Water Management Models. A guide to software.* Prentice Hall, Englewood Cliffs, N.J.

Zairi, A. El Amami H., Slatni A., Pereira L.S., Rodrigues, P.N., and Machado, T. (2002) Coping with drought: evaluation of deficit irrigation strategies for cereals and field horticultural crops in Central Tunisia. (This issue).

IRRIGATION DEMAND MANAGEMENT TO COPE WITH DROUGHT AND WATER SCARCITY

L. S. PEREIRA
Agricultural Engineering Research Center, Institute of Agronomy, Technical University of Lisbon, Portugal

Abstract: The use of water for agricultural production in water scarcity regions requires innovative research and appropriate transfer of technologies. This paper discusses some of these aspects relative to irrigation demand management. First, some concepts relative to water scarcity are discussed concerning aridity, drought, desertification and water shortage, and respective policies to cope with the respective water stressed regimes. Conceptual approaches on irrigation performances, water use and water savings are reviewed as well as issues on improvement of irrigation methods and respective performances. Main focus is placed on deficit irrigation strategies, including limitations on the applicability of related practices.

Key words: water scarcity concepts, reduced demand, irrigation performance, irrigation methods, irrigation scheduling, deficit irrigation.

1. Introduction: water scarcity concepts and water management implications

Water is becoming scarce not only in arid and drought prone areas, but also in regions where rainfall is abundant. Water scarcity concerns the quantity of resource available and the quality of water because degraded water resources become unavailable for more stringent water uses. The sustainable use of water - resource conservation, environmental friendliness, appropriateness of technologies, economic viability, and social acceptability of development issues - is therefore a priority for agriculture in water scarce regions.

Water scarcity may be due to different causes, natural or man induced, and has different duration. Natural causes relate to aridity and drought. *Aridity* is a nature produced permanent imbalance in the water availability consisting in low average annual precipitation, with high spatial and temporal variability, resulting in overall low moisture and low carrying capacity of the ecosystems. *Drought* is a nature produced but temporary imbalance of water availability, consisting of a persistent lower-than-average precipitation, of uncertain frequency, duration and severity, of unpredictable or difficult to predict occurrence, resulting in diminished water resources availability and carrying capacity of the ecosystems, largely affecting human activities and nature. Droughts are both a hazard and a disaster. A hazard because it is a natural accident of unpredictable occurrence but of recognisable recurrence. A disaster, because a drought corresponds to the failure of the precipitation regime, causing the disruption of the water supply to the natural and agricultural ecosystems as well as to the human activities.

G. Rossi et al. (eds.), Tools for Drought Mitigation in Mediterranean Regions, 19-33.

Man induced water scarcity refers to desertification and water shortage. *Desertification* is a man-induced permanent imbalance in the availability of water, which is combined with damaged soil, inappropriate land use, mining of groundwater, increased flash flooding, loss of riparian ecosystems and a deterioration of the carrying capacity of the ecosystems. Soil erosion and salinity are commonly associated with desertification. Desertification occurs in arid, semi-arid and sub-humid climates and may be influenced by climate changes. Drought strongly aggravates the process of desertification when increasing the pressure on the diminished surface and groundwater resources. Different definitions are used for desertification but main focus on land degradation, so giving to water a less important role. However, when dealing with water scarce situations it seems more appropriate to define desertification in relation to the water imbalance produced by the misuse of water and soil resources, so calling attention to the fact that abuse and misuse of land and water resources definitely are a cause for desertification. *Water shortage* is also man-induced but temporary water imbalance caused by less appropriate water management at regional or local scales, including groundwater over-exploitation, reduced reservoir capacities, disturbed and reduced land use and altered carrying capacity of the ecosystems. Degraded water quality is often associated with water shortages and, like drought, aggravates related impacts.

Policies and practices of irrigation water management under water scarcity must focus on specific objectives according to the causes and characteristics of water scarcity. Main characteristics associated with the water scarcity regimes may be summarised as described below, so requiring specific approaches to make water use sustainable under those water scarcity regimes.

Aridity is very often associated with high pressure on natural resources, a strong competition for water, aggravating the fact that water is the limiting resource for agriculture, frequent soil salinisation due to poor management of irrigation, and vulnerable and fragile ecosystems. Therefore, the sustainable use of water resources under aridity, implies:

- the effective adoption and implementation of integrated land and water resources planning
- the improvement of water and irrigation systems to achieve an increased service performance which would induce efficient water use and production,
- the adoption of water allocation policies favouring conservation and efficient use,
- valuing the water as an economic, social and environmental good, including for nature conservation,
- measures for augmenting the available water resource, including waste water and drainage water re-use, conjunctive use of water from different origins, and the use of low quality and saline water,
- the adoption of appropriate water and irrigation technologies that favour efficient water use and contribute to avoid water wastes and losses, and
- the users participation in water resources planning and systems management.

Drought requires measures and policies which are common with aridity such as those to avoid water wastes, reduce demand and make water use more efficient. But the peculiar characteristics of droughts imply specific measures:

- because they are unpredictable, preparedness measures are paramount to cope with droughts,
- as they have pervasive long term effects and their severity may be very high, appropriate reactive measures are required,
- the break in the natural water supply implies changes in water allocation and delivery policies, so in the management of water and irrigation systems,
- consequently, it is required that farmers be able to adopt reduced demand practices, and
- farmers incomes may be drastically reduced, so financial measures are also required to support farmers to cope with droughts.

Other aspects relative to drought mitigation are included in this publication, particularly under an integrated approach (Rossi, 2002 a).

Desertification and water shortage, because they are man induced and are associated to problems such as soil erosion, land degradation, mainly through salinisation, over exploitation of soil and water resources, and water quality degradation, require that policies and measures be oriented to solve the existing problems. Therefore, combating desertification and water shortage includes:

- re-establishing the environmental balance in the use of natural resources,
- restoring the soil quality,
- minimising water wastes,
- combating soil and water salinisation,
- controlling water withdrawals, and
- managing the water quality.

Coping with water scarcity requires measures and policies of water management that may be grouped into two main areas: demand and supply management. This paper refers only to demand management. This implies improved water use by the crops, avoiding irrigation water losses and higher irrigation systems performances. These aspects are first introduced and then the technical issues for demand control and reduction. Main attention is paid to deficit irrigation and the use of models to establish and evaluate deficit irrigation schedules, which application is described in various case studies relative to Portugal, Tunisia, Jordan, Syria and Italy presented in various papers in this publication.

2. Water use, water losses and irrigation performances

The term efficiency is very often used to express irrigation systems performance. It is commonly applied to each irrigation sub-system: storage, conveyance, distribution off- and on-farm, and on-farm application sub-systems. It can be defined by the ratio between the water depth delivered by the sub-system under consideration and the water depth supplied to that sub-system. Often, it is said that improving irrigation efficiencies is paramount under water scarce situations because this represents conditions of near optimal use of the water. This is true because this implies that the non-consumed and/or the lost fraction of the total water used are minimised. However, it is also commonly said that improving efficiencies would represent water savings or, in other words, low efficiencies mean high water losses. This is not fully true because the non-consumed fraction is not necessarily lost and may constitute a beneficial use such as the leaching fraction when irrigating in saline environments.

Along these lines, Allen *et al.* (1997) and Burt *et al.* (1997) proposed new concepts to clearly distinguish between consumptive an non consumptive uses, beneficial and non beneficial uses, and reusable and non reusable fractions of the non consumed water diverted into the irrigation system or subsystem. The beneficial water uses include the consumptive uses such as crop evapotranspiration, evaporation for climate control, and water incorporated in the product, and non-consumptive uses mainly leaching for salt control.

The non-beneficial uses may also be consumptive and non-consumptive. The first refer to evaporation of excess soil water, evapotranspiration from weeds and phreatophyte plants, evaporation from sprinkler irrigation, and canal and reservoir evaporation. The latter concern deep percolation and runoff due to excess irrigation and canal spills.

The non-consumptive water uses may be reusable or not. They are reusable when water is not degraded and returns to fresh surface or ground waters. This is the case of deep percolation and leaching add to non degraded water tables, or runoff and canal spills reaching usable water bodies. On the contrary, leaching, deep percolation and drainage water added to saline groundwater is not reusable, such as also the case for runoff and canal spills going to saline water bodies.

It can be said that improving irrigation efficiencies is of great importance under water scarcity conditions because high efficiencies correspond to increase the beneficial uses of the water for agricultural production. However, this improvement has to be complemented by:

- *controlling the non beneficial consumptive uses*, particularly those associated to soil evaporation, and evapotranspiration by phreatophytes and weeds receiving seepage and excess irrigation water,
- *minimising the non reusable fraction of the diverted water*, thus avoiding percolation to saline water tables or the disposal of drainage water into saline water bodies which would degrade the water quality and would make it non reusable, and
- *reducing the non beneficial but reusable fraction* by controlling deep percolation, seepage from canals, runoff return flows and canal excess water spills, which negatively impact operation and management costs and may be the cause for water-logging, competition by weeds, loss of nutrients and agro-chemicals, contamination of water bodies used for human consumption, and yield and income losses.

In general, these objectives may be achieved through multiple issues, which are commonly grouped into demand management and supply management practices and policy measures. When water scarcity conditions prevail, those are often named reduced demand management and limited supply management, and both refer to reducing the non-beneficial water use and to the augmentation of water use benefits under limited water availability.

3. Demand management

3.1. GENERAL ASPECTS

Demand management for irrigation under water scarcity includes practices and management decisions of mltiple nature: agronomic, economic, and technical, as summarised in Table I. Objectives concern a reduction of irrigation requirements, the adoption of practices leading to water savings in irrigation, both reducing the demand for water at the farm, and an increase in yields and income per unit of water used.

Agronomic and economic decisions and farming practices are dealt in several papers (Bucks *et al.*, 1990; Pereira, 1989 a, b; Tarjuelo and de Juan, 1999). Thus, only irrigation practices referring to demand management are discussed herein.

Table I. On farm management of water under scarcity

Objective	Technology
Reduced demand	Select low demand crop varieties / crop patterns
	Adopt deficit irrigation
Water saving / conservation	Improve irrigation systems uniformity and management
	Reuse water spills and runoff
	Surface mulch for controlling evaporation from soil
	Soil tillage for augmenting the soil water reserve
Higher yields per unit of water	Adopt best farming practices
	Avoid crop stress at critical periods
Higher farmer incomes	Select cash crops

Often, issues for irrigation demand management mainly refer to irrigation scheduling giving a minor role to the irrigation methods. However, a conjunctive approach is required (Pereira, 1996; 1999). Irrigation scheduling is the farmers decision process relative to "when" to irrigate and "how much" water to apply to a crop (Martin *et al.*, 1990). The irrigation method concerns "how" that desired water depth is applied to the field. The crop growth phase, its sensitivity to water stress, the climatic demand by the atmosphere, and the water availability in the soil determine when to apply an irrigation. The frequency of irrigations depends upon the irrigation method, *i.e.* on the depths of water, which are typically associated with the on-farm irrigation system. Both the irrigation method and the irrigation scheduling are inter-related (Smith *et al.*, 1996). This is particularly evident when supplemental irrigation is considered (Oweis *et al.*, 1999).

Irrigation scheduling requires knowledge on: the crop water requirements and yield responses to water (Allen *et al.*, 1998); constraints associated to each irrigation method and irrigation equipment (reviewed by Pereira and Trout, 1999), limitations to the water supply system, and financial and economic implications of the irrigation practice. To improve the irrigation method requires the consideration of the factors influencing the hydraulic processes, the water infiltration and the uniformity of water application to the entire field (Burt *et al.*, 1997; Pereira, 1999).

3.2. DEMAND MANAGEMENT: IMPROVED IRRIGATION PERFORMANCES

3.2.1. Concepts

Several performance indicators are currently used in on-farm irrigation. Relative to the uniformity of water application to the entire field, the main one is the *distribution uniformity* DU [%], which is the ratio between the average infiltrated depth in the low quarter of the field and the average infiltrated depth in the entire field (Burt *et al.*, 1997; Pereira, 1999). The main efficiency indicator is the *application efficiency* E_a [%], defined by the ratio between the average water depth added to root zone storage and the average depth of water applied to the field

An analysis of factors influencing the distribution uniformity and the application efficiency is given in Pereira (1999). It shows that the application efficiency depends upon the distribution uniformity. The latter characterises the irrigation system and E_a is an indicator of the system management. Thus, in general, the distribution uniformity values observed in field evaluations are the upper limits of the application efficiencies when keeping the system variables unchanged.

Non-consumptive water uses are largely dependent upon DU. When a system has high distribution uniformity the deep percolation and runoff are generally lower than for those having low DU. Similarly, less water volumes are required to assure appropriate leaching. In addition, yields may result higher because the differences between application depths in the fraction areas of the field being over- or under-irrigated are smaller. Useful relations between irrigation uniformity and crop yields have been made available, *e.g.* by Warrick and Yates (1987). Finally, the potential E_a is higher when the distribution uniformity is larger.

Non-consumptive water uses also depend upon the application efficiency. The primary effects result from the timeliness of irrigation. E_a tends to be lower when irrigation is practiced when the soil water deficit is small, which induces percolation (and runoff), and higher when that soil water deficit is large enough to provide for the root zone storage of the applied water. Due to its importance for optimal water use at field level, an indicator for the uniformity of distribution of the applied water is considered by Reca *et al.* (2001) in an water allocation optimization model aiming at deficit irrigation.

It can be therefore concluded that under water scarcity, drought in particular, it is important that farm irrigation systems should have high distribution uniformity and systems are managed to maximise the irrigation efficiency.

3.2.2. Surface irrigation

Several surface irrigation methods are used. The main ones are basin irrigation, furrow irrigation and border irrigation. In traditional systems, the water control is carried out manually according to the ability of the irrigator, so making it difficult to control "how much" water is applied and over-irrigation often occurs. In modernised systems, some form of control of discharge such as siphons, gated pipes, lay-flat tubes or gates, and some form of automation are used. Thus, in these systems, it is possible to control "how much" water should be applied at a given irrigation.

In surface irrigation, the uniformity mainly depends upon the system variables, which can often be modified/adapted by the irrigator but generally on a limited extent only. These system variables include: (a) the unit inflow rate, (b) the furrow, border, or basin length, (c) the hydraulic roughness coefficient, (d) the longitudinal slope and the levelling precision, (e) the soil infiltration characteristics, and (f) the furrow, border, or basin form and or width. The irrigator also controls the management variable time of cut-off, but this one also depends on some system variables that determine the advance time. The application efficiency is dependent on the same variables as DU and on the farmer management variables time of cut-off and the soil water deficit at time of irrigation.

The farmer ability plays a major role in controlling the management variables but his capability to achieve higher performances is definitely limited by the system characteristics and, often, by off-farm delivery decisions. When surface irrigated areas are supplied from collective irrigation canal systems, farm irrigation scheduling depends upon the delivery schedule, *e.g.* rate, duration and frequency, which are dictated by the system operational policies. To improve the on-farm irrigation systems should go together with the improvement of the delivery system to allow more flexibility in selecting the appropriate inflow rates and supply time.

Pitts *et al.* (1996) present an interesting analysis of field assessment of irrigation performances. The importance of uniformity in surface irrigation is well evidenced in literature. Sousa *et al.* (1995) have shown that increasing DU from 65 to 83% leads to an yield difference of 8%, while improving DU from 44 to 76% produce an yield increase of 22% for level basin irrigated maize. For both situations the reduction in demand is 160 and 390 mm, respectively. The role of precision-levelling is well analysed by Clemmens *et al.* (1999) in a study relative to Egypt. An example relative to North China shows that one main factor to improve irrigation performances is the adoption of more even surface slopes (Li and Calejo, 1998). 180 mm to 320 mm seasonal water savings could then be achieved. As shown by Fernando *et al.* (1998), the savings resulting from improving irrigation scheduling would be very limited if the farm systems would not be improved.

Surface irrigation systems are, in general, not able to apply small irrigation depths but only large ones. Because of system and delivery scheduling constraints, irrigation scheduling has to be simple. The use of easy irrigation calendars, or the adoption of scheduling charts produced by irrigation scheduling simulation models to take into consideration the actual climatic demand, are in general useful.

3.2.3. Sprinkler Irrigation

Main sprinkler systems are: the set systems, which can apply small to large water depths and are the best adapted for small farms; travelling guns, which are not suitable for applying very small or large depths, nor to irrigate heavy soils, sensitive crops, or under windy climates; and (c) continuous move laterals, which are designed to apply small and frequent irrigations in very large fields.

The irrigation uniformity essentially depends on variables characterising the system that are set at the design phase. System variables include: (a) pressure available at the sprinkler, (b) pressure variation in operating set, (c) sprinkler spacings, (d) sprinkler discharge, (e) wetted diameter, (f) sprinkler water distribution pattern, and (g) provisions to cope with wind speed. Maintenance is the main management variable influencing DU. The efficiency depends upon the same system variables as DU and on (h) soil infiltration characteristics and (i) application rate of the sprinkler. The main management variables concerning E_a are the duration and the frequency of the irrigation events in relation to the soil water deficit at time of irrigation.

More frequent problems in operating sprinkler systems concern inadequate operating pressure and excessive sprinkler spacings. Irrigators are often not in control of depths applied and low uniformities occur quite frequently (Pitts *et al.*, 1996). Based on field evaluations, Mantovani *et al.* (1995) show that, when the price of water is low, farmers optimise yields do not taking care on the water use. For a low DU (40 %) farmers use 2.25 times the required application depth, ratio that decreases to 1.25 when DU is high (85 %). On the contrary, if water is expensive, farmers under-irrigate for low uniformities, so accepting lower than potential yields, and only fully irrigate when systems can achieve high DU. This example shows the importance of achieving high DU when water is scarce. Then it is useful to adopt a target DU for design (Keller and Bliesner, 1990, Seginer, 1987), or to use DU to optimise crop patterns (Tarjuelo *et al.*, 1996).

3.2.4. Microirrigation

Microirrigation mainly include drip, micro-sprinkling, and sub-irrigation systems. Microirrigation uniformity, as for sprinkler DU, mainly depends upon the system variables, which include: (a) pressure at emitters, (b) pressure variation in the operating set, (c) flow regime of the emitter, (d) emitter susceptibility to variations in discharge, (e) emitter coefficient of manufacturing variation, and (f) the filtering capabilities of the system. Since these system variables are fixed at the design stage, the farmer can do very few to achieve high distribution uniformities but good maintenance. The application efficiency, mainly depends upon the same system variables as DU and upon management variables related to the duration of the irrigation and irrigation frequency. Therefore, the farmer may improve the application efficiency when adopting appropriate irrigation schedules but performances are limited by the system constraints.

Results of field evaluations show that irrigation performances are often lower then expected. Pitts *et al.* (1996), referring to the evaluation of 174 micro-irrigation systems in USA, found an average DU = 70%, with seventy-five percent of cases having DU below 85%. The cases with low DU were mainly due to inappropriate filtration and poor selection of emitters manufacturing characteristics. Uniformity in micro-irrigation affects the water saving capabilities of the systems and, mainly, crop yields. An excellent review by Bralts *et al.* (1987) underlines the usefulness of uniformity for design. Santos (1996) shows that the best yield for tomato (near 102 ton/ha) was achieved for 470 mm of applied water when uniformity was 90%, while the maximum yield decreased to 85 ton/ha using 500 mm when uniformity was only 60%. An extensive analysis by Ayars *et al.* (1999) shows the benefits of subsurface drip applied to several crops in maximising yields and reducing water demand relatively to other irrigation methods.

3.3. DEFICIT IRRIGATION

Deficit irrigation, as defined by English *et al.* (1990) and English and Raja (1996), is an optimising strategy under which crops are deliberately allowed to sustain some degree of water deficit and yield reduction having as limit the farmer revenues. The adoption of deficit irrigation requires appropriate knowledge of crop ET, crop responses to water deficits, including the identification of critical crop growth periods, and the economic impacts of yield reduction strategies (Viljoen and de Jager, 1997; Bates *et al.*, 1997). Deficit irrigation can only be properly applied when irrigation systems are designed and managed in such a manner that system performances are high enough to favour water savings (Reca *et al.*, 2001). When water savings have to be practiced in poorly performing systems then under-irrigation is adopted but not controlled deficit irrigation because water is poorly distributed along the irrigation field.

Deficit irrigation implies the adoption of appropriate irrigation schedules, which are built upon validated irrigation scheduling simulation models or based on extensive field trials. Appropriate examples of field trials oriented to the supplemental irrigation of wheat for the Near East are presented by Oweis *et al.* (1998), Zhang *et al.* (1998); and Zhang and Oweis (1999) referring to research performed at ICARDA. When strategies for deficit irrigation are derived from multi-factorial field trials, the optimal irrigation schedules are often based on the concept of *water productivity* (WP), defined by the ratio between the crop yield and the depth of water used by the crop, including both the rainfall and the irrigation applied depths. The WP indicator is useful to identify the best irrigation scheduling strategies for supplemental irrigation of cereals (Zhang and Oweis, 1999), to analyse the water saving performance of irrigation systems (Ayars *et al.*, 1999), and to compare different irrigation systems for deficit irrigation (Howell *et al.*, 1995; Scheneider and Howell, 1995).

The generation of irrigation scheduling strategies for deficit irrigation is commonly performed using simulation models after these have been calibrated or validated for the local conditions (*e.g.* Fernando *et al.*, 1998). These models must include appropriate yield-water functions or crop growth and yield sub-models to evaluate the yield impacts of water deficits (Pereira *et al.*, 1995). Applications of the model ISAREG to several Mediterranean countries to develop and evaluate deficit irrigation strategies are presented in this book.

Results of these applications show that deficit irrigation is relatively easy to be applied when supplemental irrigation of winter cereals is considered (Rodrigues *et al.*, 2002; Zairi *et al.*, 2002; Oweis *et al.*, 2002), *i.e.* when a part of the crop water requirements is satisfied by rainfall. These practices are adopted since long time in the Mediterranean area for wheat and their importance is confirmed by research developed at ICARDA (*e.g.* Oweis *et al.*, 1998). Similarly, in South Africa, supplemental deficit irrigation of maize is also generally feasible (Bates *et al.*, 1997). However, under extreme water scarcity, there is no evidence in literature that severe deficit irrigation could be economically feasible.

For summer crops, on the contrary, deficit irrigation is applicable only when adopting a mild deficit. In other words, when the crop growth and yield mainly depend upon irrigation, with only a very small fraction of the crop requirements being satisfied by natural rain, severe irrigation deficits are not feasible (Rodrigues *et al.*, 2002; Zairi *et al.*, 2002). Under these conditions, the best economical solution under drought is to reduce the cropped area and to apply there full irrigation or just a mild deficit. However, results vary with the prevailing economic conditions - production costs, irrigation costs and marketable yield prices. Results obtained by Zand-Parsa and Sepaskhah (2001) for maize in Iran are different from those for Portugal (Rodrigues *et al.*, 2002). Results for sunflower in Alentejo, Portugal, differ from those for maize in the same location due to differences in commodities prices and subsidies.

For summer crops, deficit irrigation may be feasible when severe stress is avoided. This is the case for cowpea (Zirska *et al.*, 1985) and grain sorghum (Scheneider and Howell, 1995) in USA, and chickpea and lentil in Syria (Zhang *et al.*, 2000). Water productivity is then highly increased but there is no evidence that high deficits could be acceptable, nor that relatively high water savings could lead to favourable economical results. Anyway, these crops are known to be less affected by water stress than maize and tomato, the main summer crops considered in the studies by Rodrigues *et al.* (2002) and Zairi *et al.* (2002).

Perennial tree crops respond differently because under drought the main farmers objective is to provide for the subsistence of the crop until the extreme water scarcity has ended. However, deficit irrigation may be feasible for citrus as referred by Castel and Buj (1990) for a Mediterranean environment. However, both fruit weight and number of fruits are smaller when water deficits are allowed. As reported by Gonzalez-Altozano and Castel (2000) citrus yield and quality are affected by large water deficits particularly during certain phenological phases such as during autumn. Positive responses to deficit irrigation are reported *e.g.* by Mitchell *et al.* (1989) for pear trees under conditions where rain contributed significantly to the crop ET, and by Ruiz-Sanchez *et al.* (2000) for apricots when mild deficits are adopted. Results from Alba *et al.* (2002) and Shatanawi *et al.* (2002) follow these trends but accepting larger yield losses than for the most common regulated deficits.

3.4. OTHER REDUCED DEMAND STRATEGIES

To cope with drought and other water scarcity conditions, reduced demand may assume other facets than deficit irrigation and improving irrigation systems that may lead to water savings. These alternative strategies (Pereira, 1989 b; Tarjuelo and de Juan, 1999) include: (a) reducing the irrigated cropped area, (b) changing crop patterns to less water demanding crops, (c) adopting drought oriented cropping practices, (d) using soil management practices that favour water conservation, and (e) selecting crop varieties less sensible to water stress such as drought resistant varieties. The adoption of these measures and practices require that adequate lead time exists to provide for timeliness in their adoption. Paulo *et al.* (2002) show that using the SPI for drought identification may contribute to improve the drought warning lead time, and Rossi (2002 b) analyses how drought watch systems could be improved to provide timely warning on droughts.

For summer crops, when available water is insufficient to adopt deficit irrigation as referred above, farmers are generally constrained to reduce the irrigated area in proportion to water availability. This reduction depends upon the capability of the irrigation systems to provide for water savings and to avoid wastage of irrigation water. High irrigation system performances are then required. To limit economic losses, irrigation scheduling shall be carefully practised to favour maximisation of yields and incomes, so to reduce the negative impacts of restricting the cropped area. Tree and vine crops are given priority for irrigation aiming at sustain their production capabilities.

When farmers need to reduce the irrigated area, usually they adopt different crop patterns (Viljoen and de Jager, 1997). They may shift to cash crops aiming at maximising the economic productivity of the water, *i.e.* to get the maximal income per unit water applied, or to crops that are less demanding than those usually cropped, *e.g.* replacing maize by sunflower, so that may easily maximise water productivity. Changes in crop patterns are favoured when farm systems are flexible enough to accommodate for different cropping techniques.

In many situations, cropping techniques may be modified to cope with limited water availability. Changes include selecting crop varieties having a short length of the crop season, anticipating planting to better use the soil water and the limited rain (Eastham *et al.*, 1999), or reducing plant density to allow for more soil water available for each plant (Zirska *et al.*, 1985). Generally these practices are selected in alternative to reducing the cropped area and are associated with crops that may take advantage of the limited rainfall. They do not apply to crops cultivated during the dry season when water is highly scarce.

Soil management practices may assume an important role in water conservation, so reducing the crop demand. Several practices may be considered having different objectives such as to control soil evaporation, to increase water infiltration into the soil, to augment the volume of soil exploitable by the plant roots, or to control runoff. Practices may relate to short term such as mulching that favours infiltration and reduces soil evaporation, medium term like conservation tillage that generally provide for augmenting the available soil water, and to long term as terracing and water harvesting. Impacts of these practices in irrigation management are limited but may contribute to successfully achieve production depending upon the prevailing climatic, environmental and economic conditions of the area. A good recent review on these aspects is presented by Unger and Howell (1999).

The adoption of crop varieties less sensible to water stress still is an issue that presents difficulties in the practice. Mechanisms that favour resistance to water stress oppose to those that provide for high yields. Adopting such drought resistant varieties under rainfed or, at the limit under supplemental irrigation is often viable. Adopting low-yield but water stress resistant varieties under irrigation is more difficult unless they could provide for high water productivity and better economically valorise the water applied. This dependency upon the economic factors that also influence the viability of deficit irrigation is both a limit to immediate adoption of such varieties and a positive long term trend to further developments in deficit irrigation when the water productivity of those varieties would compensate for the relative low yield.

4. Conclusions

Management of water under scarcity in irrigated agriculture includes multiple facets. These relate to the water stress regime that is the cause for water scarcity, and to the nature of prevailing problems. To build appropriate irrigation water management policies to cope with water scarcity, appropriate objectives have to be identified aiming at reducing the non-beneficial water uses, particularly those which are consumptive or produce non-reusable return flows.

Reduced demand can be achieved when adopting improved farm irrigation systems and deficit irrigation. Other techniques are also available but generally they are less effective in irrigated agriculture or they imply relatively high reductions in the farm revenue. The improvement of irrigation systems aiming at providing for water savings, controlling water wastes and reducing the irrigation demand in general is closely related with higher irrigation uniformities and application efficiencies. This implies better design, appropriate selection of irrigation equipment, careful maintenance, and extended use of field evaluation to support advice to farmers. When better uniformities are attained, conditions favour the application of deficit irrigation. Otherwise, in poor performing systems, under-irrigation could be practised but crop responses would be highly influenced by the non-uniformity of water application.

Deficit irrigation generally requires the use of simulation and economic models to generate and evaluate alternative irrigation schedules. Results for several Mediterranean countries show that deficit irrigation is generally feasible for supplemental irrigation of cereals but less for summer crops. Results analysed indicate that the feasibility of deficit irrigation is related with the use of rainfall by the crop during the growth season.

Acknowledgements

This research is funded by the research contract ERBIC-18CT-970169 with the European Union, and the Contract Interreg IIC 99.09.5942.1 with the Portuguese Ministry of Agriculture.

References

Alba, I., Rodrigues, P.N., and Pereira, L.S. (2002) Irrigation scheduling simulation for citrus in Sicily to cope with water scarcity. (This issue).

Allen, R.G., Willardson, L.S. and Frederiksen, H.D. (1997) Water use definitions and their use for assessing the impacts of water conservation. In: J.M. de Jager, L.P. Vermes, and R. Ragab (Eds.) *Sustainable Irrigation in Areas of Water Scarcity and Drought* (Proc. ICID Workshop, Oxford), British Nat. Com. ICID, Oxford, pp. 72-81.

Allen, R.G., Pereira, L.S., Raes, D. and Smith, M. (1998) *Crop Evapotranspiration. Guidelines for Computing Crop Water Requirements.* FAO Irrig. Drain. Pap. 56, FAO, Rome, 300p.

Ayars, J.E., Phene, C.J., Hutmacher, R.B., Davis, K.R., Schoneman, R.A., Vail, S.S. and Mead, R.M. (1999) Subsurface drip irrigation of row crops: a review of 15 years research at the Water Management Research Laboratory. *Agric. Water Manag.* 42: 1-27.

Bates, J.H.F., Grové, B. and Oosthuizen, I.K. (1997) An economic analysis of irrigation scheduling with limited water supplies. In: J.M. de Jager, L.P. Vermes, and R. Ragab (Eds.) *Sustainable Irrigation in Areas of Water Scarcity and Drought* (Proc. ICID Workshop, Oxford), British Nat. Com. ICID, Oxford, pp. 6-13.

Bralts, V.F., Edwards, D.M. and Wu, I-Pai (1987) Drip irrigation design and evaluation based on the statistical uniformity concept. In: D. Hillel (Ed.) *Advances in Irrigation* (vol. 4), Academic Press, Orlando, pp. 67-117.

Bucks, D.A., Sammis, T.W. and Dickey, J.L. (1990) Irrigation for arid areas. In: G.J. Hoffman, T.A. Howell, and K.H. Solomon (Eds.) *Management of Farm Irrigation Systems.* ASAE, St. Joseph, MI, pp. 499-548.

Burt, C.M., Clemmens, A.J., Strelkoff, T.S., Solomon, K.H., Bliesner, R.D., Hardy, L.A., Howell, T.A. and Eisenhauer, D.E. (1997) Irrigation performance measures: efficiency and uniformity. *J. Irrig. Drain. Engng.* 123: 423-442.

Castel, J.R. and Buj, A. (1990) Response of *Salustiana* oranges to high frequency deficit irrigation. *Irrig. Sci.* 11: 121-127.

Clemmens, A.J., El Haddad, Z. and Strelkoff, T.S. (1999) Assessing the potential for modern surface irrigation in Egypt. *Transactions of the ASAE.* 42(4): 995-1008.

Eastham, J., Gregory, P.J., Williamson, D.R. and Watson, G.D. (1999) The influence of early sowing of wheat and lupin crops on evapotranspiration and evaporation from the soil surface in a Mediterranean climate. *Agric. Water Manag.* 42: 205-218.

English, M.J. and Raja, S.N. (1996) Perspectives on deficit irrigation. *Agric. Water Manag.* 32: 71-83.

English, M.J., Musick, J.T. and Murty, V.V.N. (1990) Deficit irrigation. In: G.J. Hoffman, T.A. Howell and K.H. Solomon (Eds.) *Management of Farm Irrigation Systems.* ASAE, St. Joseph, MI, pp. 631-663.

Fernando, R.M., Pereira, L.S., Liu, Y., Li, Y. and Cai, L. (1998) Reduced demand irrigation scheduling under constraint of the irrigation method. In: L.S. Pereira and J.W. Gowing (Eds.) *Water and the Environment: Innovation Issues in Irrigation and Drainage*, E &FN Spon, London, pp. 407-414.

Gonzalez-Altozano, P., and Castel, J.R., (2000) Effects of regulated deficit irrigation on "Clementina de Nules" citrus trees growth, yield and fruit quality. In: M.I. Ferreira and H.G. Jones (Eds.) *Proceedings of the 3rd International Symposium on Irrigation of Horticultural Crops.* ISHS, Acta Horticulturae Nr. 537, Leuven, vol. II, pp. 749-758.

Howell, T.A., Yazar, A., Scheneider, A.D., Dusek, D.A. and Copeland, K.S. (1995) Yield and water use efficiency of corn in response to LEPA irrigation. *Transactions of the ASAE* 38(6): 1737-1747.

Keller, J. and Bliesner, R.D. (1990) *Sprinkler and Trickle Irrigation.* Van Nostrand Reinhold, New York.

Li, Y. and Calejo, M.J. (1998) Surface irrigation. In: L.S. Pereira, A. Musy, R.J. Liang, M. Hann, (Eds.) *Water and Soil Management for Sustainable Agriculture in the North China Plain.* ISA, Lisboa: 167-235.

Mantovani, E.C., Villalobos, F.J., Orgaz, F. and Fereres, E. (1995) Modelling the effects of sprinkler irrigation uniformity on crop yield. *Agric. Water Manag.* 27: 243-257.

Martin, D.L., Stegman, E.C. and Fereres, E. (1990) Irrigation scheduling principles. *In*: G.J. Hoffman, T.A. Howell and K.H. Solomon (Eds.) *Management of Farm Irrigation Systems*. ASAE, St. Joseph, MI, pp. 155–203.

Mitchell, P.D., van der Ende, B., Jerie, P.H., and Chalmers, D.J., (1989) Responses of "barlett" pear to withholding irrigation, regulated deficit irrigation and tree spacing. *J. Amer. Soc. Hort. Sci.* 114(1): 15-19.

Oweis, T., Pala, M. and Ryan, J. (1998) Stabilizing rainfed wheat yields with supplemental irrigation and nitrogen in a Mediterranean climate. *Agronomy Journal* 90(5):672-681.

Oweis, T., Hachum, A. and Kijne, J. (1999) *Water harvesting and supplemental irrigation for improved water use efficiency in dry areas*. SWIM Paper 7, ICARDA and IWMI, Colombo.

Oweis, T., Rodrigues, P.N. and Pereira, L.S. (2002) Simulation of supplemental irrigation strategies for wheat in Near East to cope with water scarcity. (This issue).

Paulo, A.A., Pereira, L.S. and Matias, P.G. (2002) Analysis of local and regional droughts in southern Portugal using the theory of runs and the standardised precipitation index. (This issue).

Pereira, L.S. (1989 a) Mitigation of droughts. 1. Agricultural. *ICID Bulletin*, 38: 1-15.

Pereira, L.S. (1989 b) Mitigation of droughts. 2. Irrigation. *ICID Bulletin*, 38: 16-34.

Pereira, L.S. (1996). Inter-relationships between irrigation scheduling methods and the on-farm irrigation systems. In: M. Smith *et al.* (Eds.) *Irrigation Scheduling: From Theory to Practice*. FAO Water Reports 8, ICID and FAO, Rome, pp. 91-104.

Pereira, L.S. (1999) Higher performances through combined improvements in irrigation methods and scheduling: a discussion. *Agric. Water Manage.* 40 (2): 153-169.

Pereira, L.S. and Trout, T.J. (1999) Irrigation Methods. In: H.N. van Lier, L.S. Pereira, and F.R. Steiner (Eds.) *CIGR Handbook of Agricultural Engineering, vol. I: Land and Water Engineering*, ASAE, St. Joseph, MI, pp. 297-379.

Pereira, L.S., van den Broek, B.J., Kabat, P. and Allen, R.G. (Eds.) (1995) *Crop-Water Simulation Models in Practice*. Wageningen Press, Wageningen.

Pitts, D., Peterson, K., Gilbert, G. and Fastenau, R. (1996) Field assessment of irrigation system performance. *Applied Engineering in Agriculture*. 12 (3): 307-313.

Reca, J., Roldán, J., Alcaide, M., López, R. and Camacho, E. (2001) Optimization model for water allocation in deficit irrigation systems. II. Aplication to the Bémbezar irrigation system. *Agric. Water Manag.* 48: 267-268.

Rodrigues, P.N, Machado, T., Pereira, L.S., Teixeira, J.L, El Amami, H. and Zairi, A. (2002) Feasibility of deficit irrigation with center-pivot to cope with limited water supply in Alentejo, Portugal. (This issue).

Rossi, G. (2002 a) An integrated approach to drought mitigation in Mediterranean regions. (This issue).

Rossi, G. (2002 b) Requisites for a drought watch system. (This issue).

Ruiz-Sanchez, M.C., Torrecillas, A., Perez-Pastor, A., and Domingo, R., (2000) Regulated deficit irrigation in Apricot trees. In: M.I. Ferreira and H.G. Jones (Eds.) *Proceedings of the 3rd International Symposium on Irrigation of Horticultural Crops*. ISHS, Acta Horticulturae Nr. 537, Leuven, vol. II, pp. 759-766.

Santos, F.L. (1996) Quality and maximum profit of industrial tomato as affected by distribution uniformity of drip irrigation system. *Irrig. Drain. Syst.* 10: 281-294.

Scheneider, A.D. and Howell, T.A. (1995) Grain sorghum response to sprinkler irrigation application methods and system capacity. *Transactions of the ASAE* 38(6): 1693-1697.

Seginer, I. (1987) Spatial water distribution in sprinkler irrigation. In: D. Hillel (Ed.) *Advances in Irrigation* (vol. 4), Academic Press, Orlando: 119-168.

Shatanawi, M., Al-Zu'bi, Y. and Al-Jayoussi, O. (2002) Irrigation management dynamics in the Jordan valley under drought conditions. (This issue).

Smith, M., Pereira, L.S. Berengena, J., Itier, B., Goussard, J., Ragab, R., Tollefson, L., and Van Hoffwegen, P. (Eds.) (1996) *Irrigation Scheduling: From Theory to Practice*. FAO Water Report 8, ICID and FAO, Rome.

Sousa, P. L., Dedrick, A. R., Clemmens, A. J. and Pereira L.S. (1995) Effect of furrow elevation differences on level-basin performance. *Transactions of the ASAE* 38 (1): 153-158.

Tarjuelo, J. M. and de Juan, J.A. (1999) Crop water management. In: H.N. van Lier, L.S. Pereira and F.R. Steiner (Eds.) *CIGR Handbook of Agricultural Engineering, vol. I: Land and Water Engineering*, ASAE, St. Joseph, MI, pp. 380-429.

Tarjuelo, J. M., de Juan, J. A., Valiente, M. and Garcia, P. (1996) Model for optimal crop patterns within the farm based on crop water production functions and irrigation uniformity. II. A case study of irrigation scheduling in Albacete, Spain. *Agric. Water Manag.* 31: 145-163.

Unger, P.W., and Howell, T.A., (1999) Agricultural water conservation - A global perspective. In: M.B. Kirkham (Ed.) *Water Use in Crop Production*. Food Production Press, The Haworth Press, New York, pp.1:36.

Viljoen, M.P. and de Jager, J.M. (1997) Optimal production strategies and financial compensation for sustainable irrigation farming during drought on the Vaalharts schema in South Africa. In: J.M. de Jager, L.P. Vermes, and R. Ragab (Eds.) *Sustainable Irrigation in Areas of Water Scarcity and Drought* (Proc. ICID Workshop, Oxford), British Nat. Com. ICID, Oxford, pp. 1-5.

Warrick, A.W. and Yates, S.R. (1987) Crop yield as influenced by irrigation uniformity. In: D. Hillel (Ed.) *Advances in Irrigation* (vol. 4), Academic Press, Orlando, pp. 169-180.

Zairi, A. El Amami H., Slatni A., Pereira L.S., Rodrigues, P.N., and Machado, T. (2002) Coping with drought: evaluation of deficit irrigation strategies for cereals and field horticultural crops in Central Tunisia. (This issue).

Zand-Parsa, Sh. and Sepaskhah, A.R. (2001) Optimal applied water and nitrogen for corn. *Agric. Water Manag.* 52(1):73-85.

Zhang, H. and Oweis, T. (1999) Water-yield relations and optimal irrigation scheduling of wheat in the Mediterranean region. *Agric. Water Manag.* 38: 195-211.

Zhang, H., Oweis, T.Y., Garabet, S. and Pala, M. (1998) Water-use efficiency and transpiration efficiency of wheat under rain-fed conditions and supplemental irrigation in a Mediterranean type environment. *Plant and Soil*, 201: 295-305.

Zhang, H., Pala, M., Oweis, T. and Harris, H. (2000) Water use and water use efficiency of chickpea and lentil in a Mediterranean environment. *Aust. J. Agric. Res.* 51: 295-304.

Zirska, L.H., Hall, A.E. and Hoover, R.M. (1985) Irrigation management methods for reducing water use of cowpea (*Vigna unguiculata* [L.] Walp.) and lima bean (*Phaseolus lunatus* L.) while maintaininig seed yields at maximum levels. *Irrig. Sci.* 6: 223-239.

Part 2

Drought identification and monitoring in Mediterranean regions

AT-SITE AND REGIONAL DROUGHT IDENTIFICATION BY REDIM MODEL

G. ROSSI and A. CANCELLIERE
Department of Civil and Environmental Engineering, University of Catania, Italy

Abstract: Identifying and characterising droughts is of major interest for assessing their impacts and for defining appropriate mitigation measures able to reduce their worst effects. Among the several methodologies proposed for analysing droughts, the run method is one of the most suitable, especially when a regional droughts have to be studied. In the present paper, the software package REDIM for at-site and regional drought analysis is presented. The package is a component of a more general Decision Support System to aid decision makers in the water resources management under drought conditions.

Key words: drought identification, run method, regional drought, stationarity

1. Introduction

Drought definition requires the preliminary identification of the component of the hydrological cycle of primary interest for the analysis and of the type of human activity affected by water deficit. As a consequence, different types of drought have been extensively investigated with reference to precipitation deficit from conventionally normal conditions (meteorological drought), to streamflow and/or groundwater flow deficit (hydrological drought), to soil moisture deficit (agricultural drought).

Human actions on the natural hydrological cycle (e.g. building of dams for streamflow regulation by reservoirs, long-distance water transfers, etc.) can shift in time and space the drought impact, complicating the task of a drought identification oriented to assess the water shortages for the different categories of uses, including the protection of water ecosystems.

Since the sixties, several drought indicators have been proposed generally based on a specific definition of drought. For example Foley (1957) and Herbst *et al.* (1966) used precipitation data to characterise meteorological drought, Palmer (1965) included also evapotranspiration, runoff and soil moisture to define agricultural drought while Russel *et al.* (1971) applied a balance between available water resources and municipal demands to characterise urban water supply shortage.

Yevievich (1967) proposed to apply the run method to a stationary time series of a selected hydrological variable with the aim of obtaining an objective identification of drought. An extension of run method allows to evaluate regional drought by including a threshold for deficit-affected areas (Tase, 1977; Rossi, 1980; Santos, 1983). More recently, new indices have been proposed with the purpose of taking into account water storage in the form of snow-pack (Surface Water Supply Index, by Shafer and Deszman, 1981 used in Colorado, USA) or to consider the different time-scales at which the drought phenomenon affects the different categories of users (Standard Precipitation Index, SPI, by McKee *et al.*, 1993). The current trend is to use several

G. Rossi et al. (eds.), Tools for Drought Mitigation in Mediterranean Regions, 37-54.

different indices with the objective to provide information to the general public about drought onset and evolution within the framework of Drought Monitoring Systems developed at national or multi-national basis.

In the present paper the investigations carried out within the EU Project "A DSS for mitigation of drought impacts in the Mediterranean regions" regarding the implementation of drought identification and characterisation methodologies are presented and a description of the developed software package is reported.

2. Drought identification criteria and indices

A general classification of the methods for drought identification stems from the answers to the following basic questions:

a) Which component(s) of the hydrologic cycle is (are) considered?
b) Which time-scale of the hydrological time series is chosen for drought identification?
c) Which approach is adopted to define onset and termination of a drought period?
d) Which method is used to assess the areal coverage of drought over a region?

Regarding the hydrologic component most of the proposed methods are based on the analysis of a single component, such as only precipitation, streamflow, etc. Other methods consider several components as is the case of the Palmer index where a water balance is carried out including precipitation, evapotranspiration, runoff and soil moisture.

Regarding time-scale a distinction can be made between methods based on a single time-interval, such as month, trimester, calendar or water year, and methods that cumulate the variable values of consecutive months, as is the case of SPI index, in order to assess the different features of the drought evolution of interest for different water uses.

Identification of drought onset and termination can be based on the value that the variable under analysis takes in a single interval without memory of the past, e.g. run method (Yevjevich, 1967) or by taking explicitly into account the memory of the past intervals (e.g. Herbst *et al.* method, 1966; Palmer, 1965; SPI). Some other methods identify drought end based on the recovery concept, i.e. that the drought finishes only when the surplus exceeds a significant part of the previous deficit (Correja *et al.*, 1987).

Drought is intrinsically a spatially distributed phenomenon and as such, characterisation of its areal features represents an important step for the assessment of its severity and effects (Rossi *et al.*, 1992). Most of the methods do not take explicitly into account the spatial characterisation of droughts, which is usually carried out at a second stage, by spatially interpolating drought characteristics previously computed at a given number of sites. Such interpolated values can then be represented graphically for example by isolines maps. A few methods however attempt to incorporate spatial information within drought characteristic, for example by deriving the relationship between a fixed value of the drought index and the percentage of the area of the region interested by the chosen value (Rossi, 1980), or by computing regional drought characteristics expressing areal coverage and areal weighted deficit as in the extended run method (Yevjevich, 1967, Tase, 1976)

Among the several indices proposed for describing droughts, a few have found generalised application. One of the simplest drought indices is the percent of normal precipitation related to a month, a season or a year. It is computed by dividing the actual precipitation by the long-term mean (typically 30-40 years), and expressing it as a percentage. Such an index has been computed by Italian Hydrographic Service since its foundation (1919), publishing the isolines of the precipitation for current year expressed as a percentage of the mean of previous values.

Palmer Drought Severity Index (PDSI) or its slightly modified version (Palmer Hydrological Drought Index or PHDI) are very popular indices in the United States for description of time and space variability of drought events, particularly oriented to drought monitoring and agricultural planning. They are based on a monthly water balance considering a simple two-layer soil and express moisture conditions in a standardised form on a scale ranging from -6 to 6. Droughts are then classified according to the value taken by the index.

The SPI index is based on the consideration that each component of water resources systems react to a deficit in precipitation over different time scales (McKee *et al.*, 1993). For example soil moisture is affected by precipitation anomalies in very short time, while streamflow, stored volumes in reservoirs and groundwater are affected by deficit over a large time scale. So the index is computed on cumulated values of precipitation over various time periods (k=3, 6, 9, 12, 24, 48 months). In a first stage a probability distribution (generally a gamma function) is fitted to the frequency curve of cumulated precipitation over k months for each individual moth (e.g. January, February, etc.); then the value of the standardised normal variable corresponding to the computed probability for actual cumulated precipitation is adopted as SPI. Droughts are classified according to the values of the index, which typically ranges between –2 and 2.

The above methods, despite their usefulness for drought monitoring, are not well suited when the purpose is to identify and characterise historical droughts either at a site or at a regional scale. These tasks however are of primary interest when an assessment of drought impacts is needed, especially in order to select best mitigation alternatives. The run method on the other hand allows an objective at site and regional drought identification and characterisation and therefore it represents an ideal candidate methodology for an analysis oriented to define best drought mitigation alternatives.

3. Drought identification by run method

3.1. OVERVIEW

Use of run analysis has been proposed as an objective method for identifying drought periods and for evaluating the statistical properties of drought. According to this method a drought period coincides with a "negative run", defined as a consecutive number of intervals where a selected hydrological variable remains below a chosen truncation level or threshold.

Such a threshold can be a fixed value in the case of a non periodic (e.g. annual) time series or a seasonally varying truncation level in the case of a stationary periodic series. The truncation level in each time interval is somewhat arbitrary and it must be selected on the basis of the objective of the study. Usually it is assumed equal to the long-period mean (or median) of the variable of interest, while other possible choices include a fraction of the mean (Clausen and Pearson, 1995), a value corresponding to a given non-exceedence probability (Zelenhasic and Salvai, 1987, and Correia *et al.*, 1987), or a level defined as one standard deviation below the mean (Ben-Zvi, 1987). In any case, the threshold should be chosen in such a way to be considered representative of the water demand level (Yevjevich, 1983, Rossi *et al.*, 1992).

The advantage of using run method for drought definition consists in the possibility of deriving the drought characteristics (such as duration, cumulative deficit) analytically or by data generation once the stochastic properties of the basic variable are known. This possibility is not limited to relatively simple cases where time dependence of consecutive values can be neglected but also when a Markov chain structure is assumed for the underlying variable (Cancelliere *et al.*, 1998).

Although the above definition of drought can be applied either to continuous or discrete time hydrological series, in practice one usually deals with series aggregated at discrete time steps. Thus, the identification of drought characteristics by run method requires the choice of the aggregation time scale to be used in the analysis. In the theoretical discussion of Dracup *et al.*, 1980, three main considerations for selecting a suitable discrete time interval are presented.

First the selection should depend almost entirely on the purpose for which the study is intended. For example, short time scales (e.g. day or 7-days interval) could be unsuitable for the identification of long drought periods, as occasional values above the threshold might cause an interruption of the drought. Zelenhasic and Salvai, (1987), tried to overcome the interruptions resulting from using daily time intervals in streamflow series by ignoring the minor deficits and the substantial surpluses occurring in a short period of time. Rodrigues *et al.*, (1993), suggest the introduction of the recovery time concept in which a certain surplus should be accumulated to recover from drought conditions. An analysis of drought identification at different time scales is carried out in Cancelliere *et al.*, (1995) and a methodology, based on the use of moving averages, is proposed to overcome some problems occurring when a short time scale is adopted. Also Tallaksen *et al.* (1997) proposed to use a moving average procedure and a method based on the sequent peak algorithm for pooling depended droughts and reducing the number of minor droughts. However it should be pointed out that all these methods are rather arbitrary and lack objectivity.

The second consideration by Dracup *et al.*, (1980) deals with the sample size adequacy of the historical drought record in order to obtain reliable drought analysis. This may contradict the first consideration since a long time scale could reduce sensibly the number of drought events available for analysis. Dracup *et al.*, 1980, suggest a tradeoff between these two considerations.

As a third consideration, short intervals generally tend to exhibit serial correlation that causes serious problems in carrying out frequency analyses, since a record of independent events is usually assumed for these types of analyses. From the previous discussion and from the work of Rodrigues *et al.* (1993), it can be inferred that a seasonal time scale of 2-4 months is the most suitable. In addition, seasonal time scales can be easily related to agricultural activities and management plans, since these are usually carried out on a seasonal basis. Drought identification at a seasonal time scale is usually carried out by assuming seasonally varying truncation levels.

The method of run can be easily extended to the analysis of regional droughts, by considering also an area threshold, i.e. a value of the area affected by deficit above which a regional drought occurs.

3.2. PRELIMINARY ANALYSES

Since the method of run is based on the differences between the observed hydrological variable and a constant (or periodic) threshold, it follows that results can be affected adversely by the presence of non stationarities in the series, such as trends. Therefore, a correct analysis must be preceded by testing the series for the presence of trends and for randomness. Here, three tests are considered, namely the t-Student test for linear trends, the Kendall τ-test and the turning point test for randomness.

Testing for the presence of a linear trend can be carried out by fitting a linear regression to the available data, using the time as independent variable. A slope coefficient significantly different than zero will indicate the presence of a linear trend. The *t-Student test* allows to verify statistically the difference from zero of the slope coefficient, by testing whether the correlation coefficient is significantly different than zero. In particular, under the null hypothesis that the population correlation coefficient ρ = 0, the test statistic t_c is computed as:

$$t_c = \frac{r \cdot \sqrt{n-2}}{\sqrt{1-r^2}} \tag{1}$$

where r is the sample correlation coefficient and n the sample size. The null hypothesis is rejected on the basis of a two-tailed test at significance level α if $|t_c| \geq t_{\alpha/2}(n-2)$

where $t_{\alpha/2}(n-2)$ is the value drawn from a Student's t distribution with *n-2* degrees of freedom and exceedence probability $\alpha/2$

The *Kendall* τ*-test* (also termed rank correlation test) is applied by computing for a series x_t ($t=1, 2, .., n$) the number p of pairs of observations (x_t,x_j) (with $j>t$) for which $x_j>x_t$ (Kendall and Stuart, 1973; Kottegoda, 1980). Under the null hypothesis of random series, the statistic:

$$z_c = \frac{\tau}{\sqrt{\operatorname{var}\tau}} \tag{2}$$

$$\text{where: } \tau = \frac{4p}{n(n-1)} - 1 \tag{3}$$

$$var\,\tau = \frac{2(2n+5)}{9n(n-1)} \tag{4}$$

converges rapidly to a standard normal distribution as n increases. Therefore, the null hypothesis is rejected with a significance level α if $|z_c|>z_{t_{\alpha/2}}$, where $z_{t_{\alpha/2}}$ is a standard normal variable corresponding to an exceedence probability $(\alpha/2)$.

The turning point test is based on counting in a series x_t $(t=1,2 \ldots.n)$ the number of turning points m , i.e. for each $t=2,3$ ..., $n-1$, the values for which x_t is larger than x_{t-1} and x_{t+1} or x_t is smaller than x_{t-1} and x_{t+1} (Kottegoda, 1980).

Under the null hypothesis the series is assumed random and it can be shown that the statistic:

$$z_c = \frac{[m - E(m)]}{\sqrt{\operatorname{var}(m)}} \tag{5}$$

$$\text{with: } E(m) = \frac{2 \cdot (n-2)}{3} \tag{6}$$

$$var(m) = \frac{(16 \cdot n - 29)}{90} \tag{7}$$

converges to a standard normal variable. Then, the null hypothesis is rejected with a significance level α if $|z_c|>z_{t_{\alpha/2}}$, where $z_{t_{\alpha/2}}$ is a standard normal variable corresponding to an exceedence probability $(\alpha/2)$.

3.3. DROUGHT ANALYSIS AT A SITE

Given a discrete time series $h(i)$, $i=1, 2, \ldots\, n$ and a truncation level h_o, it is possible to identify positive and negative deviations according to the sign of the difference between the observed values of the variable and the fixed threshold.

Then, a run is defined as a sequence of intervals characterised by deviations with the same sign preceded and followed by at least one interval with deviation of the opposite sign. Negative deviations are usually referred as deficit, whereas the corresponding run is termed as negative run or drought.

By analysing a given drought *s*, it is possible to identify the following characteristics (Figure 1):

- duration *L(s)*, defined as the number of consecutive intervals where the variable remains below the threshold;
- cumulated deficit *D(s)*, defined as the sum of the negative deviations, extended to the whole drought duration;
- intensity of drought *ID(s)*, defined as the ratio between cumulated deficit and duration.

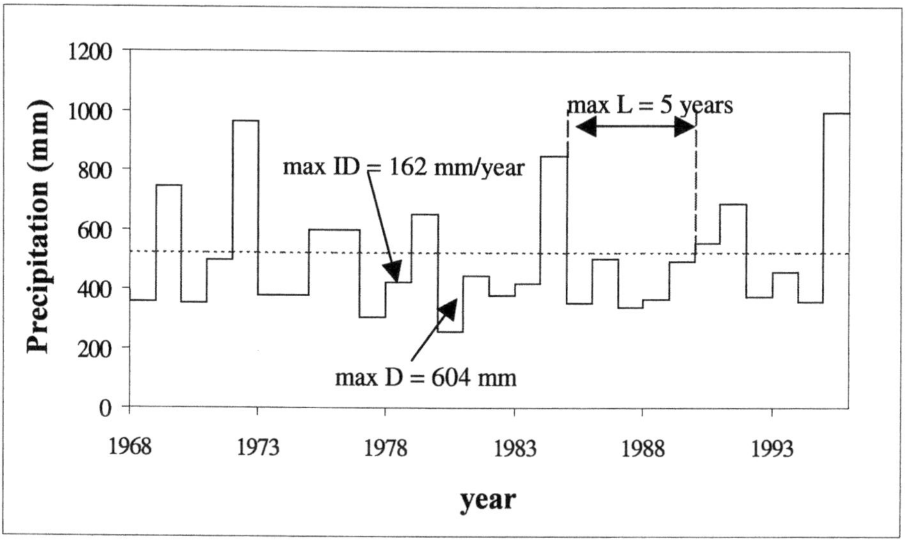

Fig. 1. Droughts identified on hydrological series and their characteristics

More precisely, by letting *d(i)* be the deficit in the interval *i*:

$$d(i) = [h_0 - h(i)]\, I[h(i)]) \quad \text{where} \quad I[h(i)] = 1 \quad \text{if} \quad h(i) < h_o,\ 0 \text{ otherwise} \tag{8}$$

The duration of the drought *s* is given by:

$$L(s) = i_f - i_i + 1 \tag{9}$$

where i_f and i_i are such that $d(i) > 0$ for $i_i \leq i \leq i_f$ and $d(i_i - 1) = 0$, $d(i_f + 1) = 0$

The cumulated deficit can be expressed by:

$$D(s) = \sum_{i=i_i}^{i_f} d(i) \qquad (10)$$

and finally the drought intensity is given by:

$$ID(s) = D(s)/L(s) \qquad (11)$$

Once the above characteristics have been computed for each drought, it is possible to characterise the whole series by considering the average, the maximum or the minimum of each characteristic and the total number of droughts.

If the time series is aggregated at a yearly time scale, a constant threshold is usually assumed. It can be given by the mean or median or computed as a function of the mean h_m and of the standard deviation s of the series (e.g. $h_o = h_m - \alpha s$ where α is an empirical coefficient or the value of a random variable $\alpha(0,1)$ corresponding to a fixed probability level). Alternatively the threshold can be computed as a sample quantile of the series corresponding to a frequency level F or to a quantile of the distribution fitted to the series. If the series is aggregated at a time scale shorter than the year, say season or month, it is necessary to take into account the periodicity of the investigated series and to adopt a variable threshold, for the selected time scale.

The main advantages of defining drought in terms of run is the relative objectivity of the identification and characterization process, once the threshold level is defined. Furthermore, it is possible to derive theoretical probability distribution functions of the drought characteristics, on the basis of the stochastic properties of the underlying series (Cancelliere *et al.*, 1998).

3.4. REGIONAL DROUGHT ANALYSIS

The above analysis can be extended to the case of regional droughts, i.e. droughts which affect large regions, by considering several series of the variable of interest and selecting, besides the truncation level at each site, an additional threshold, which represents the value of the area affected by deficit above which a regional drought is considered to occur. Thus, for a given interval i, two indices are computed, representative of the areal extension and amount of the deficits over the investigated region. Figure 2 depicts a scheme of the regional extension of the methodology. In particular, once the threshold levels $h_0(k)$ for each site $k=1,..,K$ are defined, it is possible to identify for each interval i sites which present deficit, i.e. sites for which the difference between the threshold and the observed value of the variable is positive:

$$h_0(k) - h(i,k) \qquad (12)$$

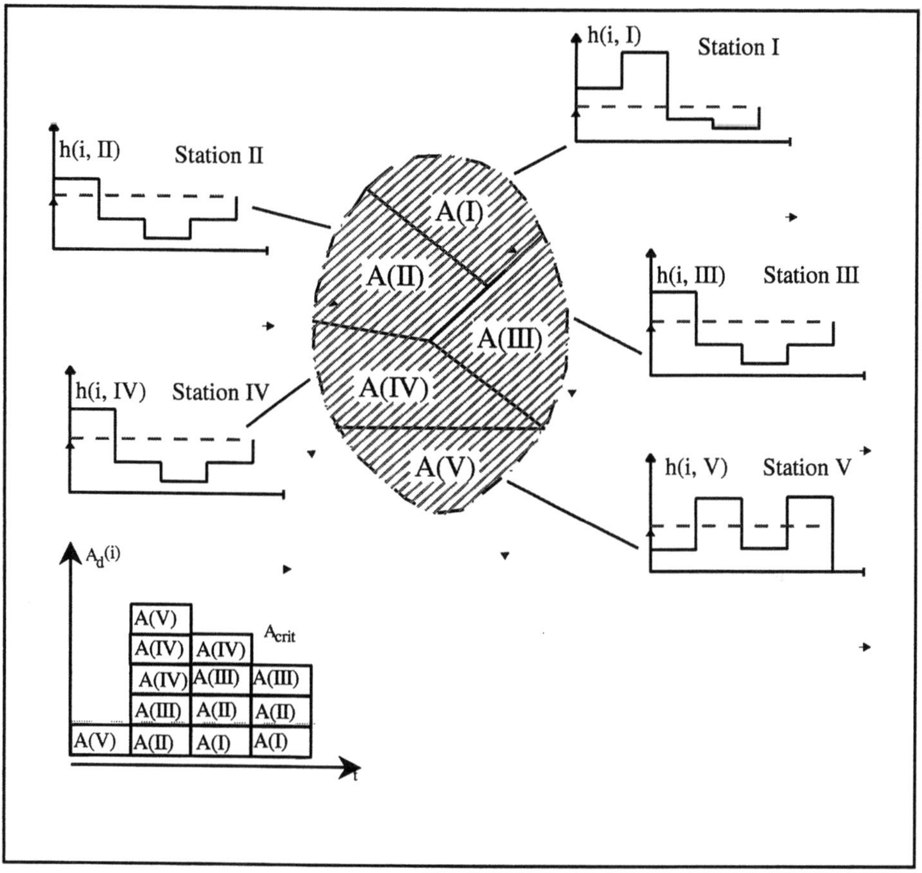

Fig. 2. Regional drought identification

Then, it is assumed that the deficit at each site is extended to an influence area around the observation station, which for example can be estimated by Thiessen polygons method (Figure 2). Such area *S(k)* is usually expressed in terms of the total area under investigation as:

$$A(k)=S(k)/S_{tot} \tag{13}$$

where the total area S_{tot} is obviously:

$$S_{tot}=\sum_{k=1}^{K}S(k) \tag{14}$$

By fixing the area threshold A_{crit}, again expressed as a fraction of the total area, two indices can be computed, namely the areal coverage of deficit $A_d(i)$:

$$A_d(i) = \sum_{k=1}^{K} I[h(i,k)]\, A(k) \tag{15}$$

where:

$$I[h(i,k)] = 1 \quad if \quad h(i,k) < h_0\,(k)$$
$$I[h(i,k)] = 0 \quad if \quad h(i,k) \geq h_0\,(k)$$

and the areal deficit $d(i)$ in the interval i:

$$\begin{aligned} d(i) &= \sum_{k=1}^{K} [h_0(k) - h(i,k)] \cdot I[h(i,k)] A(k) \quad && if \quad A_d(i) \geq A_{crit} \\ d(i) &= 0 && if \quad A_d(i) < A_{crit} \end{aligned} \tag{16}$$

The $A_d(i)$ index is a measure of the area affected by deficit, expressed as a fraction of the total area and it ranges between 0 and 1. The second index provides some insight on the total amount of the deficit in the area, being basically a sum of the deficits at each sites, weighted by the corresponding influence areas.

For each drought r, regional drought duration is defined as:

$$L(r) = i_f(r) - i_i(r)+1 \tag{17}$$

where i_f and i_i are such that $d(i) > 0$ for $i_i(r) \leq i \leq i_f(r)$ and $d(i_i(r) - 1) = 0$, $d(i_f(r) + 1) = 0$

The cumulated areal deficit is computed as:

$$D(r) = \sum_{i=i_i}^{i_f} d(i) \tag{18}$$

while the regional drought intensity is given by:

$$ID(r) = D(r)/L(r) \tag{19}$$

Finally, the mean areal coverage of drought can be computed as:

$$AD(r) = \sum_{i=i_i}^{i_f} Ad(i) / L(r) \tag{20}$$

As in the case of at site analysis, the above characteristics can be further analysed by taking the average and by considering their minimum and maximum values, in order to characterise the whole period of observations. A flow-chart depicting the regional drought identification and characterisation process is reported in Figure 3.

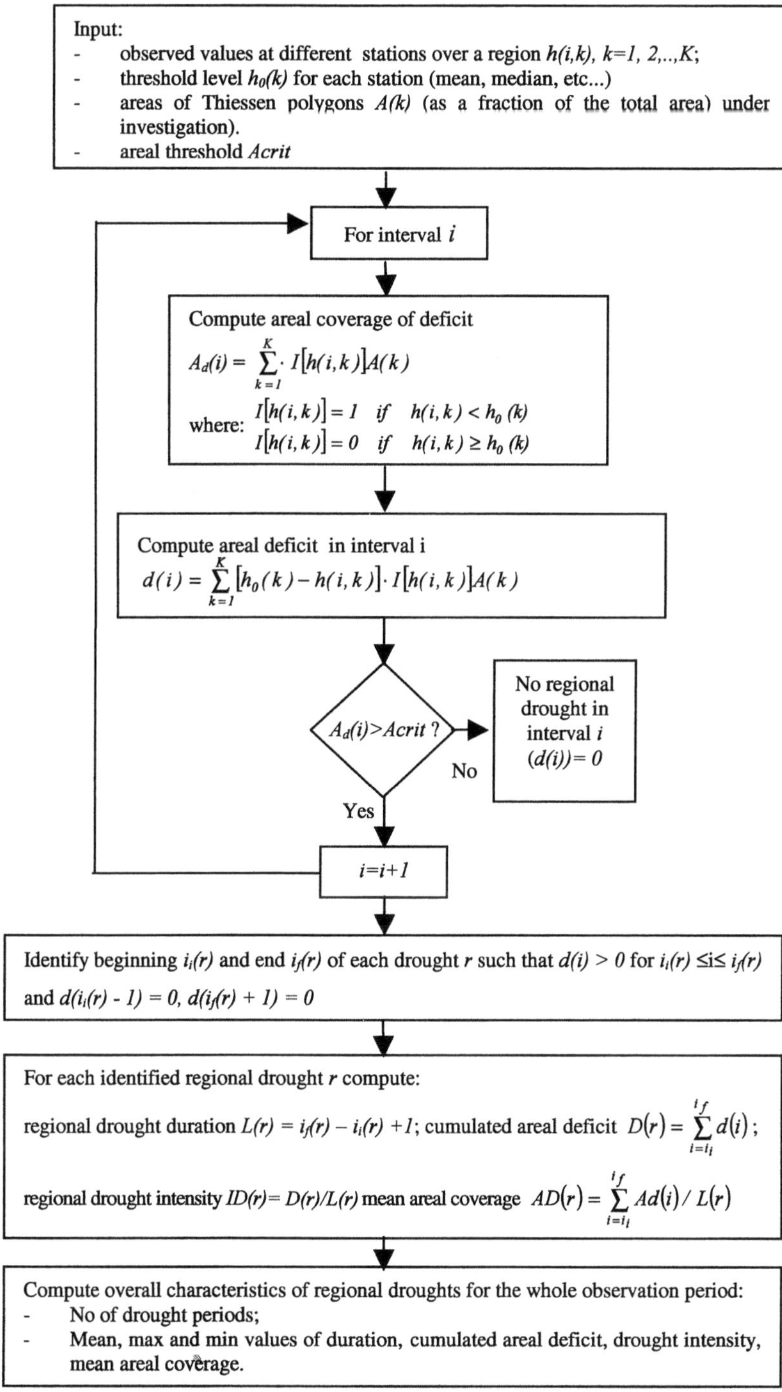

Fig. 3. Regional drought identification and characterization procedure

4. REDIM software package

4.1. OVERVIEW

The package REDIM is an user friendly software which allows to perform drought analysis on hydrological series both at a site and over a region using the method of run. In addition, it allows to test statistically for stationarity in a time series, whose absence would lead to misleading drought analyses. The software, written in Visual Basic, runs under Windows 95/98/NT platform, and is structured as a sequence of dialog boxes which guide the user throughout the analysis of stationarity and drought identification and characterisation processes. The main features of Redim can be listed as follows:

- Data files are in MS Access format, which allows flexibility and compatibility with existing databases, as well as Windows based programs such as spreadsheets, word processors, etc.;
- The Data Import function allows to create automatically MS Access database files from existing ASCIII text files, thus avoiding tedious database creation processes;
- Different aggregation time scales can be used (monthly, three monthly and yearly), with the possibility to select the initial month of aggregation in order to take into account water years;
- Testing for stationarity in hydrological series is carried out by means of three statistical tests: Student's t-test for linear trend, Kendall τ-test or rank correlation test and turning point test;
- Testing for stationarity at trimester time scale is performed separately for each of the four trimesters, in order to identify different behaviours (e.g. increasing or decreasing trends) in the different seasons;
- Identification and characterisation of drought is performed by means of run method, and different threshold levels can be selected, namely the mean, the median, the average minus the standard deviation, or sample quantile at a fixed frequency;
- Graphical output of results allows to easily identify droughts on a given time series, and/or region.

Text reports can be produced either in ASCIII form, or MsWord format, to facilitate the merging of the outputs with other existing documents. Reports contain all the details of the performed analysis (e.g. input series, type of analysis, selected thresholds, etc.), tabular results, as well as graphics already imbedded in the document, in order to simplify the preparation of the outputs in final form.

In developing the software, particular attention has been devoted to its easy of use. All information are entered through a sequence of dialog boxes. Figure 4 shows as an example the dialog box that appears at start-up, from which the type of analysis to be carried out and the typology of hydrological data to be analysed are selected. Note that regional drought analysis can be performed only with precipitation data.

In what follows, a brief description of the software is given. For further details, the user is referred to the reference manual (DICA, 2000).

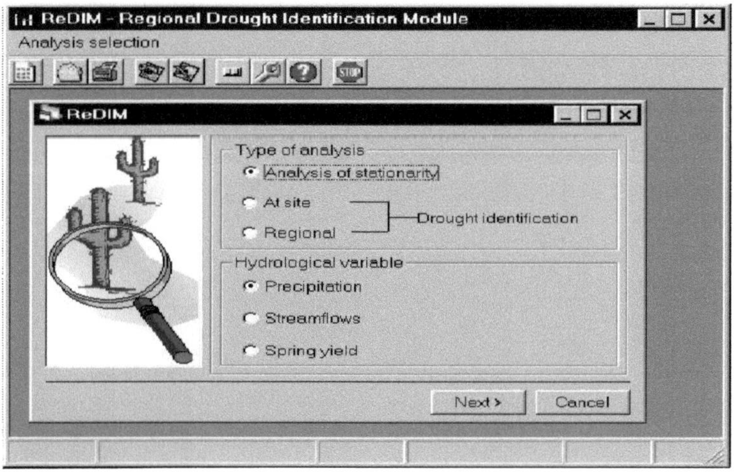

Fig. 4. Main dialog box

If stationarity analysis is chosen, after selecting the time scale and the database file containing the data, the dialog box depicted in Figure 5 appears, with the results of the three tests, namely linear trend, turning point and Kendall's τ tests.

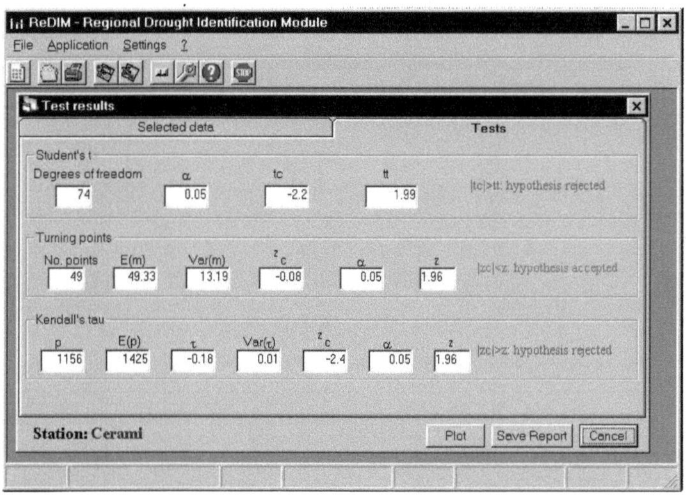

Fig. 5. Tests results dialog box

For each test, the intermediate results, as well as the test outcome are reported. Note that for each test a green label will appear if the hypothesis is accepted, otherwise the label colour will be red.

Drought analysis is performed in a similar fashion, by choosing first the time scale and the data file. Then the threshold must be selected, as shown in Figure 6.

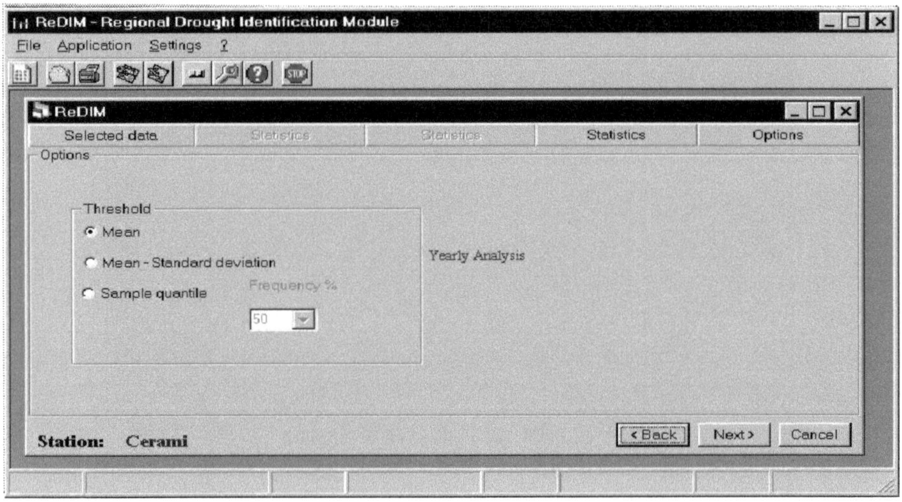

Fig. 6. Drought analysis options dialog box

At site drought analysis results are reported on tabular format (fig. 7), as well as in graphical format.

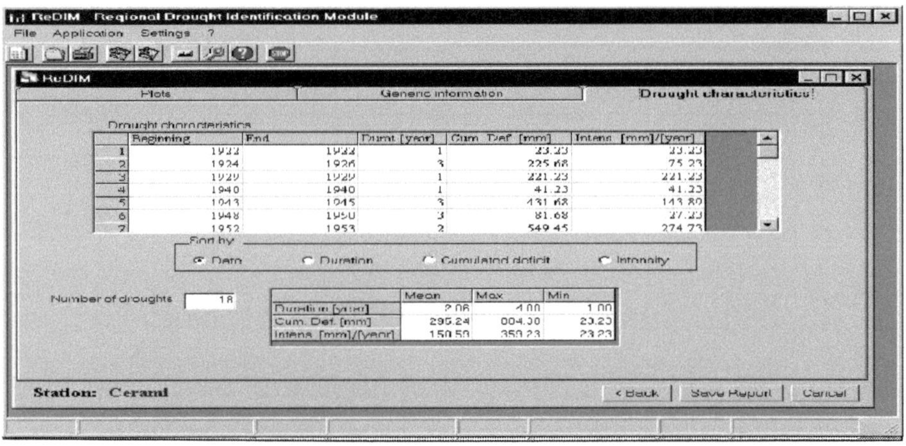

Fig. 7. At site drought analysis results

The initial steps for performing regional drought analysis are basically the same as in the case of at site analysis. After selecting the type of analysis in the main dialog box and the aggregation time scale, the dialog box in fig. 8 appears, from which the user can select the stations to be analysed following the regional drought identification methodology:

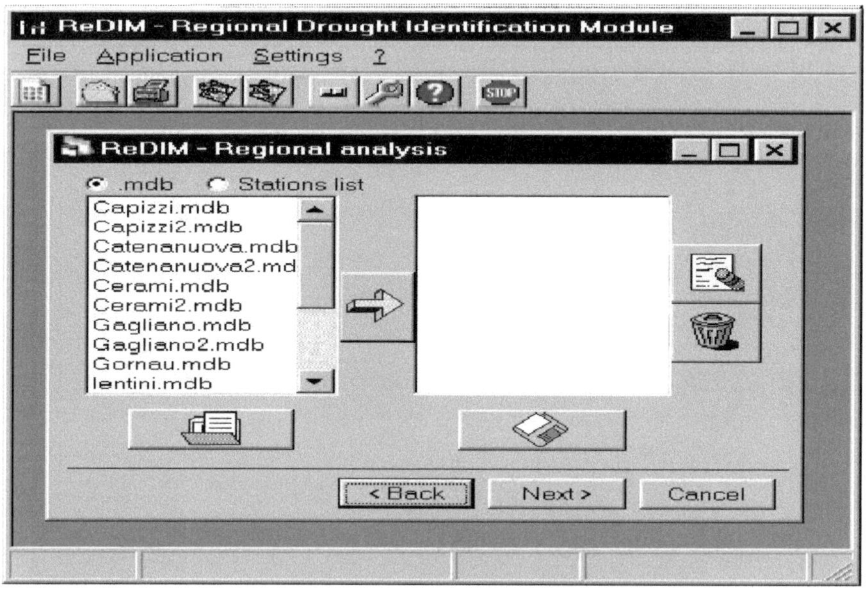

Fig. 8. Selection of precipitation stations for regional drought analysis

The next dialog box prompts for the total area of the region of interest and for the influence areas of each station, expressed as a percentage of the total area. In the case of regional drought analysis the user must provide the areal threshold expressed as a percentage of the total area, besides the deficit threshold, as shown in figure 9.

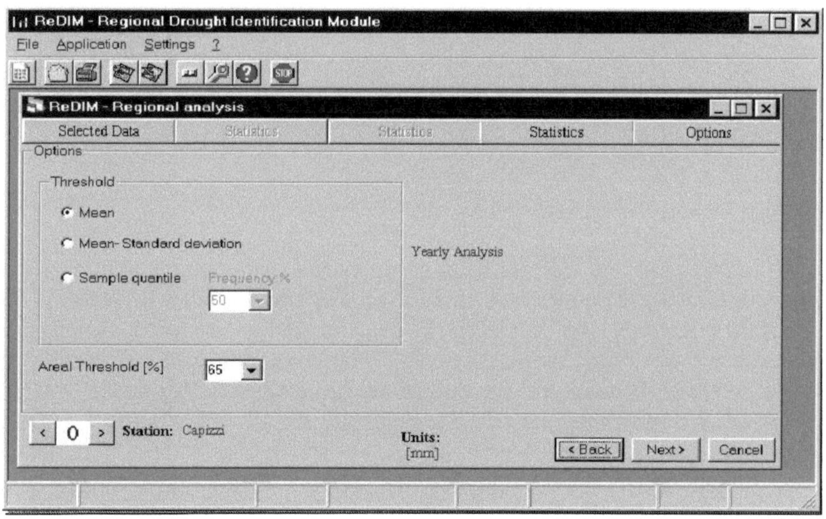

Fig. 9. Regional drought analysis options dialog box

The regional drought analysis results, whose dialog box is depicted in fig. 10, include also the mean areal coverage of drought (eq. 13). The results can also be represented in graphical format, as shown in Figure 11, which depicts the time series of areal deficit (top) with the relative threshold (65%) and the regional deficit (bottom), computed only for intervals where the areal deficit is greater than the threshold. Droughts can be easily identified as a consecutive series of red labels preceded and followed by at least one green label.

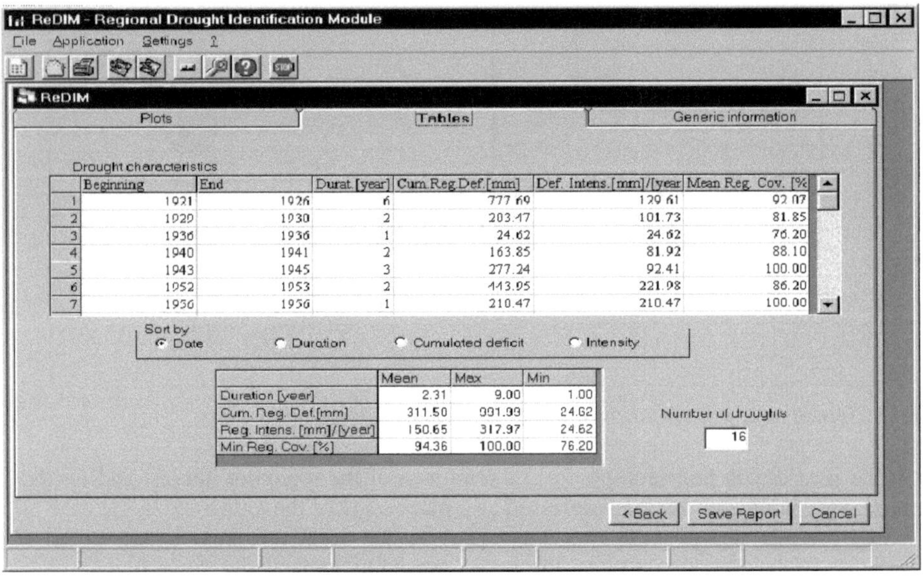

Fig. 10. Regional drought analysis results

For further details concerning REDIM model, the user is referred to the reference manual (DICA, 2000).

5. Conclusions

Drought identification and characterisation represents an important step toward a correct management of water resources. In particular, they provide basic information for assessing drought effects and for selecting the best mitigation strategies. Among the several proposed methodologies, the run method, despite some limitations, is one of the most efficient methods because of its objectivity in identifying droughts and its suitability to assess drought characteristics at a regional scale.

A software implementation of the method, with an user friendly interface has been presented as an useful tool for scientists, water managers and practitioners to better analyse the drought characteristics in a given region and to help identify the most appropriate mitigation alternatives to cope with droughts. The software has been developed with the aim to provide a common methodological ground for drought identification and characterization in different climatic and hydrologic contexts in the Mediterranean region.

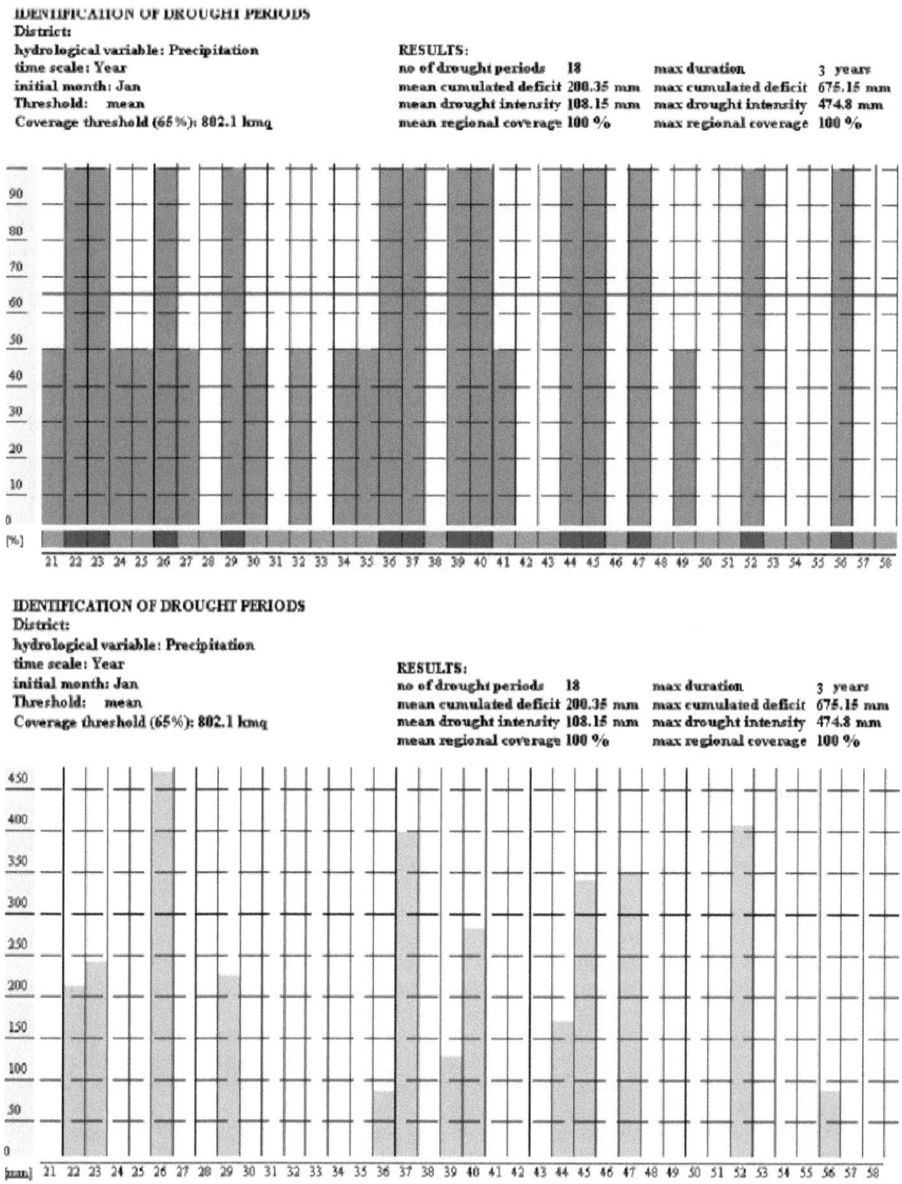

Fig. 11. Example of graphical output of regional drought identification: areal coverage of deficit (above) and cumulated areal deficit (below)

Acknowledgments

The research has been initiated with the support of CNR, Group for Prevention from Hydrogeological Disasters and carried out within the European Community INCO-DC project "A DSS for drought mitigation in Mediterranean regions" (DSS-DROUGHT, IC18-CT97-0169). Special thanks are due to V. Cozzubbo and R. Musumeci for the software implementation of some parts of the model.

References

Ben-Zvi, A. (1987) Indices of hydrological drought in Israel. *J. Hydrol.* 92: 179-191.

Cancelliere, A., Ancarani, A. and Rossi, G. (1995) Identification of drought periods on streamflow series at different time scales. In: N.X. Tsiourtis (Ed.) *Water Resources Management under Drought or Water Shortage Conditions*, Balkema, Rotterdam, pp. 49-57.

Cancelliere, A., Ancarani, A. and Rossi, G. (1998) Distribuzioni di probabilità delle caratteristiche di siccità. *Atti del XXVI Conv. di Idr. e Costr. Idrauliche*, Catania 9-12 settembre, CUECM, Catania, pp. 327-340.

Clausen, B. and Pearson, C.P. (1995) Regional frequency analysis of annual maximum streamflow drought. *J. Hydrol.* 173: 111-130.

Correia, F.N., Santos, M.A. and Rodrigues, R. (1987) Engineering risk in regional drought studies. In: L. Duckstein and E.J. Plate (Eds.) *Engineering Reliability and Risk in Water Resources*, Martinus Nijhoff Publ., Dordrecht, pp.61-86.

DICA (2000) *Regional Drought Identification Module REDIM.* Department of Civil and Environmental Engineering, University of Catania, Catania.

Dracup, J. A., Lee, K. S. and Paulson, E. D. (1980) On the definition of droughts, *Water Resour. Res.* 16(2): 297-302.

Foley, J.C. (1957) Drought in Australia, Bureau of Meteorology, Malbourne, Bulletin, no. 43, 281 p.

Herbst, P. H, Bredekamp, D. B. and Barker, H. M. G. (1966) A technique for the evaluation of drought from rainfall data. *J. of Hydrology* 4(3): 264-272.

Kendall, M., Stuart, A. (1973) *The advanced theory of statistics*, Vol. 2, Charles Griffin, London.

Kottegoda, N. T. (1980) *Stochastic Water Resources Technology*, The Macmillan Press, London.

McKee, T.B., Doesken, N.J. and Kleist, J. (1993) The relationship of drought frequency and duration to time scales. 8th Conference on Applied Climatology, Anaheim, California, pp. 179-184.

Palmer, W. C. (1965) *Meteorological drought.* U.S. Weather Bureau, Research Paper 45, Washington, D.C.

Rodrigues, R., Santos, M. A and Correia, F. N. (1993) Appropriate time resolution for stochastic drought analysis. In: J. D. Salas, R. Harboe, and J. Marco-Segura (Eds.) *Stochastic Hydrology and its Use in Water Resources Systems Simulation and Optimization*, Kluwer, Netherland, pp. 247-265.

Rossi, G. (1980) Analisi della siccità su una regione. In: E. Guggino, G. Rossi and E. Todini (Eds.) *Piene e siccità. Atti del Corso su Eventi idrologici estremi*, Erice, CLUP, Milano, pp. 361-387.

Rossi G., Benedini, M., Tsakiris, G. and Giakoumakis, S. (1992) On regional drought estimation and analysis. *Water Resour. Manage.* 6: 249-277.

Russel, C.S., Arey, D.G. and Kates, R.W. (1971) *Drought and water supply: Implications of the Massachusetts experiences for municipal planning.* John Hopkins Press, Baltimore, Maryland.

Santos, M.A. (1983) Regional droughts: a stochastic characterization. estimation and analysis. *J. Hydrol.* 66: 183-211.

Shafer, B.A. and Dezman, L.E. (1982) Development of a surface water supply index (SWSI) to assess the severity of drought conditions in snowpack runoff areas. *Proceedings of the Western Snow Conference.*

Tallaksen, L.M., Madsen, H. and Clausen, B. (1997) On the definition and modelling of streamflow drought duration and deficit volume. *Hydrol. Sci. J.* 42(1): 15-33.

Tase, N. (1976) Area-deficit-intensity characteristics of droughts. *Hydrology paper* n. 87, Colorado State University, Colorado.

Yevjevich, V. (1967) An objective approach to definitions and investigations of continental hydrologic drought. *Hydrology paper* n. 23, Colorado State University, Colorado.

Yevjevich, V., Da Cunha, L. and Vlachos, E. (Eds.) (1983) *Coping with Droughts.* Water Resources Publications, Littleton, Colorado.

Zelenhasic, E. and Salvai, A. (1987) A method of streamflow drought analysis. *Water Resour. Res.* 23(1): 156-168.

ANALYSIS OF LOCAL AND REGIONAL DROUGHTS IN SOUTHERN PORTUGAL USING THE THEORY OF RUNS AND THE STANDARDISED PRECIPITATION INDEX

A. A. PAULO, L. S. PEREIRA, and P. G. MATIAS
Agricultural Engineering Research Center, Institute of Agronomy, Technical University of Lisbon, Portugal

Abstract: The identification, monitoring and characterisation of droughts are of great importance in water resources planning. Mediterranean regions are especially affected by droughts. A study on local and regional analyses of droughts is applied to a southern region of Portugal, Alentejo, aiming at contributing to a better understanding of droughts in that region. The theory of runs and the SPI are applied at a local level in order to identify local droughts. Two time scales were used, the 3-month and the 12-month period. Further, also using the same methods, drought characteristics are derived for the region by identifying the drought periods, computing the regional coverage of drought and the drought magnitude. Results show that both methods are appropriate to characterise local and regional droughts and that the SPI is very helpful in identifying initiation and ending of droughts.

Key words: droughts, regional droughts, theory of runs, Standardised Precipitation Index.

1. Introduction

The identification and characterisation of local and regional droughts has been object of a wide research at international and national levels (Santos, 1983; Santos *et al.*, 1988; Rossi and Cancelliere, 2002). Questions relative to drought definition have been reviewed, among others, by Wilhite and Glantz (1987) and Tate and Gustard (2000), showing that the definitions of droughts are particularly related to the human activities, the ecological environments and the scale of analysis, both in space and time.

Local and regional drought indices have been developed and applied by many authors. Hayes (1999) presents a compilation of several local indices and analyses the respective advantages and disadvantages. Guttman (1998) recently compared the Palmer Drought Index (PDI) with the Standardised Precipitation Index (SPI). Limitations on the use of PDI to Mediterranean regions have been shown by Cancelliere et al. (1996). Stochastic analysis including the theory of runs has been successfully applied in Mediterranean areas for characterisation of local and regional droughts (Rossi et al. 1992; Cancelliere and Rossi, 2002), particularly in Portugal (Santos, 1983; Santos et al., 1988; Henriques and Santos, 1996). However, the SPI has not been applied to the analysis of regional droughts.

The objective of this study is to perform an in-depth analysis of hydrological droughts in Alentejo, Portugal, which could be helpful to support water resources management under drought, particularly for irrigation purposes. This paper also aims to test the SPI

G. Rossi et al. (eds.), Tools for Drought Mitigation in Mediterranean Regions, 55-78.

methodology (McKee et al., 1993; Edwards, 2000) for local and regional drought analysis compared with the well known theory of runs.

2. Drought analysis methodology

2.1. DATA

This study is applied to the region of Alentejo, in Southern Portugal (Figure 1). It is a depressed region, in terms of economic and population resources. The region is characterised by extensive agriculture, mainly cereals (wheat), olive trees, vineyards, Mediterranean forests, mainly cork and green oaks, and natural pastures. Water resources are short and irrigated crops are not widely spread. It is a region frequently affected by droughts and water scarcity.

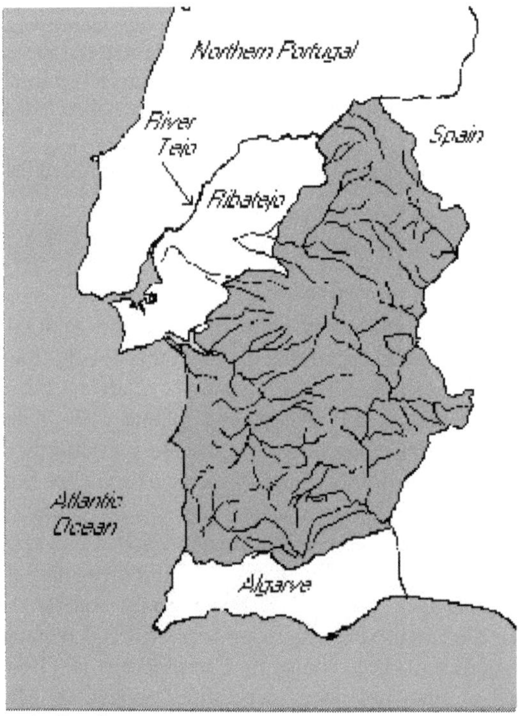

Fig. 1. Location of Alentejo Region

Monthly rainfall data from 58 stations, in the region or near its borders, are available for the period 1931/32 to 1994/95. The study is developed using the hydrological year, from October to September. The location of those rainfall stations is shown in Figure 2.

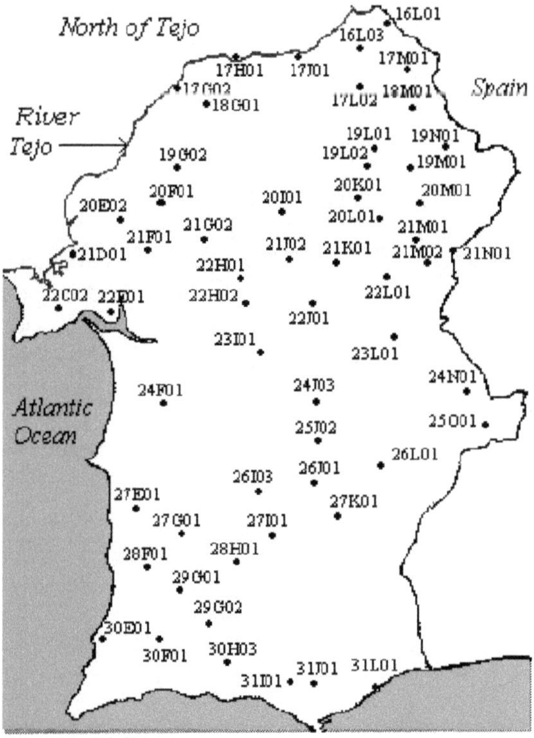

Fig. 2. Location of the rainfall stations

The identification of the rainfall stations selected after the statistical analysis is given in Table I.

Table I. Selected rainfall stations and respective areas of influence

Nr	Station Code	Station name	Area (km2)	Nr	Station Code	Station name	Area (km2)
1	16L03	Nisa	490.1	20	22J01	Evora	1142.0
2	17J01	Gavião	439.0	21	24J03	Cuba	1875.2
3	17L02	Vale do Peso	870.0	22	24N01	Amareleja	1703.5
4	17M01	Castelo de Vide	379.0	23	25J02	Beja	485.1
5	18G01	Chouto	309.1	24	26I03	Aljustrel	1447.7
6	18M01	Portalegre	527.5	25	26J01	Trindade	290.9
7	19G02	Lamarosa	121.6	26	26L01	Serpa	1439.0
8	19L02	Fronteira	885.7	27	27E01	Cer. Alentejo	1654.7
9	19M01	Monforte	1153.8	28	27G01	Reliquias	664.4
10	20I01	Pavia	1184.0	29	27I01	Castro Verde	635.5
11	20L01	Estremoz	495.0	30	27K01	Algodor	1315.1
12	21F01	Canha	74.6	31	28H01	Ald. Palheiros	805.0
13	21G02	Lavre	362.4	32	29G01	Saboia	670.0
14	21J02	Arraiolos	381.3	33	29G02	S. M. da Serra	94.0
15	21K01	Azaruja	634.8	34	30E01	Aljezur	107.5
16	21M01	Vila Viçosa	653.1	35	30F01	Monchique	56.4
17	21M02	Alandroal	897.2	36	30H03	S. B..Messines	218.2
18	22E01	Aguas de Moura	1657.0	37	31J01	S. Bras Alportel	102.1
19	22H01	Mont.-o-Novo	1072.3			Total	27293.8

To evaluate the quality of data, the randomness of the annual data sets was investigated through tests for homogeneity, absence of trends and autocorrelation. A set of non-parametric tests presented by Helsel and Hirsch (1992) were applied. The Kendall's τ autocorrelation test, the Mann-Kendall trend test and the Mann-Whitney homogeneity test were performed, using a computer program developed by Matias (1998).

The Kendall τ-test (also termed rank correlation test) evaluates the monotonic relationship between two variables (Helsel and Hirsch, 1992). Two series $x_{1i}=x_i$ $(i=1,\ldots,n-1)$ and $x_{2j}=x_j$ $(j=2,\ldots,n)$ are built from the initial time series x_t $(t=1,\ldots,n)$. Considering pairs of observations (x_{1k},x_{2k}) and (x_{1w},x_{2w}), with $w>k$, the number p of pairs of concordant observations (when $x_{1w}>x_{1k}$ and $x_{2w}>x_{2k}$), and the number m of pairs of discordant observations (when $x_{1w}>x_{1k}$ and $x_{2w}<x_{2k}$ or $x_{1w}<x_{1k}$ and $x_{2w}>x_{2k}$), are computed. Pairs with ties are not considered concordant or discordant.

Under the null hypothesis of non-existence of first order autocorrelation between the elements of x_t, the statistic z_s, corrected for continuity is defined as

$$z_S = \begin{cases} \dfrac{s-1}{\sigma_S} & ,s>0 \\ 0 & ,s=0 \\ \dfrac{s+1}{\sigma_S} & ,s<0 \end{cases} \tag{1}$$

where:

$$s = p - m \tag{2}$$

and

$$\sigma_s = \sqrt{\frac{(n-1)(n-2)(2n+3)}{18}} \tag{3}$$

When there are no ties σ_s converges rapidly to a standard normal distribution as n increases (n>10). If there are ties Equation 3 must be corrected accordingly. Therefore, the null hypothesis is rejected with a significance level α if $|z_s|>z_{\alpha/2}$, where $z_{\alpha/2}$ is the standard normal variable corresponding to an exceedance probability of $\alpha/2$. The variable of interest for the test is s (Equation 2).

The Mann-Kendall trend test (Helsel and Hirsch, 1992) is applied in the same way as the Kendall τ-test, using a bivariate sample (x_i,T_i) where x_i is the observed value at time T_i. It allows to detect if there is a significant monotonic relationship between x_i and T_i either increasing or decreasing.

The Mann-Whitney homogeneity test (Helsel and Hirsch, 1992) is a non-parametric test applied here to the mean and to the variance. Originally, it was applied in order to detect if two samples have been drawn from the same continuous population. In this specific case two series are built from the initial one, $x_{1i}=x_i$, with $i=1,2,...,n_1$, and $x_{2\ j-n_1}=x_j$, with $j=n_1+1,n_1+2,...,n_1+n_2$ (Conover, 1980; Helsel and Hirsch, 1992). The dimension n_1 is equal to the major integer less or equal to $n/2$, with $n=n_1+n_2$.

For testing the hypothesis of homogeneity of the mean, the value of u is computed by assigning to each observed value a rank. The rank-sum of the n_1 observations is u. Under the null hypothesis of homogeneity of the mean, the statistic z_u is

$$z_u = \begin{cases} \dfrac{u-1/2-\mu_u}{\sigma_u} & , u>\mu_u \\ 0 & , u=\mu \\ \dfrac{u+1/2-\mu_u}{\sigma_u} & , u<\mu_u \end{cases} \tag{4}$$

where:

$$\mu_u = \frac{n_1(n+1)}{2} \tag{5}$$

and

$$\sigma_u = \sqrt{\frac{n_1 n_2 (n+1)}{12}} \tag{6}$$

If there are no tied scores, σ_u converges to a standard normal distribution as n increases (n>20). Therefore, the null hypothesis is rejected with a significance level α if $|z_u|>z_{\alpha/2}$. When testing the homogeneity of variance the absolute differences between observed values and the respective sub-sample means are computed for each sub-sample. Then a rank is assigned to each difference. Let V be the sum of rank squares of the sub-sample with dimension n_1. Then, under the null hypothesis of homogeneity of the variance, the statistic z_v is obtained by

$$z_v = \frac{v-\mu_v}{\sigma_v} \tag{7}$$

where $\mu_v = \dfrac{n_1 n_2 (n+1)(2n+1)}{6}$ (8)

and $\sigma_v = \sqrt{\dfrac{n_1 n_2 (n+1)(2n+1)(8n+11)}{180}}$ (9)

if there are no ties. This statistic is asymptotically normal, with mean 0 and variance 1. The decision rule for this test is the same as in the previous tests.

The probability α_p is computed in all tests. It is the probability of obtaining the computed test statistic, or one even less likely, when the null hypothesis is true. It is given by

$$\alpha_p = 2[1 - P(Z \leq z)] \tag{10}$$

If $\alpha_p < \alpha$ the null hypothesis is rejected with a significant level α.

From the 58 annual time series, 43 were accepted as random, considering a significance level of 0.05. These were the rainfall stations used in further analysis. From these 43 rainfall stations, only 17 have complete records. Thus, missing monthly data for each station were estimated by linear regression using as reference station the one in the region that has a highest correlation coefficient with that to be completed.

The region of Alentejo was divided in Thiessen polygons resulting in 37 influent rainfall stations. The respective areas of influence are shown in Table I. Main descriptive statistics, referring to the 64 hydrological years for those 37 rainfall stations, are presented in Table II.

Table II. Descriptive statistics of annual precipitation (1931/32-1994/95)

Station Code	Minimum	Maximum	Median	Mean	Coefficient of Variation	Skewness Coefficient
16L03	328.6	1189.5	678.9	700.0	0.29	0.24
17J01	409.1	1479.8	808.9	826.6	0.27	0.47
17L02	411.0	1229.4	720.6	764.8	0.27	0.26
17M01	412.3	1776.4	838.4	869.5	0.32	0.62
18G01	394.2	1200.2	752.9	776.1	0.29	0.36
18M01	379.7	1863.0	769.9	851.5	0.30	1.13
19G02	337.0	1148.6	690.1	696.2	0.27	0.00
19L02	251.2	950.8	516.5	549.5	0.31	0.29
19M01	268.0	880.9	524.4	543.7	0.29	0.32
20I01	204.2	993.6	575.9	586.1	0.29	0.22
20L01	356.1	1106.4	601.4	640.9	0.30	0.43
21F01	343.8	1211.4	671.9	694.5	0.28	0.28
21G02	289.7	1064.6	615.7	647.0	0.29	0.33
21J02	237.8	1156.9	637.9	679.3	0.29	0.44
21K01	270.1	990.0	564.4	587.8	0.29	0.53
21M01	332.8	1968.2	687.2	754.2	0.37	1.53
21M02	298.7	944.2	590.6	618.8	0.29	0.15
22E01	310.8	999.8	660.9	667.4	0.28	0.01
22H01	317.4	1107.4	658.7	681.1	0.29	0.36
22J01	347.1	1034.7	602.6	639.7	0.29	0.32
24J03	224.2	932.5	549.1	562.3	0.32	0.00
24N01	242.9	816.9	488.1	530.5	0.29	0.25
25J02	193.5	913.2	530.4	560.1	0.28	0.19
26I03	214.5	1084.1	545.1	540.1	0.33	0.29
26J01	183.1	895.5	469.7	490.9	0.32	0.29
26L01	191.6	1015.5	493.7	521.8	0.33	0.45
27E01	397.8	1183.2	767.4	773.9	0.27	0.09
27G01	297.9	1121.2	672.7	688.4	0.31	0.19
27I01	182.6	889.1	525.9	534.9	0.30	0.21
27K01	200.5	866.1	466.7	503.0	0.32	0.32
28H01	245.1	949.4	589.1	609.9	0.30	0.12
29G01	136.0	1197.5	618.6	642.4	0.34	0.32
29G02	309.6	1313.5	693.4	715.6	0.33	0.47
30E01	174.0	1146.0	567.6	566.8	0.33	0.53
30F01	430.8	2046.1	1224.4	1252.1	0.30	0.25
30H03	248.4	1151.1	668.2	684.9	0.32	0.35
31J01	289.2	1532.1	792.7	835.7	0.31	0.42

2.2. IDENTIFICATION OF LOCAL DROUGHTS

2.2.1. Theory of runs

The theory of runs (TOR) is based on the choice of a critical threshold level, y_c (Guerrero-Salazar and Yevjevich, 1975). Considering a discrete time series, $x_1, x_2, \ldots, x_i, \ldots, x_n$, a negative run occurs when x_i is consecutively less than y_c, during one or more time intervals. Negative runs in rainfall time series are related to drought and the difference between y_c and x_t is referred as a deficit. A run can be determined by its length (L), its cumulated deficit (D) and its intensity (I), as exemplified in Figure 3.

Therefore, any drought, s, can be characterised by its:

- *duration, L(s)*, the number of consecutive time intervals where rainfall remains below the critical level,
- *cumulated deficit, D(s)*, the sum of consecutive deficits, and
- *intensity, I(s)*, given by the ratio *D(s)/L(s)*.

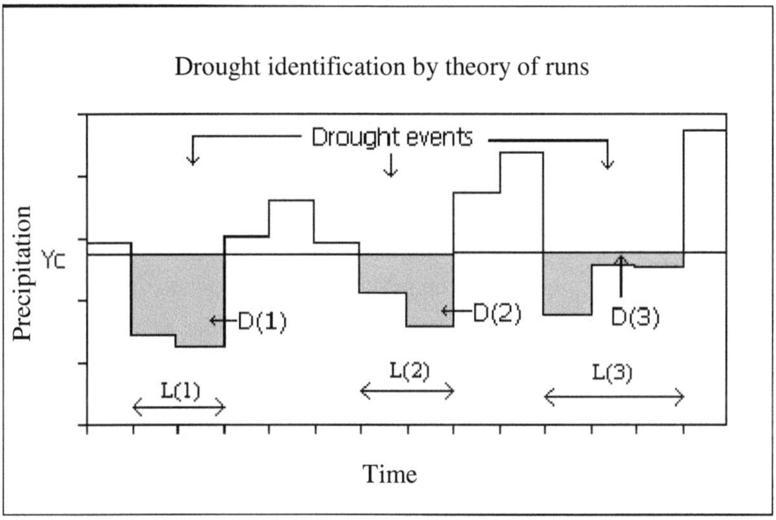

Fig. 3. Characteristics of drought events

The theory of runs can be applied in multiple time scales. In this study two time scales are used: the 12-month and the 3-month time scales. For this purpose, monthly rainfall time series are aggregated in yearly time series and in trimester time series.

The number of droughts and its characteristics are dependent on the time scale and on the choice of the critical threshold level, y_c. The critical level y_c usually adopted is a sample quantile, the sample mean or a function of the mean and of the standard deviation. When the time scale is shorter than the year, the critical level must be variable due to the seasonality of the data. In this study the threshold $y_c = \hat{\mu} - \hat{\sigma}$ is adopted after a previous analysis (Paulo et al., 2000) has shown its appropriateness.

2.2.2. Standardised Precipitation Index

The Standardised Precipitation Index (SPI) is a tool developed by McKee et al. (1993) with the purpose of defining and monitoring local droughts. It was originally conceived to identify drought periods and the severity of droughts at multiple time scales. Shorter or longer time scales may reflect lags in the response of different water resources to precipitation anomalies. As the time scale increases the SPI responds more slowly to changes in precipitation (McKee et al., 1995).

The SPI is an index based on the probability distribution of precipitation, for any time scale. The SPI calculation for a given location is based on the long-term precipitation record for the desired time period of analysis. The parameters of the probability distribution function are estimated from an historical record. If those parameters are estimated from a record referring to a different period, the SPI values may change. This is one disadvantage of the SPI. However, the SPI is less complex than the Palmer Drought Index and results better even in mountainous land or areas of frequent climatic extremes (Guttman, 1998).

The computation of the SPI, in every month, for a given time scale is performed following several steps: i) calculation of the cumulative precipitation totals for each month using a k-month time scale; ii) fitting of gamma distribution functions, to each calendar month; iii) estimation of the distribution function values for each precipitation total; iv) computation of the SPI values by transforming the gamma distribution function values into standard normal variable values.

i) Calculation of the cumulative precipitation for each month:

The precipitation total $X_{i,j}^{k}$ in a given month *j* and year *i* depends on the chosen time scale, *k*. For example, in order to compute the SPI for October in a particular year *i* with a 12-month time scale, the corresponding precipitation total is the sum of the precipitation depths from November of year *i-1* until October of year *i*. If a 3-month time scale is used, the total for October is the sum of the precipitation depths during the months August-September-October. Thus, $X_{i,j}^{k}$, precipitation total in the month *j* of year *i* and in the *k-1* past consecutive months, with $k \leq 12$, is computed by

$$X_{i,j}^{k} = \begin{cases} \sum_{l=13-k+j}^{12} w_{i-1,l} + \sum_{l=1}^{j} w_{i,l} & , j < k \\ \sum_{l=j-k+1}^{j} w_{i,l} & , j \geq k \end{cases} \tag{11}$$

where $w_{i,l}$ is precipitation in l^{th} month of year *i* [mm]

ii) *Fitting of gamma distribution functions for each calendar month*:

The two-parameter gamma distribution is defined by the probability density function

$$f(x)=\frac{1}{\beta^{\alpha}\Gamma(\alpha)}x^{\alpha-1}e^{-\frac{x}{\beta}} \quad , \; x>0 \tag{12}$$

where α is the shape parameter (α>0), β is the scale parameter (β>0) and

$$\Gamma(\alpha)=\int_0^{\infty} y^{\alpha-1}e^{-y}\,dy \tag{13}$$

is the gamma function.

The parameters α and β, of the gamma distribution can be estimated by the method of maximum likelihood using the following estimators (Clarke, 1973):

$$\hat{\alpha}=\frac{1+\sqrt{1+\frac{4}{3}A}}{4A} \tag{14}$$

$$\hat{\beta}=\frac{\overline{x}}{\hat{\alpha}} \tag{15}$$

where $A=ln\,\overline{x}-\frac{1}{n}\sum_{i=1}^{n} ln\,x_i$ (16)

and n, number of observations

iii) *Estimation of the distribution function values*:

Once the parameters α and β are estimated they can be used to calculate the distribution function value for an observed precipitation total by

$$P(X\leq x)=F(x)=\int_0^x f(t)dt=\frac{1}{\hat{\beta}^{\hat{\alpha}}\Gamma(\hat{\alpha})}\int_0^x t^{\hat{\alpha}-1}e^{-\frac{t}{\hat{\beta}}}dt \tag{17}$$

where $x=x_{i,j}^{k}$ and $z=\frac{t}{\hat{\beta}}$.

If there is a non null probability that precipitation total $X_{i,j}^{k}$ is zero, the cumulative probability becomes (Haan, 1977):

$$F(x)=q+(1-q)F^{*}(x) \tag{18}$$

where $F^{*}(x)$ is the distribution function estimated discarding the null values and q is the proportion of zeros in the precipitation series, $X_{i,j}^{k}$, $i=1,...,n$.

iv) *Computation of the SPI values*:

The value of the distribution function F(x) for any observed precipitation total is then transformed to a standard normal variable value, which is the SPI value. If the observed precipitation totals fit a normal distribution, SPI is obtained directly from

$$SPI_{i,j}^{k} = \frac{X_{i,j}^{k} - \hat{\mu}_{j}^{k}}{\hat{\sigma}_{j}^{k}} \tag{19}$$

where $\hat{\mu}_{j}^{k}$ and $\hat{\sigma}_{j}^{k}$ are respectively the mean and the standard deviation of the observed totals, with time scale k, in month j.

The SPI is a z score and represents an event departure from the mean, expressed in standard deviation units; it is a normalised index in time and space. This feature allows comparisons of SPI values between different locations and between different periods.

McKee et al. (1993) defined the criteria for a "drought event" for any of the time scales. A drought event occurs any time the SPI is continuously negative and reaches an intensity where the SPI is −1.0 or less. The event ends when the SPI becomes positive. Therefore, each drought event has a duration defined by its initiation and end, and an intensity referring to each month in the period when the event continues. The cumulated intensity of a drought can be considered as the drought magnitude, which is defined by the sum of the SPI for all the months within a drought event.

The drought intensity is arbitrarily defined according to the SPI values as listed in Table III (McKee et al, 1993). The expected time in each category is based on an analysis of a large number of rainfall stations across Colorado. The percent of time in moderate, severe and extreme droughts correspond to those expected from a normal distribution of the SPI. The 2.3% of time when the SPI values are in the extreme drought category typically correspond to the frequency expected for an "extreme" event (Wilhite, 1995).

Table III. Classification of SPI values and corresponding time in category

SPI values	Drought category	Time in category
0 to −0.99	Mild drought	24.0%
−1.00 to −1.49	Moderate drought	9.2%
−1.50 to −1.99	Severe drought	4.4%
≤ −2.00	Extreme drought	2.3%
		40.0%

As it is already mentioned before, the time scale is embedded in the computation of the monthly values of the SPI. Drought characteristics such as duration are expressed in months, whatever the time scale used. In this study, the SPI is applied in a 12 and in a 3-month time scales.

2.3. IDENTIFICATION OF REGIONAL DROUGHTS

2.3.1. Theory of runs

Regional droughts are identified and characterised using the two methods described for identification of local droughts. A regional drought is identified when a significant fraction of the total area of the region is under drought conditions or, in other words, when the sum of the areas A_j affected by drought reaches a selected critical areal threshold A_c , as explained in Figure 4. Two areal thresholds are considered in this study: 30 and 50% of the total area.

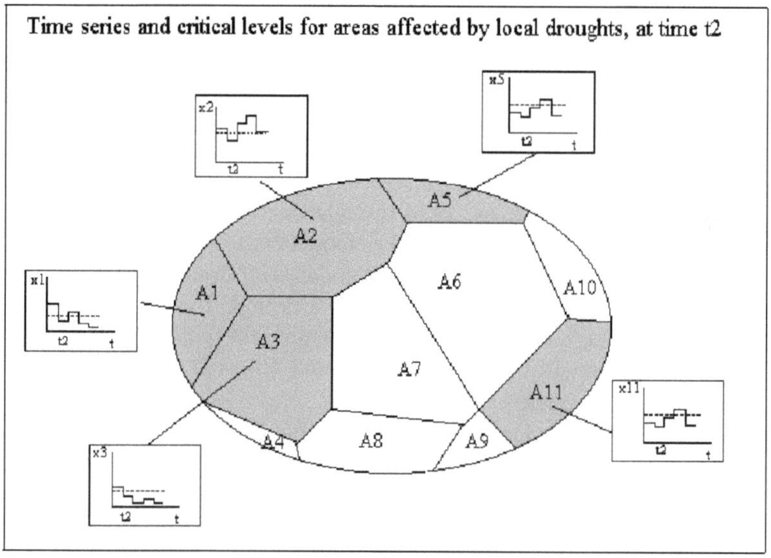

Fig. 4. Regional drought identification

Applying the theory of runs, a regional drought may be defined by its duration, cumulated areal deficit, intensity and mean regional coverage:

- the duration is the length of the consecutive time intervals in which drought affects an area that equals or exceeds a pre-defined areal threshold;
- the cumulated areal deficit is the sum of the cumulated deficits at each site, weighted by the corresponding areas of influence;
- the regional drought intensity is the ratio between the cumulated areal deficit and the duration;
- the mean regional coverage is the mean fraction of the area in the region that is affected by drought.

The identification and characterisation of both local and regional droughts using the theory of runs was accomplished with the utilisation of the software REDIM (DICA, 2000, Rossi and Cancelliere, 2002).

2.3.2. Standardised Precipitation Index

The SPI is extended to identify regional patterns of precipitation and dryness conditions, regional coverage of drought and drought severity. For this purpose a month is classified as dry if the percent area of the region affected by drought exceeds the adopted areal threshold. A regional drought event is assumed when the sum of the areas of influence of all stations affected at least by "mild drought" remains continuously above the areal threshold and, in that period, the area where SPI≤−1 exceeds the areal threshold in one or more months, the SPI being computed with a pre-selected time-scale.

The following drought characteristics are computed with the SPI:

- the regional drought severity, which is obtained for each month by combining the areal threshold and the local classifications of SPI for the time scale used. For example, if the total area having an SPI classification of "severe drought" equals or exceeds the areal threshold, then the region is classified under severe drought during that period;
- the drought duration, which is the number of consecutive months when a regional drought is classified at least as "mild drought" and that period includes one or more months where drought is classified as moderate, severe or extreme;
- the monthly regional coverage of drought under each drought category, which is the percentage of the region area affected by a drought having that severity. For example, the regional coverage of a moderate drought is the fraction of the area covered by moderate drought; it is computed from the sum of the areas of influence of all stations where −1.5<SPI≤−1;
- the regional SPI (SPI_{reg}) for each month and for each time scale, which is the weighted average of the local SPI values using the area of influence of each rainfall station as the respective weights.

3. Local droughts

The characteristics of the droughts identified by the theory of runs (TOR) and the SPI using two time scales, 12 and 3-month, are summarised in Tables IV and V.

Considering a 12-month time scale, results in Table IV show a relatively large variability of drought occurrences within the Alentejo region, When TOR is used the number of identified local droughts ranges from 5 (22H01) to 11, the average durations vary in the range 1.0 to 1.8 years (22H01), and the average cumulative deficits largely vary from 56 mm to 164 mm. Results relative to the SPI (Table V) show also a relatively large variation in the number of droughts, ranging from 10 to 18, drought duration, varying from 15.6 to 24.5 months, and drought average time in extreme and severe categories, ranging from 2.72 (30H03) to 6.60 months (26I03). Differences between both methods result from the respective methodologies. The larger number and duration of droughts with SPI indicate that the dry anomalies identified by the SPI correspond to a critical threshold closer to the mean than $y_c = \hat{\mu} - \hat{\sigma}$ used with TOR. This is expected considering the time in each drought category referred in Table III.

Results of the application of both the theory of runs and the SPI using the 3-month time scale (Tables IV and V) are in agreement but, like for the 12-month time scale, there is a trend for the SPI to identify a number of drought periods larger than that by TOR. Results concerning the local droughts identified show a large variation in the number of droughts identified: they range from 23 to 42 when the TOR is used, and from 46 to 64 when using the SPI. However, these extreme numbers refer to different locations. The drought intensity computed with TOR ranges from 16 to 43 mm/3-months, while the drought average time in severe and extreme categories given by the SPI varies from 0.63 (21M01 and 27G01) up to 1.12 (20I01). These results indicate that droughts randomly affect the region, so requiring that appropriate regional analysis be developed.

Table IV. Local drought characteristics using the theory of runs for 12-month and 3-month time scales

	12-month time scale				3-month time scale			
Station code	Nr of years under drought	Nr of droughts	Average Cumul. Deficit (mm)	Average Intensity (mm/y)	Nr of 3-month under drought	Nr of droughts	Average Cumul. Deficit (mm)	Average Intensity (mm/3-month)
16L03	11	10	70	62	40	35	20	18
17J01	10	9	94	86	40	33	32	25
17L02	11	10	77	68	40	36	29	26
17M01	13	11	66	62	43	38	30	28
18G01	10	9	87	76	37	34	28	26
18M01	7	6	106	82	44	39	31	28
19G02	12	10	104	88	40	34	31	27
19L02	10	9	76	64	46	36	22	17
19M01	10	7	86	67	45	38	19	17
20I01	11	10	70	59	46	41	23	20
20L01	12	9	56	38	42	37	21	17
21F01	12	10	86	64	41	37	23	23
21G02	11	9	90	66	38	33	29	26
21J02	7	6	114	93	43	35	35	28
21K01	9	6	100	57	41	35	19	16
21M01	8	7	95	82	26	25	35	34
21M02	11	9	85	69	29	28	33	33
22E01	12	11	87	74	47	42	21	19
22H01	9	5	148	80	40	36	27	24
22J01	12	9	67	54	45	37	23	19
24J03	11	8	113	71	27	23	28	25
24N01	10	7	75	64	29	27	26	24
25J02	9	7	98	75	33	30	24	22
26I03	12	8	101	79	31	27	24	20
26J01	10	7	96	65	30	29	19	18
26L01	9	7	94	77	28	25	26	23
27E01	13	11	84	77	30	29	29	29
27G01	11	9	104	86	33	30	23	22
27I01	10	8	107	82	32	27	27	23
27K01	7	7	81	81	29	29	18	18
28H01	11	9	100	80	32	28	27	25
29G01	9	7	138	120	42	32	29	20
29G02	9	7	123	104	26	26	23	23
30E01	8	7	118	93	28	25	24	21
30F01	11	10	164	135	27	25	47	43
30H03	11	9	102	78	28	26	25	24
31J01	9	7	145	109	29	28	27	25

The SPI methodology when compared with TOR identifies a higher number of droughts, with larger drought duration. However, the distribution of the months by the different classes of severity within each drought event reveals a relatively small number of extremely and severely dry months. Differences in results from both methods would be smaller if a higher critical threshold would be used with the theory of runs or adopting a different definition for mild droughts (Table III) such as SPI<–0.5. Nevertheless, results from both methods are coherent.

Table V. Characteristics of local droughts using the SPI for 12-month and 3-month time scales

	12-month time scale						3-month time scale					
Stat. Code	Nr. Events	Average Duration (month)	Average time (months) in classes of severity				Nr. Events	Average Duration (month)	Average time (months) in classes of severity			
			Extreme	Severe	Moderate	Mild			Extreme	Severe	Moderate	Mild
16L03	15	21.7	1.1	2.0	5.7	12.9	56	5.2	0.3	0.5	1.3	3.1
17J01	15	18.7	1.6	2.3	4.3	10.5	54	5.3	0.4	0.7	1.2	3.0
17L02	15	18.9	0.9	3.1	4.7	10.1	54	5.1	0.4	0.6	1.2	2.9
17M01	14	24.3	0.4	2.6	7.0	14.3	52	5.5	0.4	0.7	1.3	3.2
18G01	14	20.1	0.6	3.5	5.2	10.9	51	5.4	0.4	0.7	1.2	3.1
18M01	17	17.1	1.2	1.6	3.8	10.6	55	5.3	0.3	0.7	1.3	3.0
19G02	15	17.3	1.7	3	4	8.6	52	5.1	0.4	0.6	1.5	2.6
19L02	17	18.5	0.7	2.8	4.7	10.3	49	5.5	0.4	0.7	1.5	2.9
19M01	13	23.2	0.9	3.6	5.4	13.5	53	5.5	0.3	0.7	1.3	3.3
20I01	15	18.3	1.1	2.9	4.3	10.1	51	5.2	0.4	0.8	1.4	2.8
20L01	14	23.9	0.6	2.9	6.1	14.4	51	5.4	0.3	0.8	1.4	3.0
21F01	17	17.7	0.6	2.8	4.8	9.5	58	5.0	0.3	0.5	1.5	2.7
21G02	17	15.6	1.1	2.2	4.7	7.7	56	5.0	0.3	0.7	1.4	2.6
21J02	15	19.7	1.7	1.4	5.3	11.2	51	5.4	0.4	0.7	1.3	3.1
21K01	12	24.2	1.5	3.3	5.3	14.2	56	5.2	0.2	0.6	1.5	2.8
21M01	13	23.5	0.4	3.3	5.4	14.5	56	5.3	0.3	0.3	1.4	3.4
21M02	13	21.1	1.5	3.3	4.7	11.6	51	5.4	0.4	0.6	1.3	3.1
22E01	17	17.5	1.2	2.8	3.6	9.9	58	4.8	0.3	0.4	1.3	2.8
22H01	14	18.9	1.0	3.6	4.9	9.4	52	5.1	0.4	0.6	1.3	2.8
22J01	12	24.3	0.6	4.1	6.8	13.0	53	5.4	0.4	0.6	1.5	2.9
24J03	13	21.6	1.7	2.2	6.2	11.5	47	6.3	0.2	0.9	1.3	3.9
24N01	13	23.4	0.8	3.0	5.9	13.8	54	4.9	0.4	0.6	1.3	2.7
25J02	17	17.5	1.1	2.0	3.5	11.0	55	5.2	0.4	0.7	1.3	2.9
26I03	10	24.2	3.2	3.4	6.1	11.6	58	4.8	0.2	0.5	1.4	2.6
26J01	13	20.2	2.2	2.3	5.1	10.7	50	5.3	0.3	0.6	1.3	3.1
26L01	13	22.8	1.6	2.2	5.8	13.2	49	5.6	0.4	0.6	1.5	3.2
27E01	13	21.5	0.8	3.6	6.6	10.6	56	5.3	0.2	0.6	1.5	3.0
27G01	14	19.8	1.1	3.7	4.3	10.8	56	5.3	0.2	0.5	1.7	3.0
27I01	17	16.7	1.2	3.0	2.4	10.1	56	4.7	0.3	0.6	1.4	2.5
27K01	14	22.9	1.8	2.3	3.7	15.1	48	5.4	0.3	0.7	1.2	3.3
28H01	15	19.7	1.1	2.9	4.4	11.4	52	5.1	0.3	0.7	1.2	3.0
29G01	11	24.5	2.6	3.2	4.1	14.6	46	5.9	0.4	0.4	1.4	3.7
29G02	15	21.8	1.1	2.9	4.8	13.1	60	5.1	0.2	0.6	1.5	2.9
30E01	13	22.6	1.5	2.8	4.0	14.4	54	5.4	0.3	0.5	1.6	3.0
30F01	16	18.1	0.7	2.9	4.7	9.8	55	5.3	0.4	0.5	1.4	3.1
30H03	18	16.8	0.9	1.8	4.4	9.7	50	5.3	0.3	0.5	1.1	3.3
31J01	16	19.3	0.8	2.0	5.2	11.4	64	4.4	0.3	0.4	1.3	2.5

4. Regional droughts

4.1. 12-MONTH TIME SCALE

Two time-scales such as for local droughts were utilised for regional droughts: the 3-month and the 12-month time scales. With the theory of runs, two critical areal thresholds were used: 50 and 30% of the area of the region. Using the SPI, the proportion of area affected by drought was computed for each time scale and month of interest. As the SPI is obtained from moving totals, it provides a more continuous information on the variation of dryness conditions in time.

Table VI shows the main characteristics of the droughts identified by the theory of runs for the 12-month time scale covering at least 30% of the region. Results include the initiation and end of the drought, the duration, the areal coverage (% area) and the intensity. 10 droughts are identified when the areal threshold is 30% of the area. This number reduces to 7 when that threshold is 50% of the area.

Table VI. Characteristics of regional droughts using the theory of runs for the 12-month time scale

Drought number	Initiation	End	Duration (year)	Areal coverage (%)	Intensity (mm/year)
1	1931/32	1931/32	1	30	23
2	1934/35	1934/35	1	62	39
3	1943/44	1944/45	2	81	95
4	1952/53	1952/53	1	34	26
5	1956/57	1957/58	2	43	13
6	1964/65	1964/65	1	54	27
7	1980/81	1980/81	1	100	100
8	1982/83	1982/83	1	100	95
9	1991/92	1991/92	1	86	51
10	1994/95	1994/95	1	87	76
Average drought characteristics			1.2	67.7	54.5

Results in Table VI show that 5 droughts are identified for the first half of the period of 64 years under analysis, and other 5 occur in the second half. However, for the first half, droughts have longer duration and lower areal coverage than for the second half. Drought intensities tend to be higher during the last decades.

The characteristics of regional droughts identified by the SPI, for the 12-month time scale using the areal threshold of 30 and 50% are presented in Tables VII and VIII, respectively.

The initiation and ending of the drought periods are now identified to the month while with the theory of runs only the year could be identified. Results show that the number of regional droughts identified with SPI, 13 and 12 for respectively 30% and 50% areal thresholds, is larger then those identified with TOR and so is the respective duration. These differences are due to the concepts behind both methodologies and the relatively high critical threshold adopted for the theory of runs, as indicated before. In what concerns regional droughts, differences also relate to the approaches used. When the TOR is used, the areal coverage is just the average fraction of the total area in the region that is affected by drought. With the SPI, the areal coverage may be reported to different classifications of drought periods: mild, moderate or more severe drought. In the identification of regional droughts with the SPI the areal threshold is compared with the total area affected by drought, regardless its severity. Therefore, the areal coverage of moderate or more severe droughts is often lower than 30 or 50%, the areal thresholds considered. In Tables VII and VIII are presented, within each drought, the mean areal coverage of at least moderate droughts, the number of months within different categories of drought and the SPI_{reg}. The SPI_{reg} may be used as an indicator of the severity of the drought: it ranges from –0.32 to –1.20 and from –0.49 to –1.26, with areal thresholds of 30% and 50%.

Table VII. Characteristics of regional droughts when the SPI is computed with the 12-month time scale adopting the areal threshold A_c=30%

Drought Nr	Initiation	End	Duration (months)	Nr of months when drought is Moderate	Severe	Extreme	Areal cov (%) moderate drought	SPI_{reg}
1	Oct-1932	Jan-1936	40	17	2	0	30.1	-0.62
2	Oct-1938	Sep-1939	12	2	0	0	16.5	-0.32
3	Dec-1944	Apr-1946	29	9	10	5	62.0	-1.20
4	Nov-1948	Feb-1951	28	6	4	0	28.6	-0.54
5	Mar-1953	Feb-1955	24	6	1	0	20.9	-0.52
6	Jan-1957	Jan-1959	25	6	1	0	23.2	-0.56
7	Dec-1964	Oct-1965	11	8	0	0	47.2	-0.91
8	Jan-1967	Nov-1968	23	2	0	0	14.7	-0.40
9	Nov-1970	Jan-1972	15	3	0	0	14.7	-0.42
10	Feb-1973	Oct-1977	45	14	1	0	24.5	-0.67
11	Feb-1980	Feb-1984	49	6	16	4	49.9	-1.00
12	Apr-1991	Oct-1993	31	7	10	2	49.8	-0.94
13	Apr-1994	Sep-1995	18	3	3	1	46.8	-0.85
Average drought characteristics			26.9	6.8	3.7	0.9	33	-0.69
Time in category (%)			–	12	6	2	–	–

Table VIII. Characteristics of regional droughts when the SPI is computed with the 12-month time scale adopting the areal threshold A_c=50%

Drought Nr	Initiation	End	Duration (months)	Nr of months when drought is Moderate	Severe	Extreme	Areal cov (%) moderate drought	SPI_{reg}
1	Nov-1933	Dec-1935	26	10	2	0	42.1	-0.91
2	Jan-1944	Apr-1946	28	10	7	2	64.2	-1.24
3	Jan-1949	Apr-1950	16	3	4	0	45.8	-0.87
4	Apr-1954	Jan-1955	10	2	0	0	22.6	-0.54
5	Jan-1957	Nov-1958	23	2	0	0	24.5	-0.62
6	Dec-1964	Sep-1965	10	6	0	0	51.9	-1.00
7	Jan-1971	Jan-1972	13	2	0	0	16.9	-0.49
8	Feb-1973	Sep-1976	44	6	0	0	25.0	-0.69
9	Feb-1980	Oct-1982	33	3	10	1	45.6	-0.99
10	Dec-1982	Dec-1983	13	6	5	0	72.2	-1.26
11	Apr-1991	Oct-1993	31	9	4	1	49.8	-0.94
12	Aug-1993	Sep-1995	14	5	5	0	60.2	-1.12
Average drought characteristics			21.8	5.3	3.1	0.3	43.4	-0.89
Time in category (%)			–	8	5	1	–	–

The criteria used to identify the regional droughts need further discussion. Two drought events identified with the TOR having areal coverage of 100% relative to the hydrological years 1980/81 and 1982/83 (Table VI) are considered one single and long duration drought event when the SPI is used for A_c=30% (event 11 in Table VII). However when using A_c=50%, two drought events are considered (events 9 and 10 in Table VIII). Therefore a question arises: should this period be considered with a single or two drought events? As shown in Figure 5 a mild drought starts by Feb-80 with a coverage of 70% of the region, reduces coverage below 50% for only one month (Nov-82) and returns immediately to 100% coverage the next month until falling below 50% by Jan-84. When a moderate drought is considered, the area coverage starts to be above 50% by Oct-80 and falls below this threshold by Nov-81; the areal coverage becomes higher than 50% by Dec-82 and falls below this value by Nov-83. Severe drought follows a similar trend. The SPI_{reg} is always negative during the entire period except when mild drought coverage falls below 50% of the regional area, i.e. for the month Nov-82 only, and for the two final months of the period.

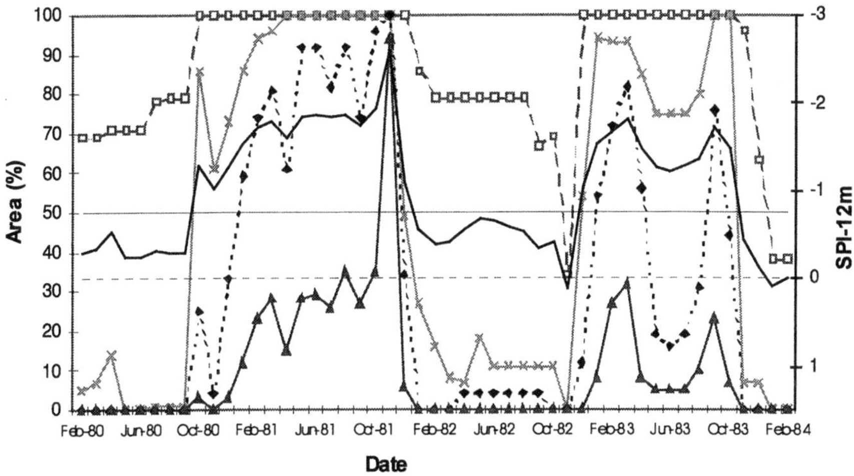

Fig. 5. Time evolution of the areal coverages having at least mild (– –□– –), moderate (·····×·····), severe (···◆···); or extremely severe (—▲—) drought, and the SPIreg (———) for the period Feb80-Feb84

It is not likely that only one month with $SPI_{reg}>0$ indicates that recovery would have occurred since the SPI_{reg} becomes smaller than –1 just a couple months later. Similarly, the indication that the areal coverage falls below the fixed threshold for one month only is probably insufficient to decide that two drought events are clearly defined before and after that period. The combined use of other drought indices such as the PDI are therefore required to appropriately validate a methodology relative to the use of the SPI in regional drought analysis.

Another example concerns the period from Nov-70 to Jan-72. Using the areal threshold of 30%, the TOR methodology does not identify any regional 12-month drought for this period. However, using the SPI this period corresponds to a regional drought (event 9 in Table VII). As shown in Figure 6, the areal coverage for a mild drought is consistently above A_c=30% during the entire period but when a moderate drought is considered the areal coverage is above that threshold for only 3 months. The fact that $SPI_{reg}<0$ is verified for the entire period could confirm that a regional drought occurred during the period despite not being confirmed when using the TOR methodology. Further research is therefore required to consistently adopt the SPI for regional analysis.

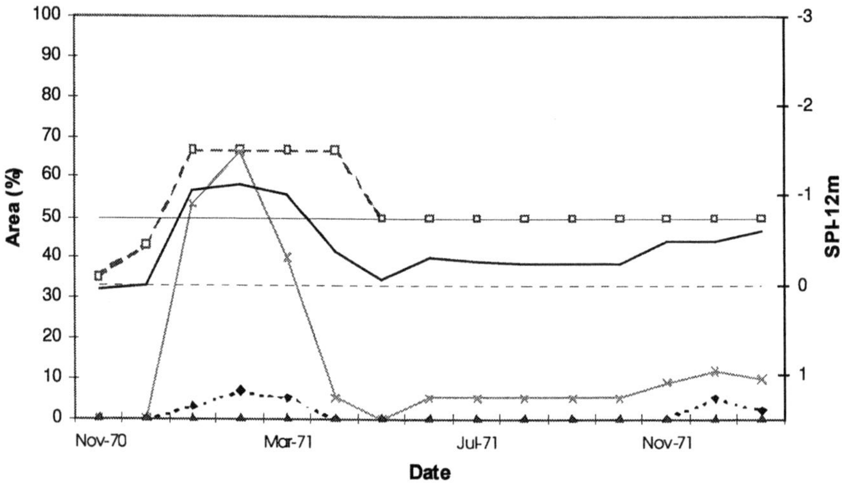

Fig. 6. Time evolution of the areal coverage relative to at least mild (--□--), moderate (—×—), severe (···◆···); and extreme (—▲—) drought categories and the SPIreg (——) for the period Nov 70-Jan72

4.2. 3-MONTH TIME SCALE

Table IX identifies the regional drought periods considering a 3-month time scale and an areal threshold of 30%, using the two methodologies. Almost all the 3-month drought periods identified with the theory of runs are as well identified by the SPI; only 3 periods identified with the SPI are not identified by the theory of runs. However, there are 34 periods identified only by the SPI. The agreement between TOR and SPI in what concerns drought identification is better for the periods Jan-Mar and Apr-Jun. The theory of runs identifies 47 trimesters affected by drought which correspond to 37 drought events having an average duration of 1.3 trimesterial periods and an average intensity of 16 mm/trimester. The SPI identifies 74 trimesterial periods under drought. The regional averages for the SPI and the number of months classified as moderate, severe and extreme, referring to the 4 trimesterial periods are given in the Table X.

Considering an areal threshold of 50%, the theory of runs identifies 25 trimesterial droughts whose average duration, intensity and regional coverage are respectively 1 trimesterial period, 24 mm/trimester and 78% of area covered. Differently, the SPI identifies 35 trimestrial periods under droughts having a mean regional coverage for moderate or more severe drought larger than 50%.

Table IX. Regional coverage of drought (in % area) computed with the SPI and the theory of runs using a 3-month time scale and an areal threshold of 30%

Year	SPI				Theory of runs		
	Regional coverage (moderate, severe and extreme droughts)				Drought periods		Regional Cov (%)
	Oct-Dec	Jan-Mar	Apr-Jun	Jul-Sep	Beginning	End	
1932/33					A-J 32/33	J-S 32/33	41
1933/34		41		61	J-S 33/34	J-S 33/34	37
1934/35	57	32		30	J-M 34/35	J-M 34/35	91
1934/35					J-S 34/35	J-S 34/35	31
1935/36	33				J-S 35/36	O-D 36/37	36
1936/37	40		59	38	A-J 36/37	A-J 36/37	59
1937/38		63	30	46	J-M 37/38	J-M 37/38	85
1941/42	43	30					
1943/44		67			J-M 43/44	J-M 43/44	66
1944/45		76	76	36	O-D 44/45	J-S 44/45	48
1946/47			64		A-J 46/47	A-J 46/47	64
1947/48				63	J-S 47/48	J-S 47/48	44
1948/49	31	31	35		J-M 48/49	J-M 48/49	77
1950/51	46						
1953/54				54	A-J 53/54	J-S 53/54	43
1954/55			56		A-J 54/55	J-S 54/55	60
1956/57	30	37			O-D 56/57	O-D 56/57	58
1957/58			48		A-J 57/58	A-J 57/58	67
1958/59	67			38			
1960/61		36	42		J-M 60/61	J-M 60/61	52
1961/62			34		A-J 61/62	A-J 61/62	34
1963/64			38		A-J 63/64	A-J 63/64	38
1964/65	43		54	34	O-D 64/65	O-D 64/65	79
1965/66				36	A-J 64/65	A-J 64/65	100
1967/68		41					
1969/70			75	59			
1970/71	98				O-D 70/71	O-D 70/71	98
1971/72	94		50		O-D 71/72	O-D 71/72	100
1972/73			50		A-J 71/72	A-J 71/72	86
1973/74	38			81	J-S 73/74	O-D 74/75	71
1974/75	92				O-D 75/76	O-D 75/76	48
1975/76	47			33	A-J 76/77	A-J 76/77	70
1976/77			56				
1979/80		55					
1980/81		97			O-D 80/81	J-M 80/81	70
1981/82			43	57	A-J 81/82	A-J 81/82	46
1982/83		83			J-M 82/83	J-M 82/83	100
1985/86	52		39	61			
1986/87	37			58	O-D 86/87	O-D 86/87	37
1987/88			62				
1988/89		46		39			
1989/90		83		45	J-M 89/90	J-M 89/90	81
1990/91			41		A-J 90/91	A-J 90/91	72
1991/92		92	41		J-M 91/92	J-M 91/92	100
1992/93		50			J-M 92/93	J-M 92/93	76
1993/94				59	J-S 93/94	J-S 93/94	36
1994/95		42	70		J-M 94/95	A-J 94/95	45
Mean	53	57	41	59			63

The number of dry months by class of regional drought severity, as computed with the SPI considering a 3-month time scale and the areal thresholds of 50 and 30% are presented in Figure 7. It can be observed that the summer period (Jul-Sep), which is already dry for the climate in the region, is the less affected by drought. The number of extremely dry months is larger for the spring period (Apr-Jun) that is the one where impacts on rainfed agriculture are higher.

Table X. Mean regional SPI, months in SPI drought classes using a 3-month time scale and a regional threshold of 30%

	3- month periods			
	Oct-Dec	Jan-Mar	Apr-Jun	Jul-Sep
SPIreg	-1.10	-1.11	-1.05	-0.89
Nr of months:				
Moderately dry	18	23	25	32
Severely dry	12	12	8	5
Extremely dry	6	4	7	0

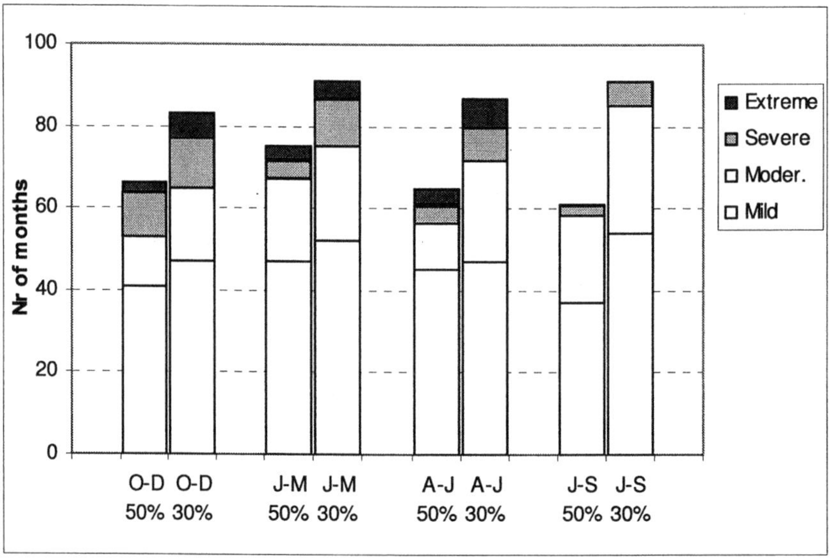

Fig. 7. Severity of regional droughts using the SPI classification for the 3-month time scale and the areal thresholds of 50 and 30%

The distribution of the regional droughts identified with the theory of runs in what concerns drought intensity is presented in Figure 8.

Differences in number of identified regional droughts using both methods are explained as for the 12-month time scale, by the assumptions behind both methods. Smaller differences would have occurred adopting a larger critical threshold y_c or defining mild droughts for SPI<–0.5. Further research is required to improve the use of the SPI for regional analysis.

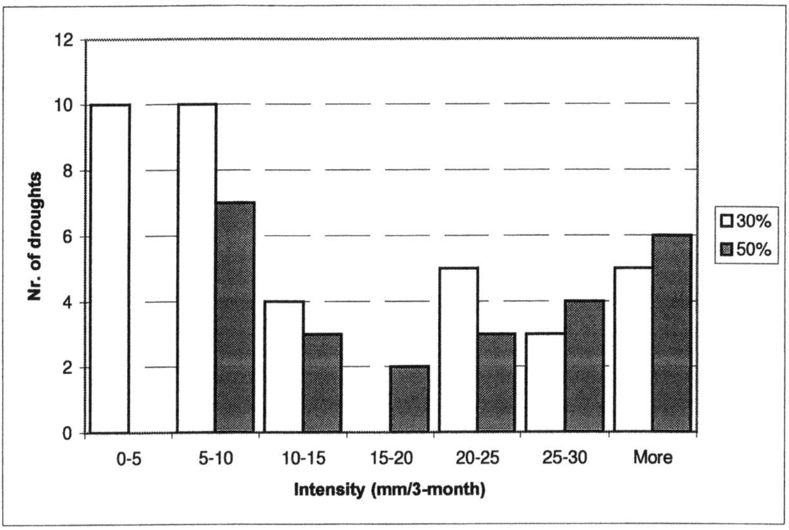

Fig. 8. Intensity of the 3-month regional droughts identified with the theory of runs and the areal thresholds of 50 and 30%

The SPI computed for a *k* time scale provides a continuous evaluation of dryness conditions in every month, which is based on the accumulated precipitation of the month of interest with the *k-1* past months. Results for both time scales and the 30% areal threshold are presented in Table XI with identification of the drought severity for each month.

Results in Table XI show not only when mild to extreme droughts occur but how severity varies in time. Results for 12-month time step show that dry periods are generally of long duration but only few among them include periods with severe or extreme droughts. Periods with moderate to severe drought are always preceded by several dry months indicating a slow initiation of droughts. This behaviour indicates that using the SPI it is possible to detect when a drought is initiating and therefore to use this information for early warning. However further research is required to effectively test the practicability of this assumption.

Comparing the time in each category used by McKee et al. (1993) to define these categories with the presented in Table XI shows that longer time under drought is observed in Alentejo, mainly for mild drought (Table XII). This indicates the need to evaluate the criteria used, preferably using other drought indices.

Analysing the results relative to the period Feb-80 to Feb-84 discussed before (Figure 5) it can be observed that it has been initiated with a mild dry sequence of 8 months and that the two periods having moderate to extremely severe droughts are sequential. When mild droughts are considered it becomes evident that a very long drought event occurred without recovery between both periods with more severe drought.

Table XI. Regional classification of drought based on SPI

Hydr.	SPI in a 12-months time scale												SPI in a 3-months time scale											
Year	O	N	D	J	F	M	A	M	J	J	A	S	O	N	D	J	F	M	A	M	J	J	A	S
1932/33	1	1	0	0	0	0	0	0	0	0	0	0						0	1	0	1	0	0	1
1933/34	0	0	0	1	1	1	0	0	0	0	0	0				0	1	1				1	1	1
1934/35	1	1	0	1	1	1	2	2	1	1	1	1	3	1	0	0	0	2	1	0	0		0	1
1935/36	1	1	1	0									2	0								0	0	1
1936/37													1	0	1						2	1	1	0
1937/38																	1	2	1	0			1	
1938/39	0	1	1	0	0	0	0	0	0	0	0	0												
1939/40																								
1940/41																								
1941/42													2	0	1	0	1							
1942/43																					0	1	0	
1943/44			0	0	0	1	2	2	1	1	1	2		0	0	2	1	2	0	0	0	0		
1944/45	2	1	2	1	2	2	2	2	3	3	3	3	0	0	1	1	2	2	3	2	1	0	0	1
1945/46	3	2	1	1	1	0	0						0	0									1	0
1946/47																					1	0	0	
1947/48																							1	1
1948/49		0	0	0	1	1	1	2	2	2	2	0	0	2	0	0	0	1	1	1	0	0		
1949/50	1	0	1	1	0	0	0	0	0	0	0	0												0
1950/51	0	0	0	0	0								1	1	0	0								
1951/52																								
1952/53						0	0	0	1	1	1	1			0	0	0	0	0	0	0	1	0	0
1953/54	0	0	0	0	0	0	0	0	0	0	0	0				0	0	0	0	0	1	2	0	1
1954/55	1	1	2	0	0								2	0	0					0	2	0	0	1
1955/56													0									1	0	
1956/57				0	0	0	1	0	0	0	0	1		0	1	2	0	0						
1957/58	1	0	0	0	0	0	0	1	0	0	0	1			0	0	0	0	1	1	2	1	0	0
1958/59	1	2	0	0									1	3							0	0	2	
1959/60																								
1960/61																0	1	1	1					
1961/62																					1	0		
1962/63																								
1963/64																					1	0		
1964/65			0	0	1	1	1	1	1	1	1	1	0	0	2	0	0		0	1	3	1	0	
1965/66	0																					1	0	
1966/67				0	0	0	0	0	0	0	0	0		0	0	2	0	0	0	0				0
1967/68	1	0	0	1	0	0	0	0	0	0	0	0	0	0	0	1	0							
1968/69	0	0																						
1969/70																			3	1				1
1970/71	0	0	0	1	1	1	0	0	0	0	0	0	2	2	2	0	0	0	0					0
1971/72	0	0	0	0									2	3	3	0				0	3	0	0	
1972/73					0	0	0	0	0	0	0	0				0	0	0	3	1	0			0
1973/74	1	1	1	2	1	0	0	0	0	0	0	0	1	1	0	0	0	0						1
1974/75	1	1	1	1	1	0	0	0	0	0	0	0	3	2	2	1	0							0
1975/76	0	0	0	0	0	1	0	1	1	1	1	0	0	2	1	1	0	0				1		
1976/77	0																		0	3	2	0		
1977/78																								
1978/79																								
1979/80					0	0	0	0	0	0	0	0				3	1	0	0	0	0		0	0
1980/81	1	1	2	2	2	2	2	2	2	2	3	2	0	0	1	3	3	2	0	0	0	0	0	
1981/82	3	3	2	0	0	0	0	0	0	0	0	0	0	1	0			0	0	1	1	1		
1982/83	0	0	1	2	2	3	2	1	1	1	2	2			0	1	2	3	1	0				
1983/84	2	0	0	0	0																			
1984/85																						0	0	1
1985/86				0	0	0	0	0	0	0	0	0	3	1	0	0				1	0	1	1	
1986/87	0	0	0	0	0	0	0	0	0	0	0	0			1	0				1	0	2		
1987/88	0																	0	3	1				
1988/89															0	1	1	0	0				1	
1989/90																		1	0	0	0	2	1	0
1990/91			0	0	0		0	0	0	0	0	0								0	2	2	0	0
1991/92	0	1	1	1	2	3	3	2	2	2	2	2	0	0	1	2	1	2	1	0				
1992/93	2	2	2	2	1	1	1	0	0	0	1	0		1	0	1	0	1	0					
1993/94	0						0	0	0	0	0	0											1	1
1994/95	0	1	1	1	2	2	2	3	2	2	2	2	0	0	0	1	1	1	1	2	2	0	0	0

0. Mild drought 1. Moderate drought 2. Severe drought 3. Extreme drought

Table XII. Time in a drought category: comparing McKee et al. (1993) data with Alentejo results

Drought category	Time in category (%)	
	McKee data (TableIII)	Alentejo (Table XI)
Mild	24.1	29.5
Moderate	9.2	11.8
Severe	4.4	6.9
Extreme	2.3	1.6
	40.0	49.7

Results also show that drought occurrence and respective severity are higher during the second half of the period under analysis, particularly in which concerns the January-March and April-June 3-month periods. However, these results can not be used as such to identify global change impacts.

5. Conclusions

The identification and characterisation of local and regional droughts in Alentejo, Portugal, using the theory of runs and the Standardised Precipitation Index has been successful, with both methods producing compatible but different results.

For determining local droughts with the theory of runs, the critical threshold $y_C = \hat{\mu} - \hat{\sigma}$ was adopted. The results of its application are coherent with those obtained using the SPI as it was originally developed, i.e. considering the occurrence of mild droughts when SPI<0. The 12-month and the 3-month time scales were used with both methods. The number and characteristics of droughts show larger differences between both methods when the 3-month time scale is used. Differences are due to the concepts and approaches used in both methods. They would be lesser if a threshold closer to the mean would be used with TOR or when the SPI for mild droughts would be defined from a lower value such as SPI<–0.5. Further research is required but adopting also other indices.

For the characterisation of regional droughts, two areal critical thresholds were used: 30% and 50% of the area covered. Both thresholds were used with TOR and SPI. Results are also different but compatible. With the SPI method, the regional classification of drought severity was based on the criterion of areal coverage in relation to the drought severity classification. Differences between results from TOR and SPI are due to the differences in supporting theories and application methodologies. Further research is needed including with application of other indices for a better interpretation of results with multiple time scales. The utilisation of other hydrological, soil moisture and cropping variables may also provide more consistent information on drought impacts on agriculture.

The SPI has shown very much appropriate to identify the months when a drought initiates and ends, when is used both at local and regional scales. For the regional analysis it was observed that moderate or more severe droughts are always preceded by several months of mild drought. This behaviour may be explored to identify when a drought is initiating and therefore be used for warning, thus to improve the lead time for adopting drought measures. However, more research is required to confirm and eventually explore this assumption.

Acknowledgements

This research is funded by the research contract ERBIC-18CT-970169 with the European Union, the Contract Interreg IIC 99.09.5942.1 with the Ministry of Agriculture, and FCT, Foundation for Science and Technology, through CEER/ISA.

References

Cancelliere, A. and Rossi, G. (2002) Droughts in Sicily and comparison of identified droughts in Mediterranean Regions. (This issue).

Cancelliere, A., Rossi, G. and Ancarani, A. (1996) Use of Palmer index as a drought indicator in Mediterranean regions. In: *Proc. IAHR Conf. From Flood to Drought*, Sun City, South Africa.

Conover, W.J. (1980) *Practical Non Parametric Statistics*. John Wiley & Sons, New York.

Clarke, R.T. (1973) *Mathematical Models in Hydrology*. FAO Irrig. Drain. Pap. Nr. 19, FAO, Rome.

Edwards, D. (2000) SPI defined. In: http://ulysses.atmos.colostate.edu/SPI.html, Colorado.

Guerrero-Salazar, P. and Yevjievich, V. (1975) Analysis of Drought Characteristics by the Theory of Runs. *Hydrology Paper Nr. 80*, Colorado State University, Fort Collins, CO.

Guttman, N.B. (1998) Comparing the Palmer drought index and the standardised precipitation index. *J. Am. Water Res. Assoc.* 34(1), pp. 113-121.

Haan, C.T. (1977) *Statistical Methods in Hydrology*. The Iowa State University Press.

Hayes, M.J. (1999) Drought Indices. In: http://enso.unl.edu/nmdc, National Drought Mitigation Centre, Lincoln, Nebraska.

Helsel, D.R. and Hirsch, R.M. (1992) *Statistical Methods in Water Resources*. Elsevier, Amsterdam.

Henriques, A.G. and Santos, M.J.J. (1996) Modelo de caracterização de secas regionais. In: *II Congresso da Água*, APRH, Lisboa, vol.II, pp.49-59.

DICA (2000) Regional Drought Identification Module REDIM. Department of Civil and Environmental Engineering, University of Catania, Catania.

Matias, P.G. (1998) *Análise de Frequência de Séries Hidrológicas Anuais*. DER/ISA.

McKee, T.B, Doesken, N.J. and Kleist, J. (1993) The relationship of drought frequency and duration to time scales. In: *8th Conference on Applied Climatology*. Am. Meteor. Soc, Boston, pp. 179-184.

McKee, T.B., Doesken, N.J. and Kleist, J. (1995) Drought monitoring with multiple time scales. *Pre-prints, 9th Conference on Applied Climatology*, 15-20 January, Dallas, TX, pp. 233-236.

Paulo, A.A.; Pereira, L.S.; Matias, P.G.; Rossi, G. and Cancelliere, A. (2000) Characterisation of droughts: Local indices and regional analysis. In: *Proc. 3rd Inter Regional Conference on Environment-Water: Water Resources Management in the 21st Century*, Budapest: 161 - 168.

Rossi, G., Benedini, M., Tsakiris, G. and Giakoumis, S. (1992) On regional drought estimation and analysis. *Water Res. Manag.* 6, pp.249-277.

Rossi, G. and Cancelliere, A. (2002) At-site and regional drought identification with REDIM model. (This issue).

Santos, M.A. (1983) Regional droughts: a stochastic characterisation. *J. of Hydrology*, 66, pp.183-211.

Santos, M.A., Correia, F. N. and Rodrigues, R. (1988) Risk assessment of regional droughts. In: F. Siccardi and R. L. Bras (Eds.) *Natural Disasters in European Mediterranean Countries*, U. S. Nat. Sci. Foundation and Nat. Res. Council, Italy, pp. 333-350.

Tate, E.L. and Gustard, A. (2000) Drought definition: a hydrological perspective. In: J.J. Vogt and F. Somma (Eds.) *Drought and Drought Mitigation in Europe*.

Wilhite, D.A. and Glantz, M.H. (1987) Understanding the drought phenomenon: the role of definitions. In: D. A. Wilhite, W. E. Easterling and D. A. Wood (Eds.) *Planning for Drought, Toward a Reduction of Societal Vulnerability*. Westview Press, Boulder, pp. 13-27.

Wilhite, D.A. (1995) *Developing a Precipitation-Based Index to Assess Climatic Conditions Across Nebraska*. Final report submitted to the Natural Resources Commission, Lincoln, Nebraska.

LOCAL AND REGIONAL DROUGHTS IN SILIANA CATCHMENT, TUNISIA

M. BERGAOUI
École Supérieur des Ingénieurs de l'Équipement Rural, Tunis

Abstract: Drought occurs when lack of rain persists and dominates a whole area and when cumulative precipitation is less than the normal. When drought occurs, necessity of management of water stored in reservoirs in order to satisfy water demands arises. In the paper, the identification and characterisation of droughts in the area of Siliana, Tunisia, during the period 1967-1997 is carried out by determining drought duration, cumulative deficit and intensity. Regional drought duration areal extent and areal cumulated deficit are also assessed.

Key words: Drought, semi-arid areas , deficit , Tunisia.

1. Introduction

The development of agriculture in the semi-arid areas requires a good knowledge of all the environmental factors that are in interaction with the cultures. Among these factors, the large variability of climatic conditions and the recurrence of severe droughts constitute the most specific characteristics of arid and semi-arid areas.

In spite of the studies devoted to improve the understanding of droughts and despite the fact that droughts are frequent phenomena in the Mediterranean area, they remain badly known from various points of view. In general terms, droughts can be defined in different ways, depending on the different meteorological, hydrological agricultural or socio-economic perspectives. From a meteorological point of view, drought is generally defined as an extended period of insufficient rainfall, normally one or more seasons, which affects a large area.

In the present paper, an analysis of meteorological droughts in the area of Siliana, Tunisia is presented. In particular the series of annual and trimester rainfall at three stations, namely Dams Lakhmess, Makthar PF and Agricultural Siliana, were used to identify and characterize droughts both at a local and regional scale by using run concepts (Yevjevich, 1967).

2. Preliminary analysis of rainfall data

The rainfall data which were collected and analysed originate from the stations of Dams Lakhmess, Makthar PF and Siliana Agricole. Table I shows the main characteristics of the selected stations, namely the location coordinates, the altitude and the corresponding areas of influence. Data were obtained mainly from the hydrological directories.

G. Rossi et al. (eds.), Tools for Drought Mitigation in Mediterranean Regions, 79-87.

Table I. Identification of the measuring sites

Station	Latitude	Longitude	Altitude (m)	Area (km^2)
Dam Lakhmess	N°:40G 01 00	E°:7G 92 00	475	127
Makthar PF	N : 39G 84 00	E : 7G 63 00	910	483
Siliana Agricole	N : 40G 09 50	E : 7G 81 00	431	430

The common period of observation for the three stations is from 1967 to 1997. Such 32 years period is believed to be long enough for the analyses carried out. The map of the investigated area with the location of the station is reported in Figure 1.

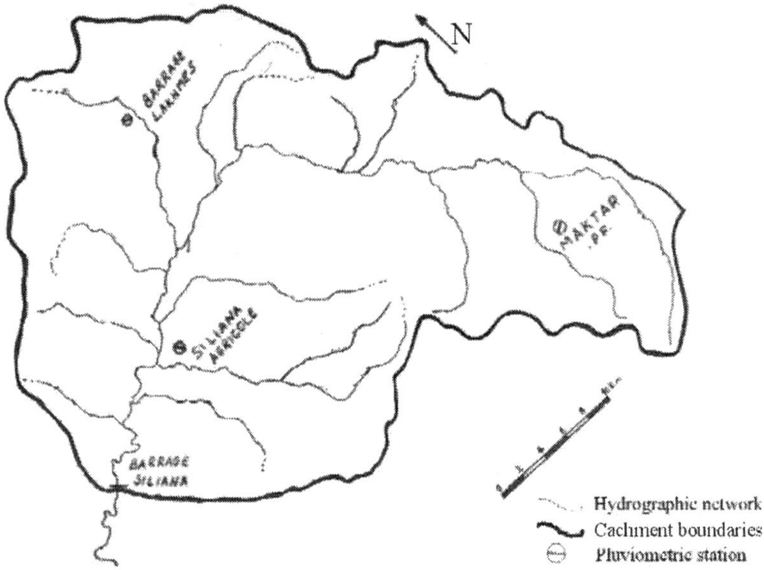

Fig. 1. Map of Siliana catchment

The basic statistical characteristics of the annual rainfall series (location and dispersion parameters) are reported in Table II.

Table II. Location and dispersion parameters (annual rainfall)

Parameters	Dams Lakhmess	Makthar PF	Siliana Agricole
Available sample (years)	32	32	32
Average(mm)	421.5	487.7	390.9
Median(mm)	370.7	483.3	367.6
Standard deviation(mm)	172.2	157.1	132.4
Minimum(mm)	115.0	100.4	103.0
Maximum(mm)	942.1	877.0	681.9
Coefficient of variation	0.4	0.32	0.34
Skewness coefficient	1.1	0.18	0.2

The location parameters (average and median) are quite different which indicates the presence of skewness in the series. The skewness coefficient differs from one station to another and varies from 0.2 to 1.1. The coefficient of variation is greater in Lakhmes than at the other stations.

The stationarity analysis of the series was carried out for each station, considering calendar, water year and trimester time scales. Three tests have been applied namely the Student t-test for linear trends, the Kendall τ-test and the turning points test for randomness, all applied to the stations of the catchment area (for details see DICA, 2000). The results of the three tests for the analysis of the series of rainfall observed at the three stations, considering calendar year, water year (starting in September) and the trimesters are summarized in Table III. Figure 2 reports the linear trend line fitted to the areal rainfall series at calendar and water year.

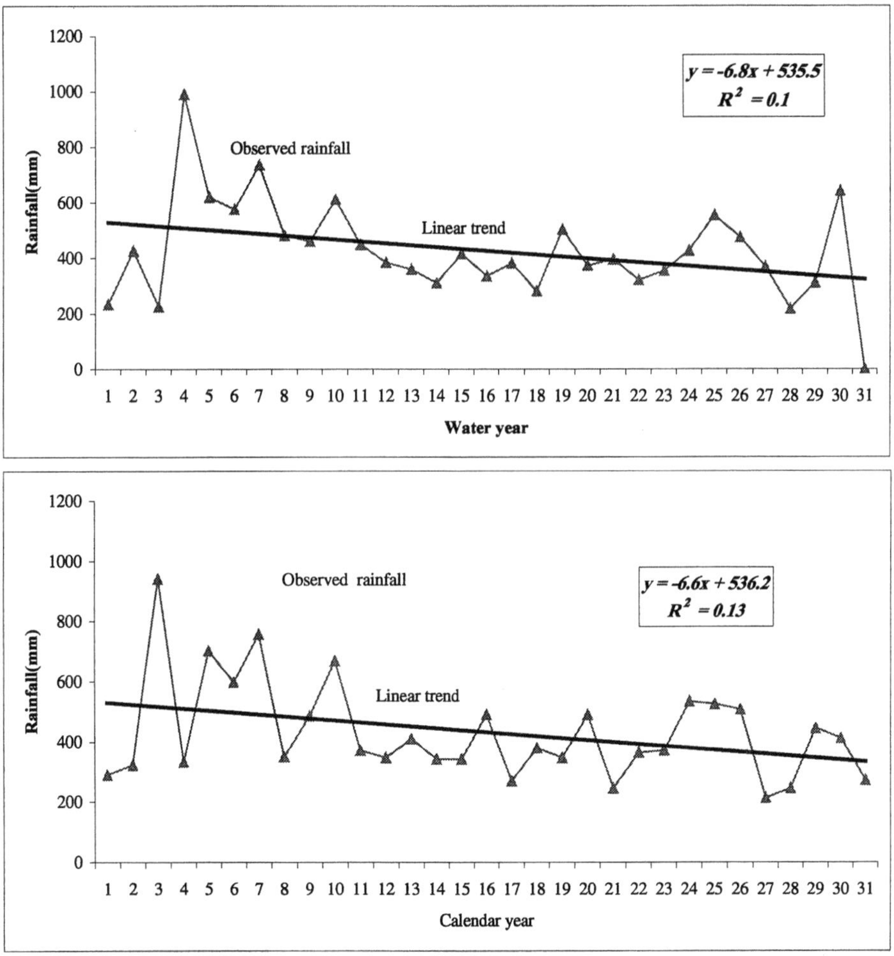

Fig. 2. Analysis of linear trend: rainfall observed at Lakhmes dam 1967-1997

Table III. Results of statistical tests on linear trend significance and series randomness

Station	Time scale	Initial month	Linear trend and Student's t test					Turning points test						Kendall's t test						
			a	b	R	t_c	Hyp	m	E(m)	var(m)	z_c	z	Hyp	p	E(p)	t	Var(t)	z_c	z	Hyp
Dams Lakhmess	Calendar year	Jan	536.15	-6.55	-0.36	-2.66	Hr	19	20.00	5.37	-0.42	1.96	Ha	203	248.0	-0.17	0.02	-1.45	1.96	Ha
	Water year	Sep	535.51	-6.83	-0.34	-2.35	Hr	19	19.33	5.19	-0.14	1.96	Ha	155	232.5	-0.32	0.02	-2.62	1.96	Hr
	Trimester	Sep	183.15	-3.39	-0.25	-1.45	Ha	18	19.33	5.19	-0.58	1.96	Ha	192	232.5	-0.16	0.02	-1.37	1.96	Ha
	Trimester	Dec	156.08	-1.80	-0.31	-1.79	Ha	20	19.33	5.19	0.29	1.96	Ha	167	232.5	-0.27	0.02	-2.22	1.96	Ha
	Trimester	Mar	157.79	-1.72	-0.23	-0.74	Ha	21	19.33	5.19	0.73	1.96	Ha	210	232.5	-0.09	0.02	-0.75	1.96	Ha
	Trimester	Jun	42.74	-0.06	-0.01	0.28	Ha	19	19.33	5.19	-0.14	1.96	Ha	222	232.5	-0.04	0.02	-0.35	1.96	Ha
Makthar PF	Calendar year	Jan	550.80	-3.16	-0.20	-0.60	Ha	19	19.33	5.19	-0.14	1.96	Ha	235	232.5	0.01	0.02	0.08	1.96	Ha
	Water year	Sep	481.14	0.89	0.05	-0.59	Ha	19	18.67	5.01	0.15	1.96	Ha	207	217.5	-0.04	0.02	-0.05	1.96	Ha
	Trimester	Sep	174.43	-1.71	-0.14	-0.97	Ha	20	18.67	5.01	0.60	1.96	Ha	220	217.5	0.01	0.02	0.09	1.96	Ha
	Trimester	Dec	163.21	-0.21	-0.03	0.06	Ha	22	18.67	5.01	1.49	1.96	Ha	224	217.5	0.03	0.02	0.23	1.96	Ha
	Trimester	Mar	151.25	-0.29	-0.05	-0.10	Ha	16	18.67	5.01	-1.18	1.96	Ha	216	217.5	0.00	0.02	-0.05	1.96	Ha
	Trimester	Jun	43.36	0.37	0.10	0.51	Ha	14	18.67	5.01	-2.07	1.96	Hr	201	217.5	-0.07	0.02	-0.58	1.96	Ha
Siliana Agricole	Calendar year	Jan	480.72	-5.04	-0.37	1.49	Ha	19	19.33	5.19	-0.14	1.96	Ha	198	232.5	-0.14	0.02	-1.16	1.96	Ha
	Water year	Sep	429.84	-2.10	-0.14	-1.37	Ha	20	18.67	5.01	0.60	1.96	Ha	196	217.5	-0.09	0.02	-0.76	1.96	Ha
	Trimester	Sep	142.33	-1.08	-0.11	-0.84	Ha	18	18.67	5.01	-0.29	1.96	Ha	202	217.5	-0.06	0.02	-0.54	1.96	Ha
	Trimester	Dec	140.15	-1.40	-0.26	-0.77	Ha	22	18.67	5.01	1.49	1.96	Ha	210	217.5	-0.02	0.02	-0.26	1.96	Ha
	Trimester	Mar	138.52	-1.29	-0.22	-1.24	Ha	17	18.67	5.01	-0.73	1.96	Ha	211	217.5	-0.02	0.02	-0.22	1.96	Ha
	Trimester	Jun	42.90	-0.01	0.00	0.02	Ha	15	18.67	5.01	-1.63	1.96	Ha	224	217.5	0.03	0.02	0.23	1.96	Ha

a: intercept; b: slope; t_c: statistic parameter of Student test; z_c: statistic parameter of Kendall test and Turning points test; R: coefficient of correlation; m: number of turning points; Ha: hypothesis accepted; Hr: hypothesis rejected

From the Table and the figure it can be inferred that the series aggregated at calendar year and water year generally present a negative linear trend. With the significance level of 5 %, the trend is significant only at the station of Lakhmes (tc>2.048). For the test of Kendall, the null hypothesis is accepted for all the stations except for water year at the Lakhmes station. By using the test of the turning point, the null hypothesis is accepted with the significance level of 5% for all the stations. The trimester series present negative trends with a significance level of 5 %. The trend is positive, but not significant, for the summer trimester (June - August) at the station of Makthar. The largest trend is related to the fall trimester. The results confirm the general decrease of the rainfall for the area of Siliana.

In addition to the above tests, the Hubert *et al.* (1989) procedure for segmentation of hydrometeorological series has also been applied. The principle of this procedure is to subdivide the series in *m* segments (*m*>1) so that the average calculated on any segment is significantly different from the average of the next segment. Given a series X_i, any series X_i , $i = i_1, i_2$ with $i_1 >= 1$ and $i_2 <= N$ where $(i_1 < i_2)$ constitutes a segment of the initial series. Any partition of the initial series in *m* segments is a segmentation of order *m* of the initial series. The selected segmentation should have, the minimum quadratic deviation *Dm*

$$D_m = \sum_{K=1}^{K=m} dk \quad with \quad dk = \sum_{i=ik-1+1}^{i=ik} (xi - \overline{X}k)^2$$

This condition is necessary but not sufficient for the determination of the optimal segmentation.

Further, it is necessary to associate a constraint in order to have averages of two contiguous segments significantly different. This constraint is satisfied by applying Scheffé test (Dagnélie, 1970). According to authors (Hubert et al., 1989) this segmentation procedure can be interpreted as a stationarity test. As shown in Table IV, a decrease and significant difference between the averages of different segments of areal rainfall series is present, that might indicate a change in the statistical properties and consequently non-stationarity.

Table IV. Results of Hubert segmentation (test of Sheffè significance level 1%)

Beginning	End	Mean (mm)	Standard deviation (mm)
1966	1968	303.2	192.9
1969	1972	715.9	169.1
1973	1997	473.4	109.9

3. Characterisation and identification of local drought

At site drought identification and characterization has been carried out by means of run concepts, using the software REDIM (DICA, 2000). The results of the analysis, applied to the data observed at the three investigated stations, are summarized in Table V. The drought characteristics are identified by using two annual time scales (water year and calendar year) and trimester time scale. Further, for each series, two threshold have been selected, namely the average and the median. From Table V, where are reported the number of drought periods, the duration, the total deficit and the intensity of each period, it can be inferred that using the calendar year the number of drought period is equal to or higher than that found while considering the water year time scale. On the other hand, deficits and intensities are generally higher when water year time scale is applied. Also, the characteristics of the drought periods, identified by taking the mean as threshold, are higher than those found for a threshold equal to the median, as expected due to the difference of the two thresholds

4. Characterisation and identification of regional drought

Also, regional drought analysis has been pursued by means of REDIM software. The characteristics of the regional drought identified in the area of Siliana are detailed in Table VI. The regional threshold was fixed at 75% of the total area. Again, three time scales were adopted, namely water year, calendar year and the trimesters. According to Table VI, characteristics of drought periods are generally higher at water year time scale, in agreement with the results of the local analysis. Thus, for the management of the water shortages and for a better use of stored water, a careful selection of the appropriate time scale is needed. From Table V and Table VI it can be inferred that the number of periods identified by the study of the regional drought is lower than that found by the identification of the local drought, as expected since the areal threshold selected, i.e. the percentage of areas affected by deficits above which drought occurs, is 75%

5. Conclusions

An analysis of rainfall series during the period 1967-1997 in Siliana catchment, Tunisia, is presented. In particular stationarity analysis of the series has been carried out by means of four statistical tests and at-site and regional drought analysis has been performed by means of run concepts. Stationarity analysis indicates a significant decrease of precipitation in the last years, both at calendar and water year time scale. The results of drought analysis seem to be affected by the time scale adopted. In particular, when water year is applied, drought characteristics duration, deficit and intensity are generally higher that those computed by applying calendar year time scale. This suggests that a careful selection of the time scale is in order for a correct water resources management in the area.

Table V. Characteristics of the identified droughts at 3 stations located in catchment of Siliana

Station	Time scale	Initial month	Threshold(mm)						No. of drought periods	Duration			Cumulated deficit			Drought intensity		
			Year		Trimester					Mean (year)	Max (year)	Min (year)	Mean (year)	Max (year)	Min (year)	Mean (year)	Max (year)	Min (year)
Dams Lakhmess	Calendar year	Jan	h(50%)	379.2					10	1.6	3	1	104.5	299.8	29.5	65.9	149.9	20.6
	Water year	Sep	h(50%)	426.5					5	3.2	7	1	357.8	518.6	200.2	178.7	426.5	65.4
	Trimester	Sep	h(50%)		125.2	123.6	131.1	36.2	26	2.6	7	1	101.3	403.3	3.8	36.0	105.5	3.8
	Calendar year	Jan	average	431.3					8	2.4	5	1	246.4	404.0	81.6	108.8	202.0	69.0
	Water year	Sep	average	434.0					5	3.6	7	1	384.8	571.2	215.2	163.6	434.0	59.8
	Trimester	Sep	average		129.3	126.5	131.8	46.3	26	2.8	8	1	110.9	434.0	2.2	33.4	108.5	2.2
Makthar	Calendar year	Jan	h(50%)	492.8					8	2.0	4	1	158.9	497.3	10.5	63.4	124.3	10.5
	Water year	Sep	h(50%)	517.1					8	1.9	5	1	216.3	682.5	0.1	100.0	196.5	0.1
	Trimester	Sep	h(50%)		124.5	149.8	144.9	37.6	28	2.1	10	1	73.1	421.1	1.2	28.7	81.3	1.2
	Calendar year	Jan	average	508.2					9	2.0	4	1	171.9	558.9	15.4	69.7	139.7	15.4
	Water year	Sep	average	506.1					8	1.5	4	1	198.3	633.1	23.2	115.7	221.6	23.2
	Trimester	Sep	average		149.3	161.4	147.4	48.0	29	2.4	11	1	100.0	543.6	9.3	36.1	87.2	9.3
Siliana Agricole	Calendar year	Jan	h(50%)	411.6					9	1.8	4	1	171.4	495.8	40.0	83.2	165.3	40.0
	Water year	Sep	h(50%)	409.6					7	2.1	5	1	209.3	325.1	57.2	106.6	163.8	57.2
	Trimester	Sep	h(50%)		114.7	116.3	116.3	35.5	28	2.1	8	1	74.3	326.2	1.7	30.6	65.2	1.7
	Calendar year	Jan	average	408.2					9	1.8	4	1	165.4	485.6	36.5	79.8	161.9	36.5
	Water year	Sep	average	403.4					7	2.0	4	1	196.7	299.2	51.0	102.7	157.6	51.0
	Trimester	Sep	average		126.8	116.3	119.3	41.0	28	2.5	8	1	85.5	349.6	13.8	35.0	69.9	12.8

Table VI. Characteristics of regional droughts in the catchme nt of Siliana.

Time scale	Initial month	Threshold	Regional Threshold	No. of Drought periods	Duration						Regional cumulated deficit			Regional drought intensity			Mean regional coverage		
					Mean		Max		Min		Mean	Max	Min	Mean	Max	Min	Mean	Max	Min
			(%)		(year)	(trim)	(year)	(trim)	(year)	(trim)	(mm)	(mm)	(mm)	(mm)	(mm)	(mm)	(%)	(%)	(%)
Year	Jan	h(50%)	75	8	1.5		3		1		132.3	293.9	25.5	75.7	147.0	25.5	96.4	100	87.8
Year	Sep	h(50%)	75	8	1.5		4		1		184.5	456.7	45.5	122.6	207.1	45.5	96.9	100	87.8
Trimester	Sep	h(50%)	75	26		1.8		7		1	69.1	282.3	5.7	34.1	69.6	5.7	96.2	100	87.8
Year	Jan	Mean.	75	8	1.6		4		1		170.0	397.2	44.1	94.3	169.4	44.1	98.1	100	87.8
Year	Sep	Mean.	75	8	1.5		4		1		176.4	432.5	37.9	117.4	202.7	38.0	97.7	100	87.8
Trimester	Sep	Mean.	75	28		2.0		8		1	84.5	322.1	10.1	39.1	78.5	10.1	96.7	100	87.8

References

Dagnélie, P. (1970) Théorie et Méthodes Statistiques. Vol. les presses agronomiques de Gembloux, 451p.

Hubert, P., Carbonnel, J.P., Chaouche, A. (1989) Segmentation des séries hydrométéorologiques. Application à des séries de précipitations et de débits de l'Afrique de l'Ouest. *J. Hydrol.*, 110(3-4): 349-367.

DICA (2000) *Regional Drought Identification Module REDIM*. Department of Civil and Environmental Engineering, University of Catania, Catania.

Yevjevich, V. (1967) An objective approach to definitions and investigation of continental hydrologic drought. *Hydrology paper* n.23, Colorado State University, Colorado.

DROUGHT IDENTIFICATION AND CHARACTERISATION IN JORDAN

A. AL-SALIHI
University of Jordan, Amman, Jordan

Abstract: An analysis of droughts in Jordan was carried out both at local and regional scale is presented. Preliminary analyses of data for stationarity and randomness have been carried out through the application of three tests, namely Student's t test, Kendall's τ test, and Turning point test. The calendar year, water year, and trimester time scales have been considered and October was assumed as the beginning month of the water year. For each series four different thresholds were used, mean, median, mean minus standard deviation, and 35% quantile. The numbers of the drought periods, for the calendar series, are almost the same for all stations. Different periods and percentages are noticed with water year series with no consistency in results between the stations.

Key words: Drought characteristics, local drought, regional drought, Jordan.

1. Introduction

The U. S. Weather Bureau defines drought, according to Dracup *et al.*, (1980) as: *"Lack of rainfall so great and so long continued to affect injuriously the plant and animal life of a place and to deplete water supplies both for domestic purposes and the operation of power plants especially in these regions where rainfall is normally sufficient for such purposes"*

This definition clearly states that the origin of drought is mainly the lack of precipitation that causes a series of consequences including agricultural and hydrological hazards associated with severity and duration of rainfall lack. A number of authors (Ben-Zvi, 1987; Dracup *et al.*, 1980) share this view. At the same time, however, they prefer to adopt an operational definition that distinguishes between meteorological, agricultural, and hydrological droughts with respect to the parameter of prime interest, which could be precipitation (Chang, 1991) (meteorological drought), soil moisture (agriculture drought), or streamflow (Cancelliere *et al.*, 1995) and groundwater levels (hydrological drought). The alternatives come from the complexity of the hydrological process that controls the distribution of rainfall via the various paths within the hydrological cycle, resulting from the interdependence between climatic (Bordi and Sutera, 1999; Rao and Al-Wagdany, 1995), hydrologic, geologic and geomorphic parameters. Since most studies, including the present one, focus on the shortage of water supply, it is preferable to adopt the operational definition of hydrological drought.

However, when dealing with a regional drought, it seems necessary to consider precipitation together with streamflow for two main reasons. Firstly, runoff records can provide satisfactory drought identification for a river basin (Tarboton, 1994), but, in many situations, they fail to detect a drought on a large region. This is evident from the

G. Rossi et al. (eds.), Tools for Drought Mitigation in Mediterranean Regions, 89-102.

results obtained by Ben-Zvi (1987). Moreover, precipitation, as a hydro-meteorological input, is inherently regional in nature. In addition, the nature of streamflow as a dependent variable, controlled by the current and antecedent precipitation as it is interact with the watershed characteristics, and the fact that most streams are regulated by dams and other hydraulic structures, affected as well by the human activities upstream, appear as serious limitation that should be considered when using streamflow as parameter for drought analysis.

Drought real-time forecasting is extremely important for drought management. The methodology suggested by Cancelliere *et al.* (1996), is based on the-Palmer Hydrologic Drought Index (PHDI). However it would require a check of the validity of empirical coefficients computed for USA in the Mediterranean regions.

In the present paper an application of REDIM model (DICA, 2000) for the identification and characterization of droughts in Jordan is presented. In particular, droughts characteristics identified on precipitation series related to two case studies are analysed and compared.

2. Drought identification

According to Yevjevich (1967) an objective identification of drought characteristics can be obtained through the application of the concept of "run" i.e. by superimposing on a selected hydrological variable a truncation level that divides the time series in wet and dry spells. In particular, for each negative spell (run) three characteristics can be computed: *duration,* defined as the time difference between the onset and the end of the drought event, *severity,* defined as the cumulative water deficiency, expressed in the same unit as the considered variable, and *drought intensity,* which is the ratio of severity and duration. Two characteristics are enough to identify the drought, since the third one can be derived from them.

The truncation level is usually selected as the average value of the variable of interest; however a number of arbitrary levels can be adopted. These include specified percentages of the average flow (Clausen and Pearson, 1995), or a given percentage of the nonexceedence probability (Zelenhasic and Salvai, 1987; Rodrigues *et al.*, 1993), or a level defined as one standard deviation below the mean (Ben-Zvi, 1987). Sometimes, it is more convenient to specify the truncation level by considering it as a descriptor of the level of water demand (Yevjevich, 1983; Rossi *et al.*, 1992).

The identification of drought characteristics requires the choice of the discrete time interval to be used in the analysis of the time series. In the theoretical discussion of Dracup *et al.* (1980), as reported in Rossi and Cancelliere, (2002), three main considerations for selecting a suitable discrete time interval are presented. First the interval should be long enough to enable the identification of long and multi-year drought periods, thus maintaining what Rodrigues *et al.* (1993), described as the "macroview trends". Zelenhasic and Salvai (1987), tried to overcome the interruptions resulting from using days as discrete intervals by ignoring the minor deficits and the substantial surpluses occurring in a short period of time. Rodrigues *et al.* (1993) suggested the introduction of the recovery time concept in which a certain surplus should be accumulated to recover from drought conditions, however these methods seem arbitrary and lack objectivity. Second, the sample size of the historical drought record should be adequate for reliable drought analysis. This may contradict the first consideration since a long discrete interval could reduce the number of drought events available for analysis. Dracup *et al.* (1980) suggested a trade-off between these two considerations. Third, short intervals generally tend to exhibit serial correlation that causes serious problems in carrying out frequency and stochastic analyses, since the latter assume a record of independent events. From the previous discussion and from the work of Rodrigues *et al.* (1993), it can be inferred that the seasonal time intervals of 2-4 months are the most suitable as discrete units. In addition, seasonal time units can most easily be related to agricultural activities and management plants, since these are usually determined on a seasonal basis. Thus drought, at any season, will be identified as negative departure from the truncation level assigned for the corresponding season.

Run method can also be extended to the identification and characterization of regional droughts. The regional drought characterisation methodology is based upon the basic concepts of point-drought and drought-affected area, the former being related to a threshold (truncation level or critical level) below which a water shortage exists, whereas the later concept is related to another threshold (critical area) above which the integration of the various areas affected by point-droughts represent a significant portion of the whole region under study.

3. Case Study

Wadi Kaferin and Wadi Hisban catchments drain from the eastern highlands near Wadi Sir, Amman and Hisban westward to the Jordan Valley, and to the Jordan River, a few kilometers north of the Dead Sea. The catchment areas, above the Kaferin Dam, and that of Wadi Hisban at the diversion weir are 161 and 60 km^2, respectively. The valleys are deeply incised in limestone and sandstone rocks, and fall from a maximum elevation of 1067m above sea level to - 150 m below sea level over a distance of 31 km and with approximately half the fall occurring over a distance of 7 km in the upper part of the catchment. The mean catchment elevation is 500 m. The average annual rainfall over Wadi Kaferin and Wadi Hisban catchments ranges from about 170 mm near Kaferin Dam to more than 525 mm near Wadi Sir in the upper part of the Kaferin catchment. Rainfall generally increases with elevation from west to east. The rate of increase is rather rapid at the western slopes and slows down near the mountaintops (Figure 1).

Two rainfall stations exist within the Wadi Kaferin catchment, at Wadi Sir and Na'ur towns. A third station at Kaferin town near the dam was operated until 1974. For the period after 1974, the data from South Shuna climatic station was used for replacing the records of Kaferin station. A good correlation has been found between the records of the stations and the missing data for the stations have been estimated accordingly using regression technique.

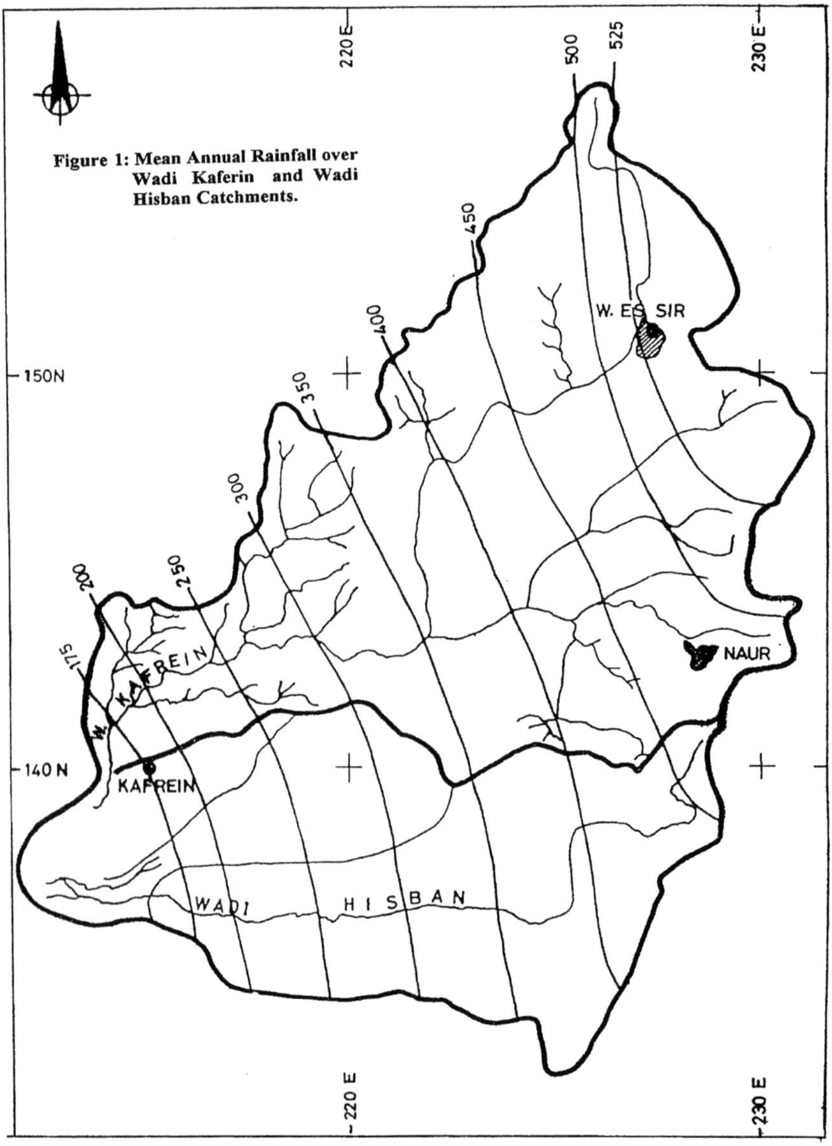

Fig. 1. Mean annual rainfall over Wadi Kaferin and Wadi Hisban catchments

Three rainfall stations affecting the catchment area of Wadi Hisban are considered. These stations are Addasiya, Madaba, and Na'ur stationthe latter already considered for Kaferin Wadi catchment. Records are available up to the year 1996. Table I gives details on the locations and other information on these rainfall stations and the catchment areas.

Table I. Annual characteristics of investigated precipitation series

Local Rainfall Station	Station ID	Location			Annual Rainfall (mm)		Catchment Area, km² (%)	
		Altitude	Latitude N	Longitude E	Mean	St. Deviation	W. Kaferin	W. Hisban
Addasiya	AN0004	750	31° 51'	35° 46'	211.0	140.0		51.00 (85)
Madaba	CC0001	785	31° 43'	35° 48'	332.3	251.0		1.20 (2)
Na'ur	AN0003	800	31° 52'	35° 50'	415.9	162.0	54.74 (34)	7.80 (13)
South Shuna	AM0007	-160	31° 54'	35° 38'	174.3	55.5	45.08 (28)	
Wadi Sir	AN0002	720	31° 57'	35° 49'	578.2	221.5	61.18 (38)	
Regional Area								
Wadi Kaferin					409.9	139.8	161.00 (100)	
Wadi Hisban					240.1	101.9		60.00 (100)

Regression analysis has been used to fill in the missing data. The analysis showed excellent correlation between Na'ur and Wadi Sir, and South Shuna stations.

The catchment areas were divided into Thiessen polygons using Wadi Sir, Na'ur, and South Shuna stations for Wadi Kaferin and Addasiya, Madaba, and Na'ur stations for Wadi Hisban. Area weighted factors were assigned to the stations. In addition, rainfall factors were developed for each station to take into account the rapid increase of rainfall with altitude and stations at the boundary of the area. A combined area-rainfall factor was then obtained by multiplying the area factor with the rainfall factor.

These factors were used to calculate the monthly area rainfall. The average area rainfall for Kaferin catchment was about 410 mm/yr and for Hisban catchment 240 mm/yr.

4. Drought identification

The drought analysis has been applied to the two regional areas, Wadi Kaferin and Wadi Hisban. Preliminary analyses of data and drought identification runs are carried out for different aggregation periods and thresholds.

Preliminary analyses of data for stationarity and randomness have been carried out on the series of each station. In particular three tests, Student's t test, Kendall's τ test, and Turning point test, are applied. The calendar year, water year, and trimester time scales are considered. In all cases, a significance level of 0.05 is used and October as the beginning month of the water year is applied.

Addasiya station and Madaba station records don't pass the randomness Kendall's τ test and the student's t test for both calendar and water year periods where the hypothesis is rejected at 0.05 levels. Moreover, trimester series for Addasiya is rejected for the randomness Kendall's τ test while the Madaba trimester series doesn't pass the turning point test. The other three stations pass all tests and for all periods, as shown in Table II.

Table II. Results of stationarity and randomness tests.

Rainfall Station	Student's t Test			Turning Point Test			Kendall's τ Test		
	Calendar Year	Water Year	Trimester	Calendar Year	Water Year	Trimester	Calendar Year	Water Year	Trimester
Addasiya	X	X	√	√	√	√	X	X	X
Madaba	X	X	√	√	√	X	X	X	√
Na'ur	√	√	√	√	√	√	√	√	√
South Shuna	√	√	√	√	√	√	√	√	√
Wadi Sir	√	√	√	√	√	√	√	√	√

X = Rejected
√ = Accepted

Drought characteristics were computed for all five stations, using annual calendar and water year, rainfall and three months rainfall beginning in October as the first month of the water year in Jordan. For each series four different thresholds were used; mean, median, mean minus standard deviation, and 35% quantile. The results are summarised in Tables III, IV, and V for calendar year, water year, and trimester respectively.

Table III. Identified drought characteristics for trimester (1973-1996). Local analysis

Rainfall Station	Thresholds	Threshold	Drought Periods			Duration			Cumulated Deficit			Drought Intensity			Date
			No.	Total	Total	Mean	Max	Min	Mean	Max	Min	Mean	Max	Min	Max Cum
		(mm)		(trims)	(%)	(trims)			(mm)			(mm/trims)			
Addasiya	**Mean**	133.77-3.27	19	60	65.2%	3.16	9	1	69.3	335.7	0.8	15.6	48.0	0.8	2°94/95-4°95/96
	Median	135.10-0	15	50	54.3%	3.33	8	1	76.4	318.9	2.1	18.5	45.6	2.1	2°94/95-4°95/96
	Mean-Std	60.40-0	5	7	7.6%	1.4	2	1	38.9	76.2	2.3	23.6	38.1	2.3	1°-2°91/92&95/96
	35% Quantile	114.16-0	13	33	35.9%	2.54	6	1	46.8	156.4	0.4	21.4	89.3	0.4	1°-4°91/92&95/96
Madaba	**Mean**	223.87-0	14	68	73.9%	4.86	10	1	173.4	433.8	13.1	44.0	165.4	6.6	2°74/75-2°76/77
	Median	164.60-0	20	52	56.5%	2.6	6	1	41.9	129.2	1.7	19.0	106.1	0.9	3°82/83-4°83/84
	Mean-Std	7.90-0	0	0	0.0%	0	0	0	0.0	0.0	0.0	0.0	0.0	0.0	-
	35% Quantile	140.36-0	17	32	34.8%	1.88	6	1	28.9	86.8	0.9	19.1	81.9	0.6	3°82/83-4°83/84
Na'ur	**Mean**	264.41-0	17	59	64.1%	3.47	7	1	127.7	511.1	13.4	36.3	91.8	6.7	2°94/95-4°95/96
	Median	259.50-0	18	52	56.5%	2.89	7	1	96.1	466.9	2.6	29.3	72.7	2.1	2°94/95-4°95/96
	Mean-Std	139.27-0	7	8	8.7%	1.14	2	1	30.1	104.3	2.8	28.7	104.3	2.8	2°1994/1995
	35% Quantile	223.54-0	15	32	34.8%	2.13	4	1	62.7	188.5	0.2	35.0	188.5	0.1	2°1994/1995
South Shuna	**Mean**	101.64-0	18	54	58.7%	3	7	1	48.5	115.8	2.5	15.6	30.3	1.3	2°94/95-1°95/96
	Median	109.50-0	17	52	56.5%	3.06	8	1	54.1	124.3	0.4	22.3	74.1	0.2	2°92/93-4°93/94
	Mean-Std	63.65-0	6	6	6.5%	1	1	1	14.9	28.3	2.0	14.9	28.3	2.0	2°1994/1995
	35% Quantile	81.72-0	14	32	34.8%	2.29	7	1	19.7	47.1	2.0	10.6	46.3	1.0	2°86/87-4°86/87
Wadi Sir	**Mean**	371.98-0	19	59	64.1%	3.11	7	1	157.0	440.8	7.9	55.3	164.9	3.9	2°94/95-4°95/96
	Median	347.20-0	19	52	56.5%	2.74	7	1	116.1	305.7	2.7	59.8	244.9	1.4	2°88/89-2°89/90
	Mean-Std	217.73-0	6	6	6.5%	1	1	1	31.1	115.4	2.7	31.1	115.4	2.7	2°1994/1995
	35% Quantile	298.74-0	18	32	34.8%	1.78	6	1	39.2	196.4	0.1	28.7	196.4	0.1	2°1994/1995

The number of drought periods for the calendar series are almost the same for the five stations and about 50% of the period is less than the mean. The period 1995 to 1996 is almost common as drought period for the five stations. Different periods and percentages are noticed with water year series with no consistency in results between the stations. But mainly two periods are pointed out which are 1983/1984 to 1985/1986 and 1993/1994 to 1995/1996 with about 60% drought period. The median seams more critical than the mean as threshold. Trimester series exhibit more drought periods.

Table IV. Identified drought characteristics for water year (1973/1974-1995/1996). Local analysis

Rainfall Station	Thresholds	Threshold (mm)	Drought Periods No.	Drought Periods Total (years)	Drought Periods Total (%)	Duration (years) Mean	Duration (years) Max	Duration (years) Min	Cumulated Deficit (mm) Mean	Cumulated Deficit (mm) Max	Cumulated Deficit (mm) Min	Drought Intensity (mm/year) Mean	Drought Intensity (mm/year) Max	Drought Intensity (mm/year) Min	Date Max Cum
Addasiya	Mean	218.30	6	12	50.0%	2	5	1	159.4	404.5	0.3	73.3	202.3	0.3	1995-96
	Median	230.35	6	12	50.0%	2	5	1	183.5	428.6	12.4	85.4	214.3	12.4	1995-96
	Mean-Std	114.61	2	3	12.5%	1.5	2	1	102.8	197.1	8.4	53.5	98.6	8.4	1995-96
	35% Quantile	185.98	4	8	33.3%	2	4	1	160.3	339.9	54.0	86.4	169.9	41.9	1995-96
Madaba	Mean	333.28	9	13	54.2%	1.44	2	1	197.9	347.1	29.5	132.7	173.5	29.5	1995-96
	Median	271.65	8	12	50.0%	1.5	2	1	126.5	223.8	57.8	83.9	111.9	57.8	1995-96
	Mean-Std	99.60	0	0	0.0%	0	0	0	0.0	0.0	0.0	0.0	0.0	0.0	-
	35% Quantile	206.23	6	8	33.3%	1.33	2	1	46.6	103.4	12.3	40.5	103.4	6.1	1995
Na'ur	Mean	415.94	8	13	54.2%	1.63	4	1	198.9	874.0	20.8	96.1	218.5	20.5	1993-96
	Median	399.55	8	12	50.0%	1.5	4	1	172.8	808.4	4.5	84.4	202.1	4.5	1993-96
	Mean-Std	251.67	2	3	12.5%	1.5	2	1	146.8	261.6	32.0	81.4	130.8	32.0	1995-96
	35% Quantile	328.90	4	8	33.3%	2	4	1	171.7	525.8	42.2	65.6	131.5	30.4	1993-96
South Shuna	Mean	176.13	7	13	54.2%	1.86	5	1	88.4	188.3	49.9	58.1	113.5	30.3	1981-85
	Median	163.10	8	12	50.0%	1.5	3	1	62.0	103.5	39.9	49.2	103.5	20.3	1995
	Mean-Std	109.01	4	4	16.7%	1	1	1	18.7	49.4	0.4	18.7	49.4	0.4	1995
	35% Quantile	141.03	7	8	33.3%	1.14	2	1	35.9	81.4	0.5	33.1	81.4	0.5	1995
Wadi Sir	Mean	580.93	8	14	58.3%	1.75	3	1	279.8	530.4	76.9	161.8	265.2	71.5	1995-96
	Median	505.60	8	12	50.0%	1.5	2	1	157.6	379.7	1.6	95.0	189.9	1.6	1995-96
	Mean-Std	360.45	3	3	12.5%	1	1	1	90.6	183.9	24.0	90.6	183.9	24.0	1995
	35% Quantile	454.85	6	8	33.3%	1.33	2	1	122.4	278.3	14.0	96.7	278.3	14.0	1995

Table V. Identified drought characteristics for water year (1973/1974-1995/1996). Local analysis

Rainfall Station	Thresholds	Threshold (mm)	Drought Periods No.	Drought Periods Total (years)	Drought Periods Total (%)	Duration (years) Mean	Duration (years) Max	Duration (years) Min	Cumulated Deficit (mm) Mean	Cumulated Deficit (mm) Max	Cumulated Deficit (mm) Min	Drought Intensity (mm/year) Mean	Drought Intensity (mm/year) Max	Drought Intensity (mm/year) Min	Date Max Cum
Addasiya	Mean	219.80	5	9	39.1%	1.8	3	1	163.2	369.0	11.8	74.0	123.0	11.8	93/94-95/96
	Median	247.50	5	12	52.2%	2.4	7	1	221.9	840.2	17.4	71.8	120.0	14.4	89/90-95/96
	Mean-Std	124.77	2	2	8.7%	1	1	1	124.8	124.8	124.8	124.8	124.8	124.8	91/92&95/96
	35% Quantile	208.84	5	8	34.8%	1.6	3	1	143.7	336.1	0.8	72.1	134.2	0.8	93/94-95/96
Madaba	Mean	341.50	6	15	65.2%	2.5	5	1	307.1	655.0	167.6	146.7	229.7	84.5	74/75-78/79
	Median	261.30	9	12	52.2%	1.33	2	1	82.4	149.5	8.4	64.4	149.5	8.4	1983/1984
	Mean-Std	82.63	0	0	0.0%	0	0	0	0.0	0.0	0.0	0.0	0.0	0.0	-
	35% Quantile	226.14	7	8	34.8%	1.14	2	1	57.5	114.3	4.6	52.7	114.3	4.6	1983/1984
Na'ur	Mean	419.66	5	16	69.6%	3.2	5	1	278.6	749.9	46.9	82.9	187.5	28.1	92/93-95/96
	Median	401.80	6	12	52.2%	2	4	1	188.0	678.5	5.4	69.1	169.6	5.4	92/93-95/96
	Mean-Std	251.15	2	4	17.4%	2	3	1	116.3	220.2	12.4	42.9	73.4	12.4	93/94-95/96
	35% Quantile	370.64	4	8	34.8%	2	3	1	209.4	578.6	28.0	84.1	192.9	28.0	93/94-95/96
South Shuna	Mean	175.40	7	13	56.5%	1.86	4	1	69.8	183.2	22.2	38.4	72.9	14.4	83/84-86/87
	Median	173.10	7	12	52.2%	1.71	4	1	65.5	174.0	19.9	37.0	70.6	18.2	83/84-86/87
	Mean-Std	119.17	3	3	13.0%	1	1	1	24.9	37.7	16.7	24.9	37.7	16.7	1983/1984
	35% Quantile	149.40	5	8	34.8%	1.6	3	1	40.9	84.7	8.0	24.2	46.9	8.0	83/84-85/86
Wadi Sir	Mean	587.86	5	15	65.2%	3	5	1	404.8	557.9	145.0	143.3	186.0	82.2	83/84-85/86
	Median	554.30	7	12	52.2%	1.71	3	1	219.7	457.2	107.0	132.5	234.8	76.8	83/84-85/86
	Mean-Std	359.57	2	2	8.7%	1	1	1	42.3	44.6	40.1	42.3	44.6	40.1	1985/1986
	35% Quantile	449.54	7	8	34.8%	1.14	2	1	54.8	134.5	2.2	52.5	134.5	2.2	1985/1986

Regional run method is applied for both regions that are Wadi Kaferin and Wadi Hisban. Two areal coverage thresholds are used, namely 50% and 75%. For each areal coverage threshold, four different thresholds are used which are mean, median, mean-standard deviation, and 35% quantile, similar to local analysis. Tables VI, VII, and VIII are summarised the results of the regional analyses for calendar year, water year, and trimester periods respectively.

The number of regional droughts is slightly less than the number of droughts identified locally in the single stations. Also the other characteristics, such as duration, cumulated deficit, and drought intensity, present slightly lower values. Comparative figures were developed as a bar chart to compare the drought duration for local and regional analyses for different intensities, as shown in Figures 2, 3, 4, and 5.

Table VI. Identified drought characteristics for calendar year. Regional analysis

Regional Area	Thresholds	Drought Periods No.	Total (years)	Total (%)	Duration Mean (years)	Max	Min	Cumulated Deficit Mean (mm)	Max	Min	Drought Intensity Mean (mm/year)	Max	Min	Mean Coverage Mean (%)	Max	Min	Date Max Cum
	Mean 50% Crit.80.5 km^2	9	14	58.3%	1.56	2	1	169.1	434.0	51.4	103.7	217.0	51.4	85.8	100.0	62.0	1995-96
	Mean 75% Crit.120.75 km^2	7	8	33.3%	1.14	2	1	159.3	307.0	52.6	139.1	307.0	52.6	100.0	100.0	100.0	1995
	Median 50% Crit.80.5 km^2	9	12	50.0%	1.33	2	1	113.3	362.8	15.6	78.2	181.4	15.6	87.1	100.0	62.0	1995-96
W. Kaferin	**Median 75%** Crit.120.75 km^2	6	7	29.2%	1.17	2	1	133.6	270.0	15.6	116.1	270.0	15.6	100.0	100.0	100.0	1995
161 km^2	**Mean-Std 50%** Crit.80.5 km^2	2	2	8.3%	1	1	1	87.7	149.4	26.1	87.7	149.4	26.1	100.0	100.0	100.0	1995
	Mean-Std 75% Crit.120.75 km^2	2	2	8.3%	1	1	1	87.7	149.4	26.1	87.7	149.4	26.1	100.0	100.0	100.0	1995
	35% Quan 50% Crit.80.5 km^2	6	7	29.2%	1.17	2	1	86.5	220.5	15.2	77.2	220.5	15.2	85.7	100.0	62.0	1995
	35% Quan 75% Crit.120.75 km^2	4	4	16.7%	1	1	1	102.9	220.5	39.3	102.9	220.5	39.3	100.0	100.0	100.0	1995
	Mean 50% Crit.30.0 km^2	6	12	50.0%	2	5	1	167.4	427.5	2.1	76.8	213.8	2.1	92.9	100.0	85.0	1995-96
	Mean 75% Crit.45.0 km^2	6	12	50.0%	2	5	1	167.4	427.5	2.1	76.8	213.8	2.1	92.9	100.0	85.0	1995-96
	Median 50% Crit.30.0 km^2	6	12	50.0%	2	5	1	183.7	441.2	12.4	85.2	220.6	12.4	92.9	100.0	85.0	1995-96
W. Hisban	**Median 75%** Crit.45.0 km^2	6	12	50.0%	2	5	1	183.7	441.2	12.4	85.2	220.6	12.4	92.9	100.0	85.0	1995-96
60 km^2	**Mean-Std 50%** Crit.30.0 km^2	2	3	12.5%	1.5	2	1	104.4	201.6	7.2	54.0	100.8	7.2	91.5	98.0	85.0	1995-96
	Mean-Std 75% Crit.45.0 km^2	2	3	12.5%	1.5	2	1	104.4	201.6	7.2	54.0	100.8	7.2	91.5	98.0	85.0	1995-96
	35% Quan 50% Crit.30.0 km^2	4	8	33.3%	2	4	1	157.9	345.0	53.5	84.8	172.5	39.9	97.1	100.0	91.5	1995-96
	35% Quan 75% Crit.45.0 km^2	4	8	33.3%	2	4	1	157.9	345.0	53.5	84.8	172.5	39.9	97.1	100.0	91.5	1995-96

Table VII. Identified drought charactestics for water year. Regional analysis

Regional Area	Thresholds	Drought Periods No.	Total (years)	Total (%)	Duration Mean (years)	Max	Min	Cumulated Deficit Mean (mm)	Max	Min	Drought Intensity Mean (mm/year)	Max	Min	Mean Coverage Mean (%)	Max	Min	Date Max Cum
	Mean 50% Crit.80.5 km^2	5	17	73.9%	3.4	5	1	275.9	468.4	81.3	83.1	117.1	46.1	89.7	100.0	82.0	92/93-95/96
	Mean 75% Crit.120.75 km^2	7	10	43.5%	1.43	3	1	152.2	253.4	36.2	111.0	209.6	36.2	100.0	100.0	100.0	88/89-90/91
	Median 50% Crit.80.5 km^2	7	11	47.8%	1.57	3	1	98.5	209.9	4.1	58.4	116.3	4.1	87.9	100.0	72.0	83/84-85/86
W. Kaferin	**Median 75%** Crit.120.75 km^2	6	6	26.1%	1	1	1	93.3	165.3	37.0	93.3	165.3	37.0	100.0	100.0	100.0	1993/1994
161 km^2	**Mean-Std 50%** Crit.80.5 km^2	2	2	8.7%	1	1	1	35.5	49.8	21.1	35.5	49.8	21.1	86.0	100.0	72.0	1993/1994
	Mean-Std 75% Crit.120.75 km^2	1	1	4.3%	1	1	1	49.8	49.8	49.8	49.8	49.8	49.8	100.0	100.0	100.0	1993/1994
	35% Quan 50% Crit.80.5 km^2	6	9	39.1%	1.5	3	1	70.2	140.2	5.6	50.7	133.1	5.6	78.7	100.0	66.0	83/84-85/86
	35% Quan 75% Crit.120.75 km^2	3	3	13.0%	1	1	1	89.4	133.1	38.8	89.4	133.1	38.8	100.0	100.0	100.0	1993/1994
	Mean 50% Crit.30.0 km^2	5	9	39.1%	1.8	3	1	172.3	413.0	12.9	79.3	137.7	12.9	99.0	100.0	95.0	93/94-95/96
	Mean 75% Crit.45.0 km^2	5	9	39.1%	1.8	3	1	172.3	413.0	12.9	79.3	137.7	12.9	99.0	100.0	95.0	93/94-95/96
	Median 50% Crit.30.0 km^2	6	11	47.8%	1.83	3	1	172.2	460.2	13.4	80.3	153.4	13.4	94.4	100.0	86.0	93/94-95/96
W. Hisban	**Median 75%** Crit.45.0 km^2	6	11	47.8%	1.83	3	1	172.2	460.2	13.4	80.3	153.4	13.4	94.4	100.0	86.0	93/94-95/96
60 km^2	**Mean-Std 50%** Crit.30.0 km^2	2	2	8.7%	1	1	1	112.6	119.2	106.1	112.6	119.2	106.1	91.5	98.0	85.0	1995/1996
	Mean-Std 75% Crit.45.0 km^2	2	2	8.7%	1	1	1	112.6	119.2	106.1	112.6	119.2	106.1	91.5	98.0	85.0	1995/1996
	35% Quan 50% Crit.30.0 km^2	5	8	34.8%	1.6	3	1	143.6	361.3	0.7	71.4	120.4	0.7	92.6	100.0	85.0	93/94-95/96
	35% Quan 75% Crit.45.0 km^2	5	8	34.8%	1.6	3	1	143.6	361.3	0.7	71.4	120.4	0.7	92.6	100.0	85.0	93/94-95/96

Table VIII. Identified drought characteristics for trimester. Regional analysis

Regional Area	Thresholds	Drought Periods Number	Total (trims)	Total (%)	Duration Mean (trims)	Max	Min	Cumulated Deficit Mean (mm)	Max	Min	Drought Intensity Mean (mm/trims)	Max	Min	Mean Coverage Mean (%)	Max	Min	Date Max Cum
W. Kaferin 161 km^2	**Mean 50%** Crit. 80.5 km^2	18	60	65.2%	3.33	7	1	115.5	373.7	9.5	36.6	120.6	4.7	90.4	100.0	72.0	86°-92°
	Mean 75% Crit. 120.75 km^2	18	40	43.5%	2.22	7	1	90.2	311.5	8.2	46.1	120.6	4.1	100.0	100.0	100.0	86°-89°
	Median 50% Crit. 80.5 km^2	21	49	53.3%	2.33	7	1	75.4	232.9	2.9	40.9	190.1	1.5	86.6	100.0	62.0	78°-84°
	Median 75% Crit. 120.75 km^2	15	27	29.3%	1.8	7	1	85.0	232.9	2.9	62.6	190.1	1.5	100.0	100.0	100.0	78°-84°
	Mean-Std 50% Crit. 80.5 km^2	6	6	6.5%	1	1	1	25.1	87.2	6.5	25.1	87.2	6.5	83.3	100.0	62.0	86°
	Mean-Std 75% Crit. 120.75 km^2	3	3	3.3%	1	1	1	36.9	87.2	11.0	36.9	87.2	11.0	100.0	100.0	100.0	86°
	35% Quan 50% Crit. 80.5 km^2	13	30	32.6%	2.31	7	1	46.8	151.7	1.6	29.5	151.7	0.8	88.9	100.0	67.0	86°
	35% Quan 75% Crit. 120.75 km^2	11	16	17.4%	1.45	4	1	38.3	151.7	3.3	32.2	151.7	3.3	100.0	100.0	100.0	86°
W. Hisban 60 km^2	**Mean 50%** Crit. 30.0 km^2	19	60	65.2%	3.16	9	1	73.8	357.7	1.5	16.8	51.1	1.5	96.1	100.0	85.0	86°-92°
	Mean 75% Crit. 45.0 km^2	19	60	65.2%	3.16	9	1	73.8	357.7	1.5	16.8	51.1	1.5	96.1	100.0	85.0	86°-92°
	Median 50% Crit. 30.0 km^2	16	46	50.0%	2.88	8	1	73.8	244.5	1.8	27.4	125.0	1.8	96.2	100.0	85.0	69°-76°
	Median 75% Crit. 45.0 km^2	16	46	50.0%	2.88	8	1	73.8	244.5	1.8	27.4	125.0	1.8	96.2	100.0	85.0	69°-76°
	Mean-Std 50% Crit. 30.0 km^2	5	7	7.6%	1.4	2	1	37.0	67.4	3.9	23.8	43.7	3.9	95.4	98.0	85.0	89°-90°
	Mean-Std 75% Crit. 45.0 km^2	5	7	7.6%	1.4	2	1	37.0	67.4	3.9	23.8	43.7	3.9	95.4	98.0	85.0	89°-90°
	35% Quan 50% Crit. 30.0 km^2	14	27	29.3%	1.93	4	1	45.0	153.7	0.5	23.6	102.0	0.5	96.5	100.0	85.0	89°-92°
	35% Quan 75%	14	27	29.3%	1.93	4	1	45.0	153.7	0.5	23.6	102.0	0.5	96.5	100.0	85.0	89°-92°

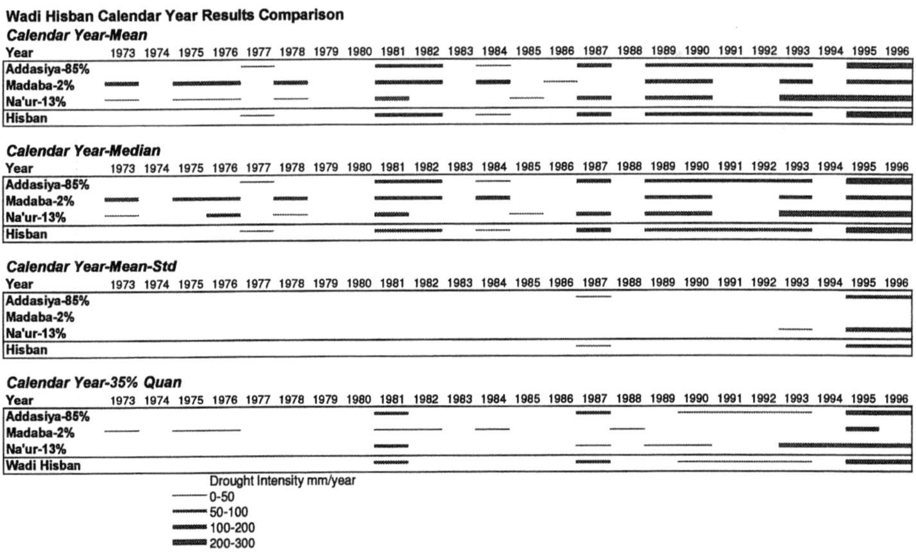

Fig. 2. Wadi Hisban calendar year results comparison

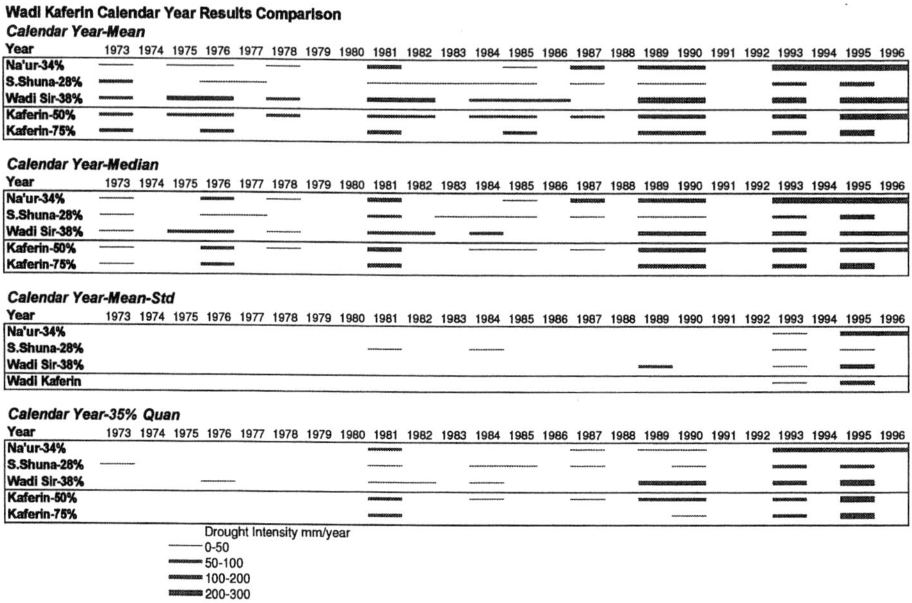

Fig. 3. Wadi Kaferin calendar year results comparison

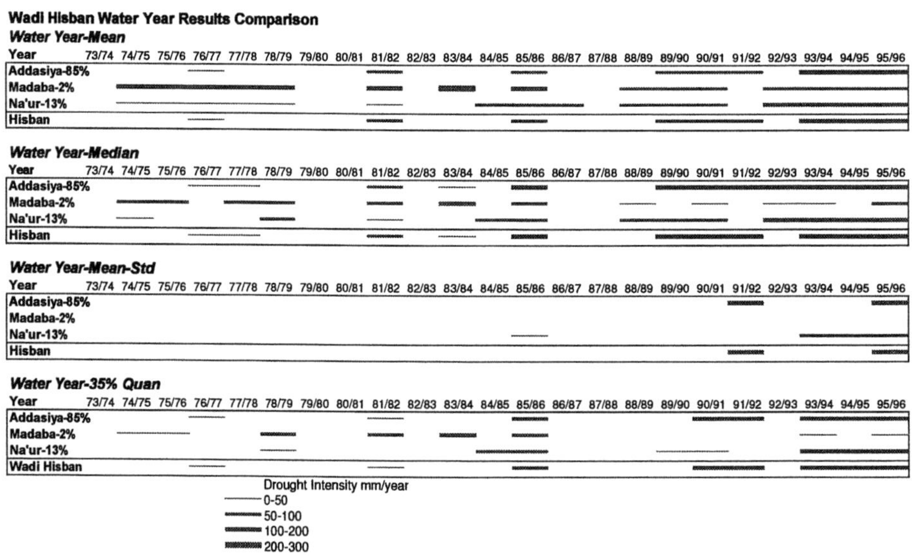

Fig. 4. Wadi Hisban water year results comparison

The comparison of results shows that the drought phenomenon is more severe on the Hisban region with respect to Kaferin reservoir. For example when the median is adopted as threshold and 50% as areal threshold, on the basis of the calendar year the droughts present mean duration and cumulated deficit (the median) of 2 years and

183.7 mm at Hisban and 1.33 years and 113.3 mm at Kaferin reservoir. The most severe drought of the period 1973-1996 is 2 years long (1995-1996) with 441.2 mm of cumulated deficit at Hisban region and the same two years with a lower cumulated deficit of 362.75 mm at Kaferin reservoir.

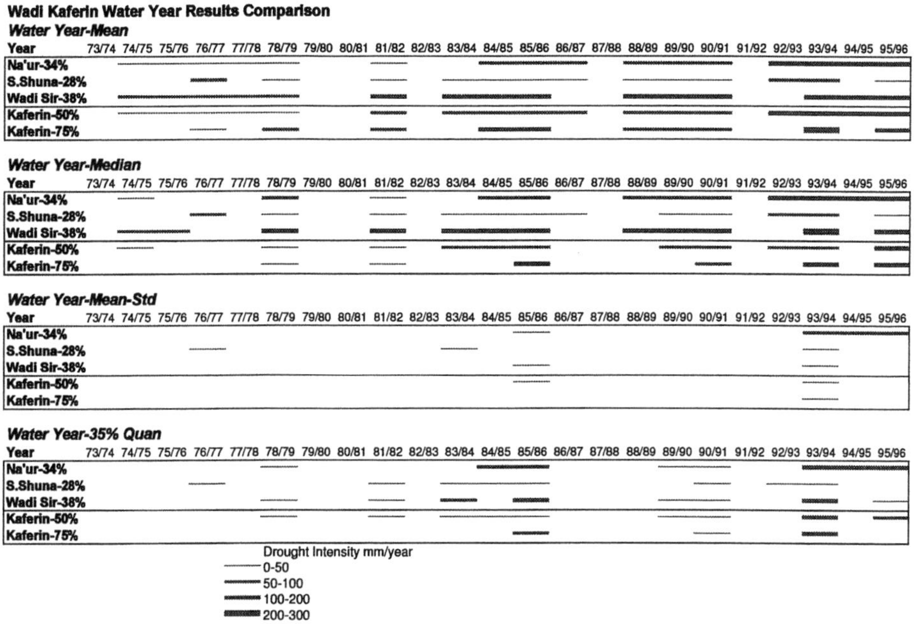

Fig. 5. Wati Kaferin water year results comparison

The computed drought characteristics present differences when either calendar or water year or even if the trimester is used. For example the worst drought of Hisban region lasts for two years (1995-1996) with cumulated deficit of 441.2 mm, as mentioned above, if computed on calendar year and three years (1993/1994-1995/1996) with cumulated deficit 460.2 mm if computed on water year basis. This will give drought intensity of 260.6 mm/year for calendar year and 153.4 mm/year for water year.

Trimester series analyses show different drought characteristics and the main drought period in between 86° and 92° trimester. The maximum cumulated deficit is 244.5 mm for Hisban region and 232.9 for Kaferin reservoir. These droughts last for seven trimesters for both regions with 34.9 mm/trimester for Hisban region and 33.3 mm/trimester for Kaferin reservoir as drought intensity.

The monthly mean precipitation deficit and temperature departure from mean were studied. A correlation coefficient of 0.575 was found and the mean deficit was estimated to increase by 5 mm for each 1°C temperature increase above mean as shown in Figure 6. The increase in temperature seams to increase the demand and decrease the supply, which enhance the drought condition.

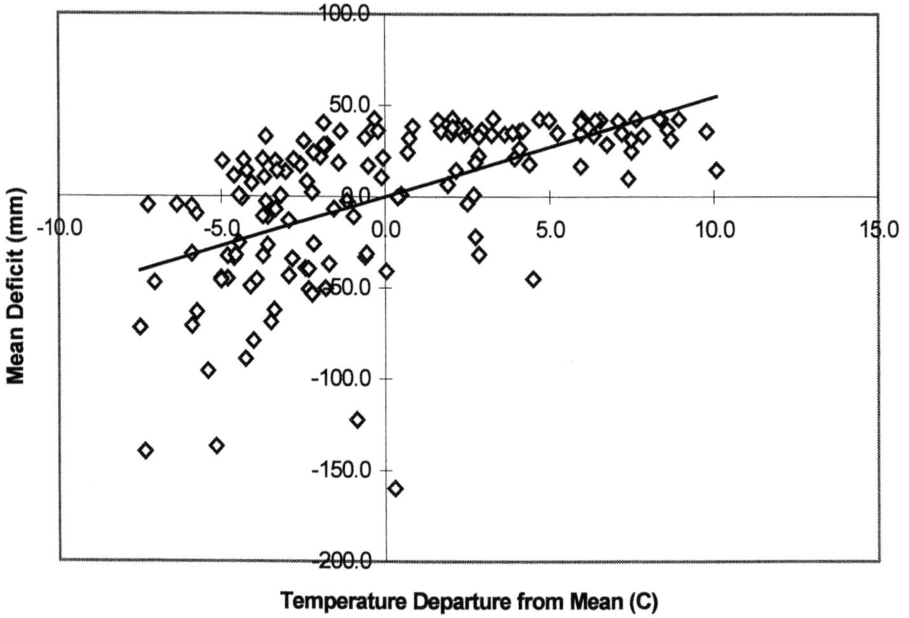

Fig. 6. Mean deficit vs. temperature departure from mean

Frequency analysis is carried out for streamflow of Wadi Hisban, Wadi Kaferin and Kaferin-Hisban. The probability of the drought characteristics is estimated by fitting several probability functions to drought duration and cumulative deficit computed on historical series. Preference can be given to probability distributions theoretically. Frequency distribution is fitted for the maximum deficit and duration. The mean monthly deficit and duration for different return periods are estimated. Pearson and Log Pearson Type III are the fitted distributions for deficit and duration. A strong positive correlation (above 0.7) between maximum monthly mean deficit and duration for the three-streamflow data was found as shown in Figure 7. This result agreed with McKee et al., (1993) findings.

5. Conclusions

An analysis of at-site and regional drought in Jordan has been presented with reference to precipitation series observed in two catchment areas (wadis). A preliminary analysis of stationarity and randomness in the series carried out by means of three tests indicates the presence of trends in two out of six investigated series, both at calendar and trimester series.

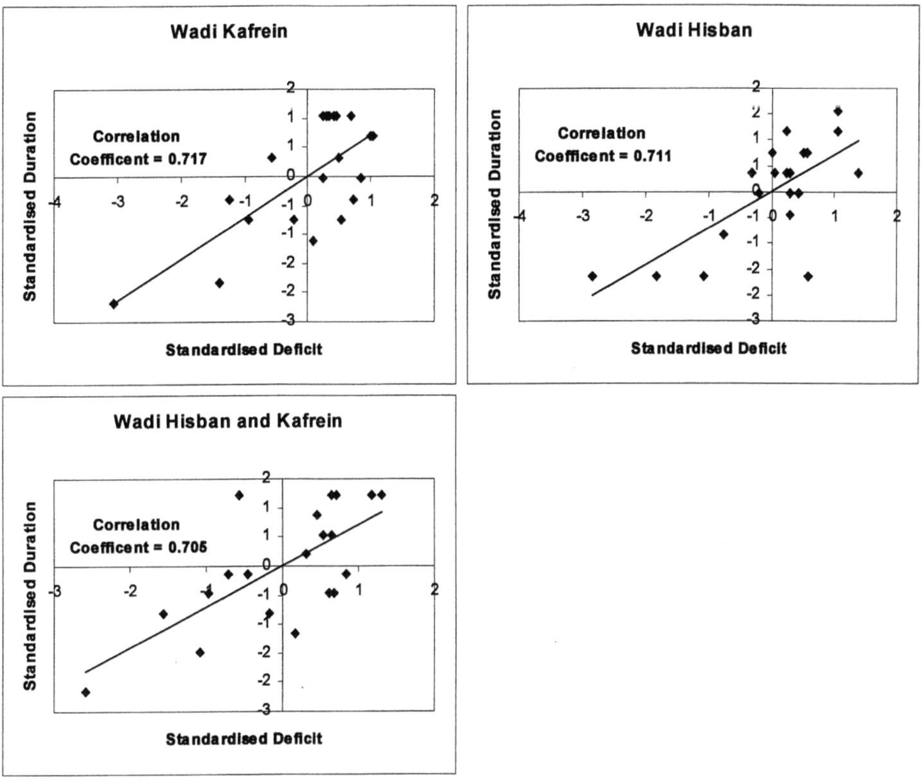

Fig. 7. Strong positive correlation between deficit and duration

Drought analysis results are variable according to the different time scales and thresholds adopted. In particular it appears that droughts identified at water year time scale are more severe and longer than those identified at calendar year. An analysis of the relationship between the series of drought duration and cumulative deficit shows that there is a strong linear dependence between these two variables. Further, the Pearson and Log Pearson Type III were identified as the most suitable distributions for streamflow deficit and duration for both wadis.

References

Ben-Zvi, A. (1987) Indices of hydrological drought in Israel. *J. Hydrol.* 92: 179-191.

Bordi, I. and Sutera, A. (1999) Identification of drought: meteorological aspects. *Proceedings of workshop on Territorial Planning and Coping with Effects of Drought*, Taormina 3-5 June, Italy.

Cancelliere, A., Ancarani, A. and Rossi, G. (1995) Identification of drought periods on streamflow series at different time scales. In: N.X. Tsiourtis (Ed.) *Water Resources Management under Drought or Water Shortage Conditions* (Proc. of the EWRA 95 Symposium, Nicosia), Cyprus, pp. 49-57.

Cancelliere, A., Rossi, G., Ancarani, A., 1996. Use of Palmer Index as a drought indicator in Mediterranean Regions. *Proceedings of the Conference From flood to drought*, SunCity, SouthAfrica, 5-7 august, 1996.

Chang, T. J. (1991) Investigation of Precipitation Droughts by Use of Kriging Method. *J. Irrig. Drain. Engng.* 117(6): 935-943.

Clausen, B. and Pearson, C.P. (1995) Regional frequency analysis of annual maximum streamflow drought. *J. Hydrol.* 173: 111-130.

DICA (2000) *Regional Drought Identification Module REDIM.* Department of Civil and Environmental Engineering, University of Catania, Catania.

Dracup, J. A., Lee, K. S. and Paulson, E.G. (1980) On the Definition of Droughts. *Water Resour. Res.* 16(2): 297-302.

Mckee, T. B., Doesken, N. J. and Kleist, J. (1993) The Relationship of Drought Frequency and Duration to Time Scales. *Proceedings of the Eighth Conference on Applied Climatology.* 17-22 January, Anaheim, California, USA, pp. 179-184.

Rao, A. R. and Al-Wagdany, A. (1995) Effects of Climatic Change in Wabash River Basin, *J. Irrig. Drain. Engng.* 121(2): 207-215.

Rodrigues, R., Santos, M.A. and Correia, F.N. (1993) Appropriate time resolution for stochastic drought analysis. In: J.D. Salas, R. Haroe and J. Marco-Sgura (Eds.) *Stochastic Hydrology and its Use in Water Resources Systems Simulation and Optimisation.* Kluwer, Netherlands, pp. 247-265.

Rossi, G., Benedini, M., Tsakiris, G. and Giakoumakis, S. (1992) On Regional drought estimation and analysis. *Water Resour. Manage.* 6: 249-277.

Tarboton, D. G. (1994) The source hydrology of severe sustained drought in the south-western United States. *J. of Hydrol.* 161: 31-69.

Yevjevich, V. (1967) An objective approach to definitions and investigations of continental hydrologic drought. *Hydrology paper* n. 23, Colorado State University, Fort Collins, Colorado.

Yevjevich, V., Da Cunha, L. and Valchos E. (Eds.) (1983) *Coping with Droughts.* Water Resources Publications, Littleton, Colorado.

Zelenhasic, E. and Salvai, A. (1987) A Method of Streamflow Drought Analysis. *Water Resour. Res.* 23(1): 156-168.

DROUGHTS IN SICILY AND COMPARISON OF IDENTIFIED DROUGHTS IN MEDITERRANEAN REGIONS

A. CANCELLIERE and G. ROSSI
Department of Civil and Environmental Engineering, University of Catania, Italy

Abstract: An analysis of at-site and regional drought characteristics of precipitation series in Sicily is presented, based on the application of the run method to two areas in Sicily as well as to the whole Sicilian territory. The drought analysis has been preceded by stationarity and randomness testing of the precipitation series. Comparison of the results of similar investigations carried out in Portugal, Tunisia and Jordan has been carried out by analysing the linear trend results and the dimensionless drought characteristics. Further, contemporaneous occurrences of deficit have been assessed by means of conditional probability concepts. Results confirm that there is a general increase in the frequency, severity and duration of droughts in the recent years over the Mediterranean, which can partly be explained in terms of a general negative trend of precipitation.

Key words: droughts, run method, stationarity, Mediterranean region

1. Introduction

Drought identification and characterisation is of major interest for any investigation oriented to face the problems of water resources planning and management under drought conditions. The computation of drought characteristics is the first step for assessment of drought impacts and for selecting the necessary measures for coping with water shortage consequent to drought.

This paper aims to present and compare the results of the drought identification and characterization analyses carried out within the first Task of the European Commission INCO-DC project "A DSS for mitigation of drought impacts in the Mediterranean regions". The paper is organized as follows: first, the results of the application of REDIM software for stationarity and randomness analysis and drought identification by means of run method (DICA, 2000) to the Italian Case-study are presented and discussed. Such results are also compared with an enlarged analysis extended to the whole Sicily region. Then, the stationarity analysis and the drought identification and characterization carried out on the different Case-studies selected by the partners of the Project, located in Portugal, Italy, Tunisia and Jordan, are presented and compared in terms of dimensionless characteristics as well as deficit contemporaneity.

For the methodological aspects, the reader is referred to Rossi and Cancelliere, (2002).

G. Rossi et al. (eds.), Tools for Drought Mitigation in Mediterranean Regions, 103-122.

2. Drought analysis in Italian case study

2.1. INVESTIGATED AREAS AND PRELIMINARY ANALYSIS OF HYDROLOGICAL SERIES

The Italian Case Study in the project consists of the irrigation system of the Catania Plain. It includes the Pozzillo reservoir on Salso river and the irrigated district which can be considered one of the most important in southern Italy. The irrigation district is mainly supplied by the Salso river inflows regulated by Pozzillo reservoir (net capacity 123 10^6 m^3). The system lies in a drought-prone area which experiences frequent periods of water scarcity due to significant reduction of precipitation and runoff.

As shown in Figure 1, two areas have been selected for the analysis, both located in the Eastern part of Sicily island. The first area is the catchment of the Salso river (tributary of Simeto river) at Pozzillo dam, with a catchment surface of 577 km^2. The second area covers the central part of Catania Plain irrigation district and has an irrigable surface of about 47000 ha.

Drought identification in both areas of the Italian case-study has been carried out through REDIM software. For each area, four precipitation gauging stations with the related polygons areas computed by Thiessen's method have been considered. In particular the stations of Capizzi, Cerami, Gagliano Castelferrato and Nicosia for Salso catchment and the stations of Catenanuova, Lentini, Motta S. Anastasia and Ramacca for Catania Plain district have been selected for the analysis. Table 1 reports the main characteristics of the investigated stations, while their location is depicted in Figure 1.

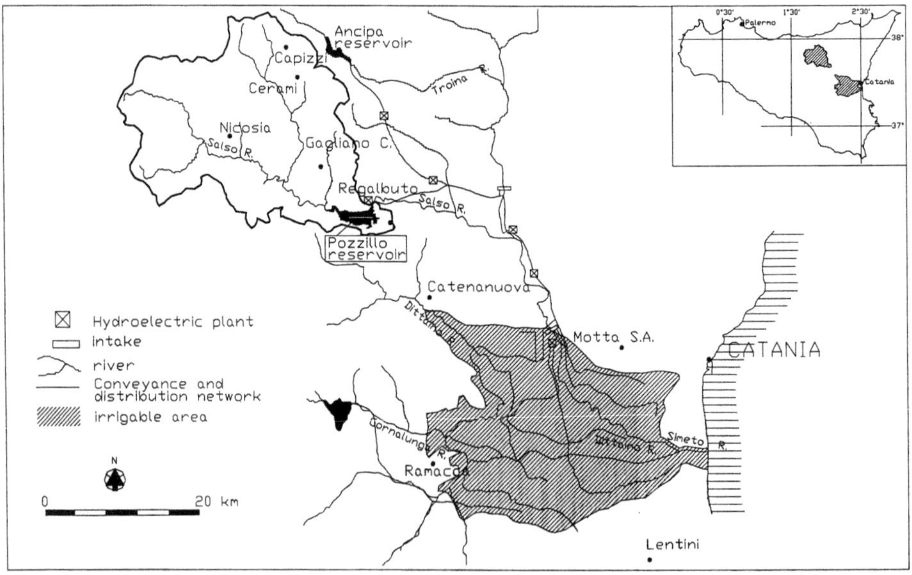

Fig. 1. Location of investigated areas for the Italian Case Study

The preliminary assessment of the stationarity and randomness during the period 1921-1996 was carried out on the series of each station, considering both calendar and water year (with September as the beginning month) and trimester time scales. Three tests were considered, namely the linear trend test, the Kendall's τ test and the turning point test. For the details, the reader is referred to Rossi and Cancelliere (2002), namely the value of Student's t for the correlation coefficients of the linear trend and the value of the z statistics of the other tests.

Table I. Characteristics of investigated stations for Italian case study

	Stations	Altitude *m a.s.l.*	Annual Mean *(mm)*	Stand. Dev. *(mm)*	Polygon area *(% of total area)*
Pozzillo Catchment	Capizzi	1089	864	202	11
	Cerami	980	697	179	18
	Gagliano	837	609	161	41
	Nicosia	650	721	212	30
Catania Plain district	Catenanuova	185	468	144	49
	Lentini	43	650	237	12
	M.S. Anastasia	254	469	168	11
	Ramacca	257	554	205	28

The results of the tests are presented in Figure 2 for the stations located within the Pozzillo reservoir catchment and in Figure 3 for the stations within Catania Plain district.

From the plots, it can be inferred that calendar and water year series present generally a negative linear trend (3 out of 4 stations) for both the areas. The trend is significant at 0.05 level only at Cerami (Pozzillo basin) and Ramacca (Catania Plain). The same stations do not pass the randomness Kendall's τ test. According to the turning point test, the hypothesis of randomness is rejected at 0.05 level only for the calendar year series at Gagliano stations.

Trimester series present generally negative trends also (significant at 0.05 level only for Ramacca station in 3 trimesters). However positive trends (always not significant) are observed for summer trimester June-August in 2 stations of Pozzillo reservoir basin and in 2 stations of Catania Plain. Further, a positive trend for winter trimester is present in 2 stations (Catenanuova and Motta) of Catania Plain and positive trends in autumn and spring trimesters occur at Nicosia station.

The results seem to confirm a general decrease of rainfall in the last decades for this part of Sicily. However, the hypothesis of different trends for precipitation series related to different seasons does not present a decisive validation.

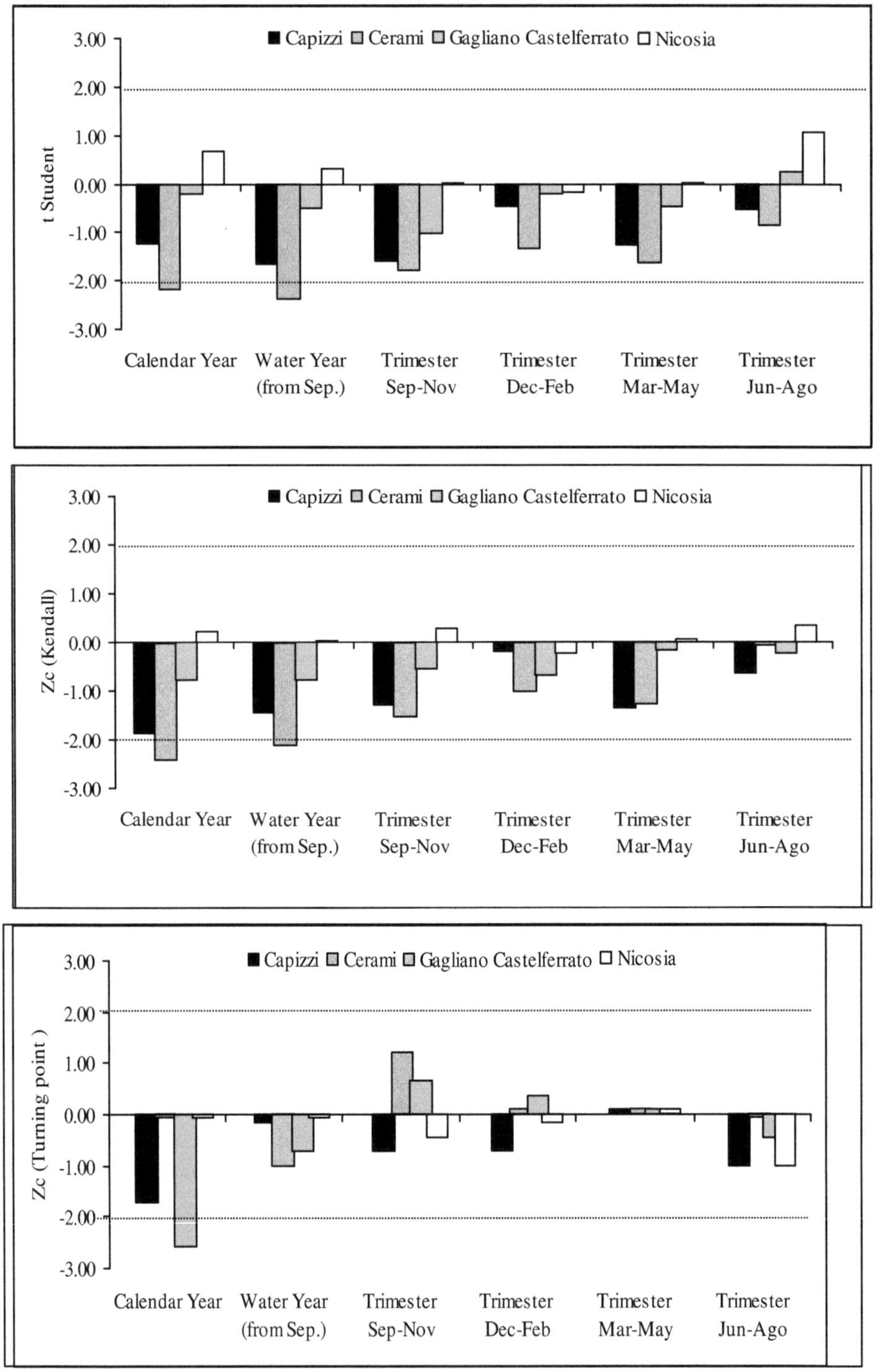

Fig. 2. Stationarity analysis results for Pozzillo catchment precipitation stations

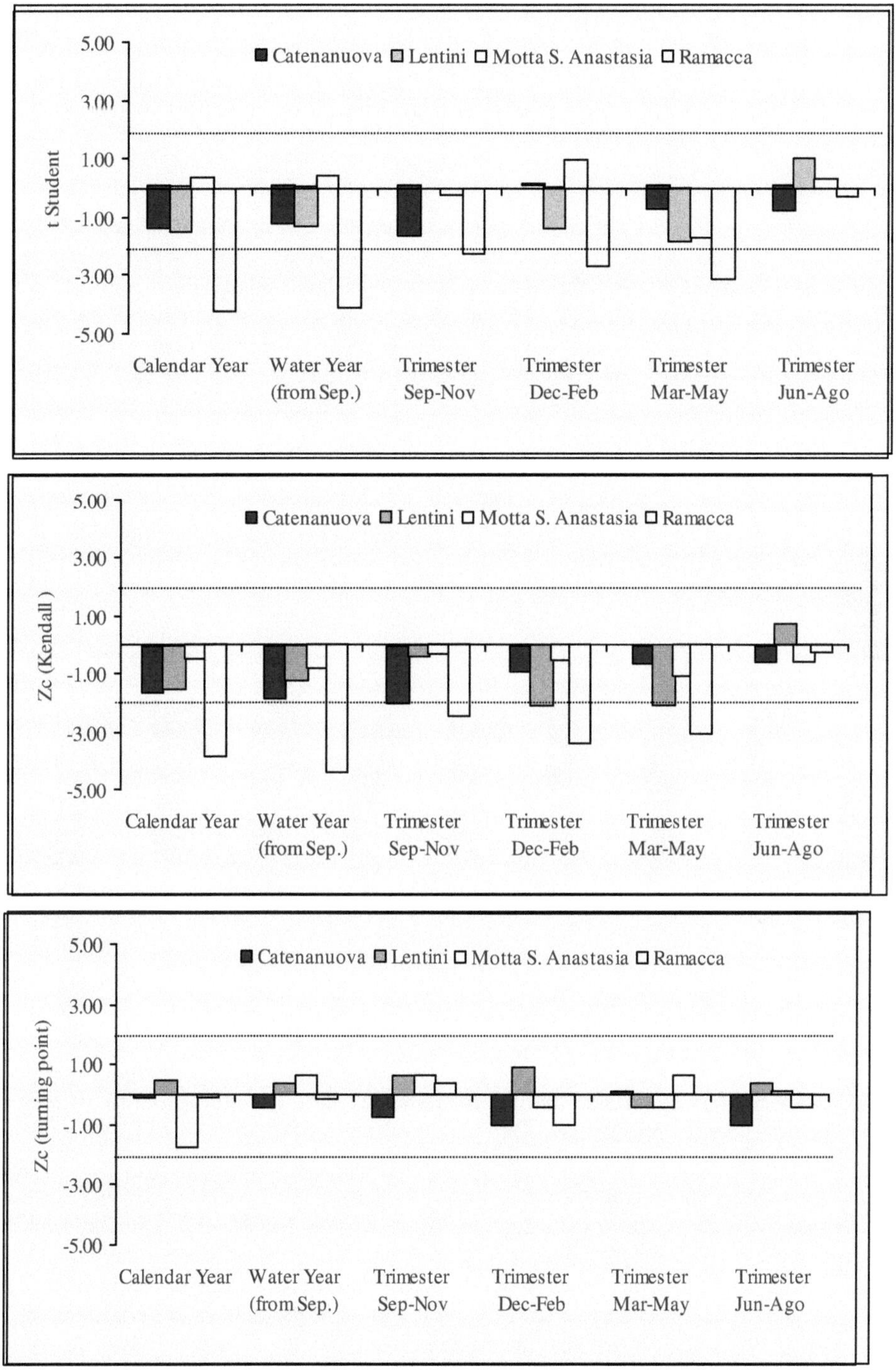

Fig. 3. *Stationarity and randomness analysis results for Catania plain precipitation stations*

2.2. AT-SITE AND REGIONAL DROUGHT ANALYSIS

Drought characteristics were computed for the two investigated areas, using annual (calendar and water year) precipitation and three-months precipitation series beginning in September. For each series two different thresholds were used, namely median and average values.

The number of drought periods and the main parameters of duration, cumulative deficit and drought intensity are listed in table IIa for the stations within Pozzillo reservoir catchment and in table IIb for the stations located in the Catania Plain district. A dimensionless analysis was also carried out in order to compare cumulated deficits identified at stations with different climatic characteristics. More specifically, Figure 4 reports the mean and max cumulated deficits divided by the mean annual precipitation value, at water year time scale and using the mean value as a threshold. It can be inferred from the figure that stations located in Pozzillo catchment present droughts with lower mean and max dimensionless cumulated deficit than those located in Catania plain. In particular the maximum dimensionless cumulated deficit is observed in Ramacca station (Catania plain), with a value of 3.3 times the mean annual precipitation. Such particularly high value can probably be explained in terms of the significant negative trend exhibited by the precipitation series at this station.

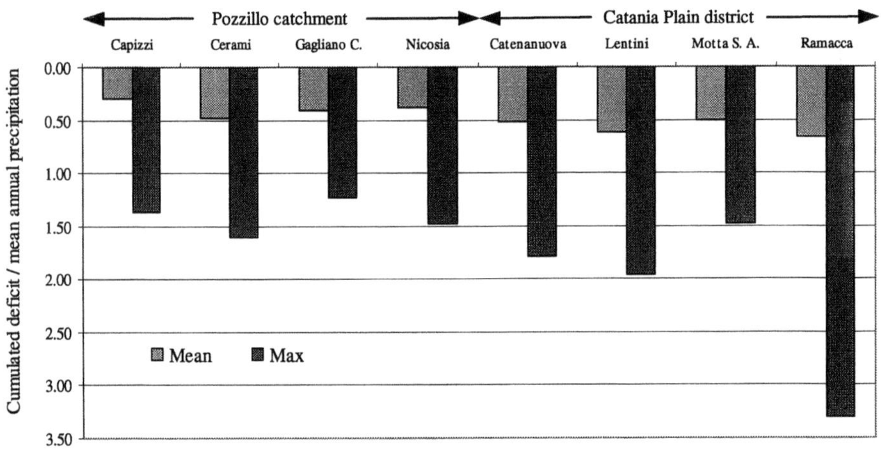

Fig. 4. Dimensionless cumulated deficits identified at the 8 analysed stations (water year time scale, threshold mean).

The characteristics of identified regional droughts in the two areas are listed in table IIIa and IIIb. Since the areal coverage threshold was fixed as 75% of the total area, the number of regional droughts is less than the number of droughts identified in the single stations. Also the other characteristics (duration, cumulated deficit and drought intensity) present generally lower values. Note that the minimum duration (not shown in the table) is always one year or one trimester.

Table II a. Characteristics of identified droughts at 4 stations located in the Pozzillo reservoir catchment

Station	Time scale		Threshold (mm)	No. of drought	Duration				Cumulated deficit			Drought intensity			
					mean		max		mean	max	min	mean		max	
					(year)	(trim.)	(year)	(trim.)	(mm)	(mm)	(mm)	(mm/yr)	(mm/trim.)	(mm/yr)	(mm/trim.)
CAPIZZI	calendar year	h(50%)	862.7	20	1.90		7		286.2	644.1	9.8	151.8		391.7	
	water year	h(50%)	855.0	22	1.73		6		248.5	1151.0	38.0	141.9		333.0	
	trimester	h(50%)	233.0, 350.0, 194.0, 42.8	77		1.97		8	110.9	530.1	2.0		55.0		233.0
	calendar year	average	864.1	20	1.90		7		288.9	648.4	11.2	153.2		393.1	
	water year	average	859.3	22	1.77		6		256.1	1177.0	42.3	142.6		337.3	
	trimester	average	249.5, 359.1, 198.1, 52.6	76		2.21		8	133.8	609.2	1.5		60.0		242.1
CERAMI	calendar year	h(50%)	699.0	18	2.11		4		298.9	811.4	25.0	151.2		361.0	
	water year	h(50%)	666.6	17	2.24		6		234.4	533.2	14.8	114.5		237.6	
	trimester	h(50%)	203.1, 257.0, 147.0, 41.6	68		2.24		6	105.3	470.7	3.6		46.7		139.0
	calendar year	average	697.2	18	2.06		4		295.2	804.3	23.2	150.6		359.2	
	water year	average	691.9	15	2.87		14		335.4	1104.4	111.9	139.8		262.9	
	trimester	average	218.9, 272.1, 148.1, 52.7	68		2.57		8	131.8	529.0	3.7		50.3		154.1
GAGLIANO C.	calendar year	h(50%)	586.8	15	2.53		11		240.0	1027.5	28.8	113.7		327.8	
	water year	h(50%)	565.0	19	2.00		6		152.0	457.6	19.0	75.7		157.5	
	trimester	h(50%)	195.0, 224.0, 124.2, 29.0	74		2.03		7	88.5	401.2	1.0		43.7		125.0
	calendar year	average	609.1	15	2.87		12		299.4	1277.1	51.1	119.6		350.1	
	water year	average	607.7	19	2.42		7		246.6	751.5	6.7	100.5		200.2	
	trimester	average	201.5, 230.7, 129.8, 45.8	74		2.34		11	108.7	453.2	0.7		46.2		131.7
NICOSIA	calendar year	h(50%)	683.0	18	2.11		7		247.8	840.2	22.0	116.6		273.6	
	water year	h(50%)	693.2	20	1.90		6		234.7	907.2	23.2	111.9		193.5	
	trimester	h(50%)	207.2, 284.8, 154.0, 33.0	74		2.05		9	103.1	541.8	0.8		49.4		148.8
	calendar year	average	721.3	17	2.53		9		354.0	1155.6	63.5	151.5		311.9	
	water year	average	718.8	21	1.95		6		272.3	1061.0	19.7	128.8		219.1	
	trimester	average	222.9, 293.8, 158.1, 44.0	72		2.39		19	128.5	1101.3	0.7		51.2		149.8

Table II b. Characteristics of the identified droughts at 4 stations located in the Catania Plain irrigation district

Station	Aggregation time scale	Threshold (mm)		No. of drought	Duration				Cumulated deficit			Drought intensity			
					mean		max		mean	max	min	mean		max	
					(year)	(trim.)	(year)	(trim.)	(mm)	(mm)	(mm)	(mm/yr)	(mm/trim.)	(mm/yr)	(mm/trim.)
CATENANUOVA	calendar year	h(50%)	453.0	18	2.11		7		211.5	714.4	14.0	103.4		239.0	
	water year	h(50%)	434.0	19	2.00		7		159.4	611.6	25.0	77.0		154.0	
	trimester	h(50%)	141.0, 155.0, 101.0, 23.0	72		2.10		8	72.8	242.0	1.0		34.9		100.0
	calendar year	average	467.7	19	2.16		7		230.8	817.1	4.7	105.1		213.7	
	water year	average	465.6	18	2.33		7		240.1	832.7	31.6	100.9		185.6	
	trimester	average	162.9, 168.5, 99.3, 34.9	69		2.57		9	104.8	360.4	4.9		39.8		113.5
IENTINI	calendar year	h(50%)	633.1	20	1.90		5		314.8	691.4	5.1	166.9		308.8	
	water year	h(50%)	626.0	20	1.90		4		333.1	680.8	23.0	188.8		500.0	
	trimester	h(50%)	207.0, 222.0, 106.0, 20.0	78		1.95		7	107.0	462.0	2.0		55.3		173.0
	calendar year	average	649.8	18	2.33		6		387.2	1232.4	21.8	176.2		311.1	
	water year	average	649.9	19	2.11		7		400.0	1270.8	46.9	205.7		523.9	
	trimester	average	237.8 260.6, 125.0, 26.5	70		2.63		11	176.8	609.7	3.0		71.8		211.7
M. S. ANASTASIA	calendar year	h(50%)	465.0	19	2.00		6		246.7	833.3	12.0	131.7		286.0	
	water year	h(50%)	435.0	21	1.81		4		164.4	581.3	32.0	90.4		193.8	
	trimester	h(50%)	157.0, 172.0, 85.0, 15.0	70		2.16		6	88.3	351.8	1.0		41.4		111.0
	calendar year	average	469.4	21	1.90		6		231.4	859.5	2.4	123.4		290.4	
	water year	average	467.4	20	2.25		7		238.4	688.4	70.8	110.3		183.9	
	trimester	average	167.5, 188.5, -90.6, 20.8	69		2.59		8	113.1	419.8	0.6		42.6		127.5
RAMACCA	calendar year	h(50%)	544.1	17	2.24		8		351.3	1254.3	12.1	145.2		331.1	
	water year	h(50%)	508.0	15	2.53		11		270.8	1321.0	4.0	89.7		187.0	
	trimester	h(50%)	166.6, 202.0, 104.0, 28.0	70		2.17		9	104.5	478.0	1.0		45.3		144.0
	calendar year	average		18	2.22		8		354.2	1311.9	3.5	143.5		341.5	
	water year	average		16	2.63		11		372.0	1835.8	12.8	118.9		217.8	
	trimester	average	183.9, 215.4, 116.4, 39.0	68		2.66		9	141.0	599.8	1.0		49.1		159.4

Table III a. Characteristics of the identified regional droughts. Pozzillo reservoir catchment (areal threshold 75%)

Time scale	Threshold	Number of drought	Duration				Regional cumulated deficit			Regional drought intensity			Mean regional coverage		
			mean		max		mean	max	min	mean	max	min	mean	max	min
			(year)	(trim.)	(year)	(trim.)	(mm)	(mm)	(mm)	(mm)	(mm)	(mm)	(%)	(%)	(%)
water year	h(50%)	15	2.00		7		258.4	755.9	46.4	149.8	355.2	46.4	97.4	100.0	81.9
calen. year	h(50%)	16	1.94		6		201.6	556.1	19.8	105.7	162.2	19.8	94.9	100.0	76.2
trimester	h(50%)	70		1.83		6	97.0	442.9	2.5	53.7	144.6	2.5	95.8	100.0	76.2
water year	average	16	2.19		8		305.8	964.9	24.6	161.2	380.0	24.6	95.4	100.0	76.2
calen. year	average	21	1.86		6		231.0	778.2	49.6	123.1	200.0	46.9	95.5	100.0	76.2
trimester	average	72		2.03		10	114.9	522.8	7.9	55.9	153.7	7.9	96.4	100.0	76.2

Table III b. Characteristics of the identified regional droughts. Catania Plain irrigation district (areal threshold 75%)

Time scale	Threshold	Number of drought	Duration				Regional cumulated deficit			Regional drought intensity			Mean regional coverage		
			mean		max		mean	max	min	mean	max	min	mean	max	min
			(year)	(trim.)	(year)	(trim.)	(mm)	(mm)	(mm)	(mm)	(mm)	(mm)	(%)	(%)	(%)
water year	h(50%)	19	1.53		4		238.9	533.8	38.7	152.4	249.7	38.7	98.2	100.0	89.5
calen. year	h(50%)	19	1.42		4		167.3	519.3	47.6	113.2	231.2	47.6	98.4	100.0	81.7
trimester	h(50%)	64		1.69		5	87.4	269.7	6.6	50.9	120.4	6.6	96.4	100.0	81.7
water year	average	20	1.50		4		241.3	565.8	21.9	154.7	259.2	21.9	98.6	100.0	89.5
calen. year	average	18	1.94		7		270.1	1003.5	37.9	134.3	176.3	37.9	98.5	100.0	90.9
trimester	average	72		1.92		6	107.3	322.5	9.3	55.3	137.9	9.3	97.5	100.0	81.7

The comparison of the two tables indicates that regional drought events are generally more severe in absolute value on Pozzillo reservoir catchment than on Catania Plain district. For example, at calendar year time scale, the droughts present mean duration and cumulated deficit respectively of 2 years and 258.4 mm at Pozzillo and 1.53 years and 238.8 mm at Catania Plain. Also the most severe drought of the period 1921-96 is 7 years long (1983-89) with 755.9 mm of cumulated deficit at Pozzillo and 4 years long (1986-89) with 553.8 mm of cumulated deficit at Catania Plain. However, such different values of cumulated deficit are due to the different climatic conditions of the two areas, since a dimensionless analysis of cumulated deficit in the two regions reported in the following indicates an opposite result, in agreement with the at-site dimensionless analysis shown previously.

The differences between the characteristics computed on calendar and water year are not negligible, thus confirming that it is preferable to characterise droughts on water year series, especially when the purpose of the analysis is to study regulation storage facilities. For example the worst drought of Pozzillo basin last from 1983 to 1989 (with cumulated deficit 755.9 mm) if computed on calendar year and from 1985-86 to 1989-90 (with cumulated deficit 551.6 mm) if computed on water year.

2.3. COMPARISON WITH DROUGHT CHARACTERISTICS OVER SICILY

The results of the regional drought analyses carried out for the two areas of interest of the Case Study have been compared with the results of an enlarged analysis of drought on the whole Sicily territory (Alecci et al., 2000). Such an analysis was developed by using the monthly precipitation series of 171 stations for the period 1921-96. With reference to the whole Sicily region (approximately 25,000 km^2), the Pozzillo catchment area represents about 2%, while the Catania plain district area is less than 2%.

A preliminary stationarity analysis on the 171 precipitation stations at the yearly and three-months time scales was carried out in order to detect linear trends. The results are shown in Figure 5 from which it can be inferred that at the five investigated time scales the percentage of stations with a negative trend far exceeds the percentage of stations with a positive trend. More than half of such negative trends are significant at 5% level for water year and winter trimester series, whereas very few negative trends are significant for summer trimester series. The greater percentage of stations exhibit negative trends in winter trimesters while the summer precipitation series present the largest number of positive trends, none of which however are significant.

The similarity of the trend analysis results for water year and winter trimester series is a direct consequence of the fact that winter trimester precipitation accounts for approximately 40% of the water year annual precipitation. Also, fall and spring trimester series, present a similar trend behaviour, probably due to the similarity of the precipitation regimes in these seasons. On the other hand, the above findings seem to indicate a different trend behaviour of the winter and summer trimesters in Sicily.

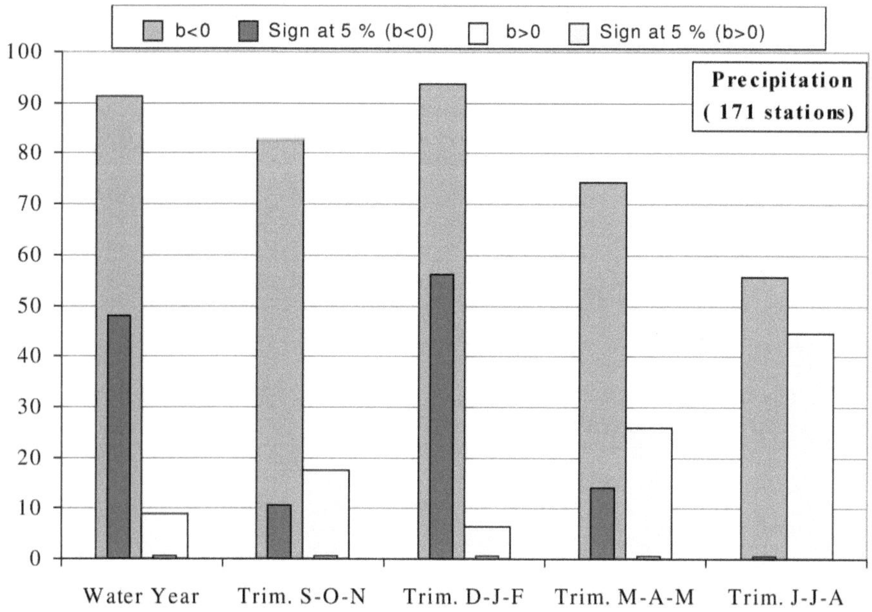

Fig. 5. Linear trend test results for the investigated precipitation series in Sicily.

Table IV reports a summary of the results of the regional drought analyses and in particular the main statistics of drought duration, cumulative deficit and drought intensity for Pozzillo catchment, Catania plain district and the whole Sicily region computed on water year precipitation series. A comparison between the identified droughts on the three areas indicates that the number of identified droughts on the two areas of the investigated Case-study are significantly larger than the corresponding number of droughts computed on the whole Sicily region. Droughts identified on the Pozzillo reservoir catchment and in Catania plain district are generally longer than those identified on the whole Sicily region and present a greater areal extension percentage, as expected with reference to the larger area of Sicily. This is evident regardless of the adopted threshold. For example, the maximum duration observed in the whole Sicily is 3 years, against maximum durations ranging between 4 and 7 years for the other two areas. Further, in no case the maximum areal extension of deficit is 100% for the whole Sicily, whereas the maximum areal extension for Catania plain and Pozzillo catchment is always 100%. In order to compare the drought characteristics of the three regions taking into account the different climatic conditions, the cumulated deficit and the intensity have been divided by the mean areal precipitation of each investigated area. Such dimensionless values are reported in Table V, along with the sum of drought durations divided by the total length of the observation period. It is confirmed that droughts identified on Pozzillo catchment and Catania plain are generally more severe in terms of dimensionless cumulated deficit than those identified on the whole Sicily (e.g. the maximum values range from 0.79 to 1.10 for Pozzillo and from 0.99 to 1.90 for Catania plain, against a maximum of 0.61-0.69 for the whole Sicily). On the other hand, the dimensionless drought intensity identified on Pozzillo catchment does not differ significantly from the one identified on the whole Sicily, while the one identified on Catania plain is slightly greater.

Table IV. Regional drought characteristics identified on water year precipitation series (areal threshold:75%)

Region	Areal rainfall (mm)	Threshold (mm)	No of droughts	Duration mean (years)	Duration max (years)	Sum of durations (years)	Cumulated deficit mean (mm)	Cumulated deficit max (mm)	Average intensity mean (mm/year)	Average intensity max (mm/year)	Areal extension of deficit mean (%)	Areal extension of deficit max (%)
Pozzillo .	705.5	median	16	1.9	6	31	202.7	558.8	106.3	162.9	94.9	100.0
catch		mean	21	1.8	6	39	230.9	778.1	123.0	200.1	95.3	100.0
Catania	528.0	median	18	1.5	4	28	179.1	520.5	108.1	208.6	98.1	100.0
plain		mean	18	1.9	7	35	270.1	1004.0	134.2	176.3	98.5	100.0
Sicily	716.0	median	12	1.4	3	17	190.7	460.1	131.0	162.7	86.7	95.6
region		mean	14	1.5	3	21	217.2	513.4	144.1	180.8	88.5	97.2

Table V. Dimensionless regional drought characteristics identified on water year precipitation series (areal threshold: 75%)

Region	Areal rainfall (mm)	Threshold (mm)	Sum of durations/period (%)	Cumulated deficit/areal rainfall mean	Cumulated deficit/areal rainfall max (mm)	Cumulated deficit/areal rainfall Total	Intensity/areal rainfall mean (%)	Intensity/areal rainfall max (%)
Pozzillo.	705.5	median	41.3	0.29	0.79	4.60	0.15	0.23
catch		mean	52.0	0.33	1.10	6.87	0.17	0.28
Catania	528.0	median	37.3	0.34	0.99	6.10	0.20	0.40
plain		mean	46.7	0.51	1.90	9.21	0.25	0.33
Sicily	716.0	median	22.7	0.25	0.61	3.05	0.17	0.22
region		mean	28.0	0.29	0.69	4.06	0.19	0.24

Figure 6 shows a comparison of identified droughts over the three areas using the mean as threshold. The graph indicates an apparent increase in frequency and duration of regional droughts over the whole Sicily beginning in the 60's, without apparent increase of drought intensity. Droughts identified in the Catania plain district somehow confirm such an increase, while no significant changes can be inferred from droughts identified on Pozzillo reservoir catchment. In any case, it appears that in the recent years the identified droughts on the three areas tend to coincide more than in the first part of the observation sample. In particular, until the 60's, only three one-year droughts are identified at the whole regional level, although many droughts are identified on Catania plain and on the Pozzillo catchment. However the graph indicates a general agreement in the identification of the severe drought that occurred at the end of the 80's which can probably be considered the most severe drought observed during the 1921-1996 period in terms of duration and cumulated deficit.

The comparison between droughts identified on a smaller scale (Pozzillo catchment and Catania plain) with those identified over a larger region (whole Sicily) seems to confirm that some kind of scaling of quantities such as cumulated deficit or areal extension of deficit with the extent of the investigated area occurs. This is in agreement with the fact that when larger areas are investigated, the areal threshold of deficit (e.g. 75%) is more difficult to be reached, and therefore a smaller number of droughts are identified.

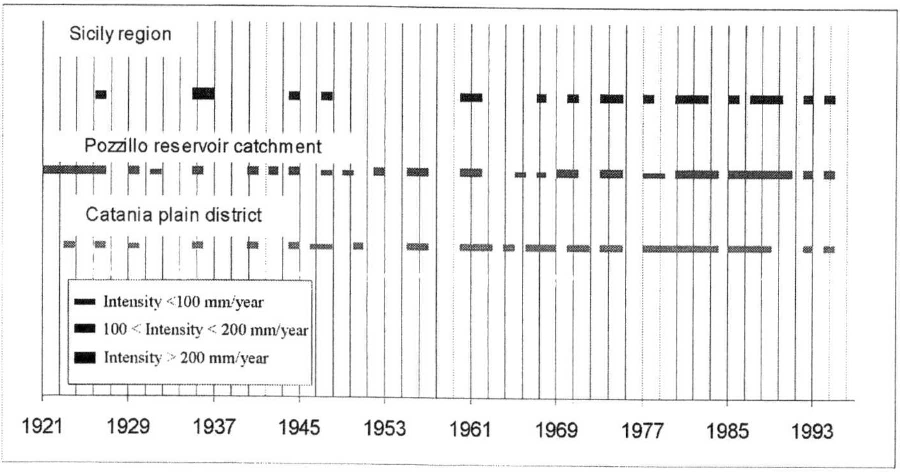

Fig. 6. Regional droughts identified on the Sicily region, the Pozzillo reservoir catchment and the Catania plain district rainfall series during 1921/22-1995/96 (water year starting in September; threshold equal to the median; areal threshold 75%)

3. Comparison of identified droughts in Mediterranean regions

3.1. ANALYSED CASE-STUDIES

The location of the investigated case studies is depicted in Figure 7. The case studies cover a range of approximately 35° of longitude and of about 5° of longitude, in the Mediterranean basin. Note that in order to limit the differences in the surfaces of the investigated areas, it was decided, for the sake of comparison , to restrict the analysis of Alentejo region only to sub-region C.

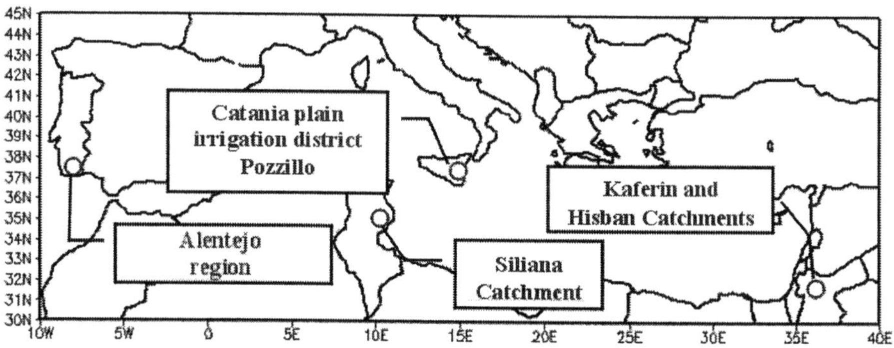

Fig. 7. Location of case studies

A companion paper (Paulo et al., 2002) reports the results of a wider analysis carried out on the whole Alentejo region (approx. 27000 km^2), using both run method and the Standardized Precipitation Index to investigate drought features of the area. The main characteristics of the investigated precipitation series are indicated in Table VI. In particular time series of monthly precipitation were available in a variable number of station within the basins or districts of all Case Studies. The longest records of precipitation are related to the stations of the two investigated areas of Sicily: 76 years from 1921 to 1996, while the shortest are the ones related to the stations within the two Jordan basins, i.e. 23 years from 1973 to 1996. The average annual value of areal precipitation ranges between the maximum of 706 mm (Alentejo sub region C, Portugal and Catania plain irrigation district, Sicily) and the minimum of 240 mm (Hisban Wadi in Jordan), with an intermediate value (440 mm) for Tunisian studied region.

Table VI. Characteristics of investigated case studies

Partner and case study	Surface (km^2)	Analysed period	No. of stations	Mean (mm)
University of Lisbon				
-Alentejo sub region C	4375	1932-1995	7	706
University of Catania				
-Salso river catchment at Pozzillo	577	1921-1996	4	528
-Catania plain irigation district	470	1921-1996	4	706
INGREF Tunis				
-Siliana catchment	1040	1967-1997	3	440
University of Jordan				
-Kaferin Wadi catchment at Kaferin	161	1973-1996	3	410
-Hisban Wadi catchment at diversion	60	1973-1996	3	240

Figure 8 shows the monthly averages and standard deviations of the areal precipitation, computed by means of Thiessen polygons on the basis of available records at the stations.

3.2. STATIONARITY ANALYSIS

A comparison of the trend characteristics of precipitation series in Mediterranean regions has been carried out by using the results of the stationarity analyses developed for each Case Study by the different partners of the INCO-DC Project (Bergaoui, 2002; Salihi, 2002; ISA, 2000). All tests were applied on water year precipitation series, starting on September, in all cases except than in Jordan case, where October was considered as a starting month.

Figure 9 reports a plot where the linear trend t-statistics for the 6 investigated areas are shown. The line segment indicates the range of the statistics, while the marker shows the mean value. It can be inferred from the plot that the linear trend t-statistic presents in some cases a wide range, with occasional positive values. Alentejo region, Siliana catchment and Kaferin catchment always exhibit negative trends and all analysed areas show at least on series with a negative trend significant at 5% level. On the other hand, none of the few positive trends are significant.

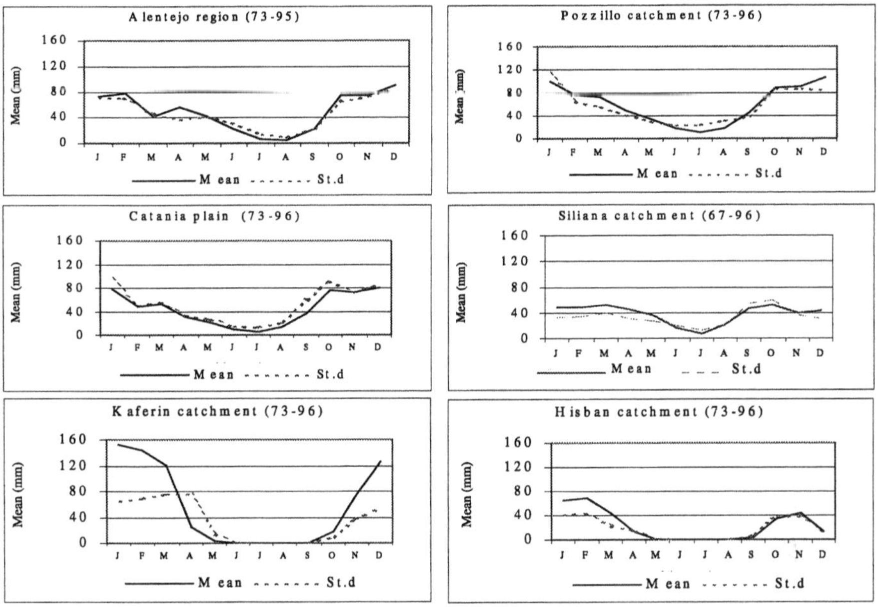

Fig. 8. *Monthly means and standard deviation of areal precipitation in investigated regions*

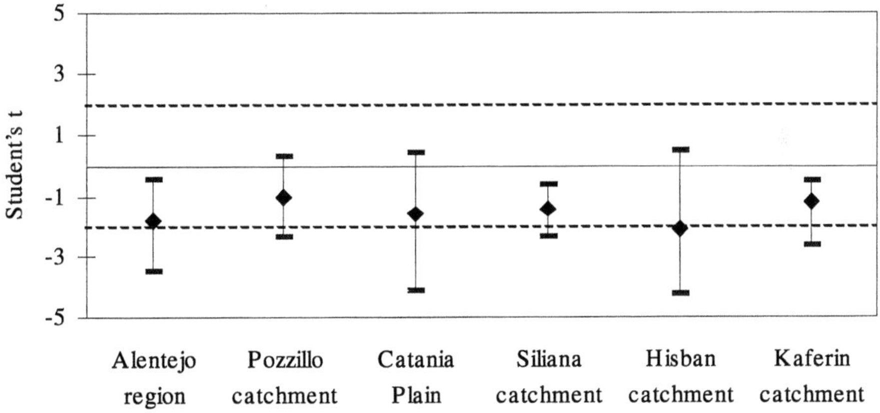

Fig. 9. *Student's t statistics (max, mean and min) for linear trend in investigated cases study*

The trend analysis results seem to confirm that a generally decreasing trend is present in all case studies.

3.3. COMPARISON OF DROUGHT OCCURRENCE AND CHARACTERISTICS

Drought characteristics in Mediterranean regions have been compared by using the results of the analyses developed in each Case Study by the different partners of the INCO-DC Project. Characteristics of regional droughts were computed by considering as variable thresholds in each station either the mean or the median and as critical area the 75% of the total surface of the investigated catchment or district. In all cases, water year precipitation series were used.

Figure 10 shows the regional drought occurrences in the 6 investigated areas adopting the mean of each series as threshold. The asterisks indicate the time span of available records for each region. It can be inferred from the figure that the last decade is characterised by more frequent and longer droughts in all case studies, the only exception being Siliana, for which an apparent decrease of drought lengths and intensity is observed.

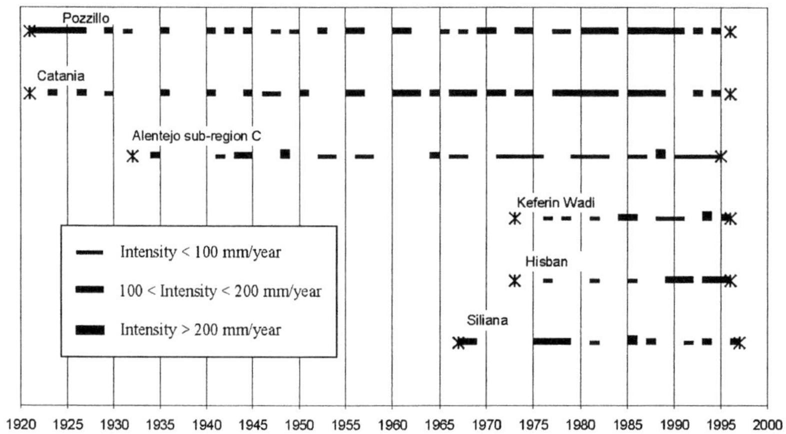

Fig. 10. Regional drought periods identified on water year precipitation, (threshold equal to the mean, areal threshold 75 %)

Table VII reports a summary of the regional drought characteristics identified on the different investigated areas on water year precipitation series. In order to compare droughts across different regions with different climatic conditions, dimensionless cumulatated deficit and drought intensity have been computed by dividing the corresponding drought characteristics by the mean areal precipitation of the investigated area.

It can be inferred from the table that longest droughts are observed in the two Sicilian case studies, although Alentejo region in Portugal exhibits on average longer droughts. The two Jordan regions (Keferin Wadi and Hisban Wadi) on the other hand are affected by shorter droughts. The table also indicates that mean dimensionless cumulated deficit is generally higher in Hisban Wadi basin in Jordan (.72) and Alentejo region in Portugal (.57).

Table VII. Regional drought characteristics identified on annual precipitation series (areal threshold 75 %; variable observation periods)

Region and analyzed period	Threshold	No of observ.	No of drought periods	Duration		Σ durations/ observ. period	Cumulated deficit / mean areal rainfall			Intensity / mean areal rainfall			Areal extension of deficit			Most severe drought years		
				mean	max		mean	max	min	mean	max	min	mean	max	min	Length	Cum. Def.	Intensity
	(mm)	(years)		(years)		(%)				(1/year)			(%)					
Alentejo region	median	64	12	2.31	5	43.8	0.56	1.37	0.20	0.25	0.41	0.10	94.0	99.0	84.0	71-75	79-82	48
Sept. 1932-Aug. 1995	mean	64	13	2.23	5	45.3	0.57	1.45	0.10	0.25	0.42	0.10	95.0	99.0	84.0	71-76	79-83	48
Pozzillo catchment	median	75	16	1.94	6	41.3	0.38	1.06	0.04	0.20	0.31	0.04	94.9	100.0	76.2	21-26	21-26, 85-89	40
Sept. 1921-Aug. 1996	mean	75	21	1.86	6	52.0	0.44	1.47	0.09	0.23	0.38	0.09	95.3	100.0	76.2	21-26,85-90	21-26, 85-90	40
Catania plain	median	75	18	1.56	4	37.3	0.25	0.74	0.07	0.15	0.30	0.07	98.1	100.0	81.7	85-88	85-88	80
Sept. 1921-Aug. 1996	mean	75	18	1.94	7	46.7	0.38	1.42	0.05	0.19	0.25	0.05	98.5	100.0	90.9	77-83	77-83	26
Siliana catchment	median	31	8	1.50	4	38.7	0.42	1.04	0.10	0.28	0.47	0.10	97.0	100.0	87.8	75-78	75-78	87
Sep. 1967-Aug. 1997	mean	31	8	1.50	4	38.7	0.40	0.98	0.09	0.27	0.46	0.09	97.7	100.0	87.8	75-78	75-78	87
Kaferin catchment	median	23	6	1	1	26.1	0.23	0.40	0.09	0.23	0.40	0.09	100.0	100.0	100.0	95-96	95-96	95-96
Oct. 1973-Sept. 1996	mean	23	7	1.43	3	43.5	0.37	0.62	0.09	0.27	0.51	0.09	100.0	100.0	100.0	88-90	88-90	93
Hisban catchment	median	23	6	1.83	3	47.8	0.72	1.92	0.06	0.33	0.64	0.06	94.4	100.0	86.0	93-95	93-95	85-86
Oct. 1973-Sept. 1996	mean	23	5	1.80	3	39.1	0.72	1.72	0.05	0.33	0.57	0.05	99.0	100.0	95.0	89-91, 93-95	89-91	93-95

Droughts identified on the two Jordan basins have the largest minimum and average areal extension, probably due to the limited extension of the investigated basins which leads to contemporaneous occurrence of deficits in all stations.

Figure 11 shows the dimensionless regional cumulated deficits identified at the six investigated areas. It can be inferred from the figure that Wadi Hisban area in Jordan exhibits highest mean and maximum dimensionless regional cumulated deficits. This is probably due to the significant trend affecting precipitation in that area (see Figure 7). The two Italian areas (Pozzillo catchment and Catania plain) show max values that are comparable with those observed at Alentejo area, though the latter area exhibits a larger mean value.

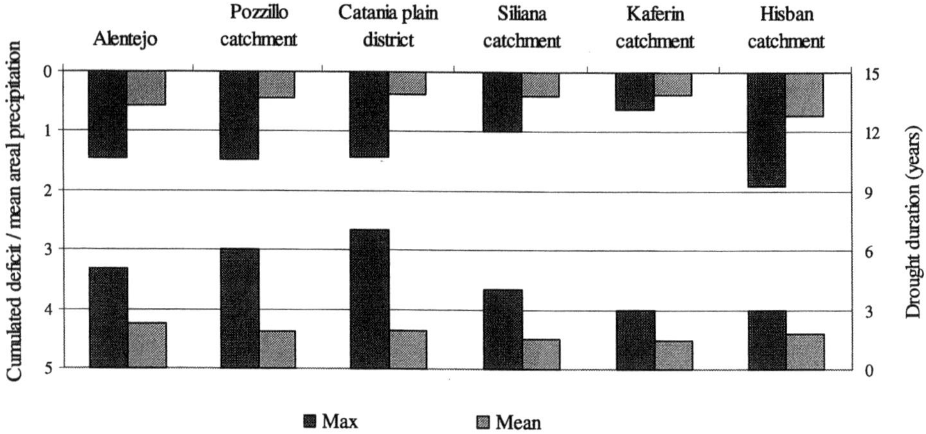

Fig. 11. Mean and maximum drought duration and dimensionless regional deficit at the six investigated areas (threshold equal to the mean, areal threshold 75 %)

In order to assess the contemporaneous occurrence of regional deficit in the investigated areas, taking into account the different lengths of precipitation series of the different case-studies, some new computations were made. In particular regional drought identification was carried out on the reduced precipitation series of the Italian and Portoguese Case-studies in order to consider a common record period 1973-1995 for all investigated areas. However, the analysis was not re-applied for Siliana catchment for which the original 1967-1996 record was considered. Analyses were carried out at water year time scale, using 75% areal threshold and the mean as deficit threshold.

Contemporaneous occurrence of deficits was assessed by developing a contingency table in which the conditional probabilities p_{ij} of observing a regional deficit in area i given a regional deficit occurs in area j are reported. In formulae:

$$p_{ij} = P[\text{regional deficit in year } t \text{ at site } i \mid \text{regional deficit in year } t \text{ at site } j]$$

High values of p_{ij} indicate contemporaneity between regional deficits at sites i and j thus suggesting a link in drought occurrences between the two sites.

The conditional probabilities can be estimated as:

$$p_{ij} = \frac{\sum_{t=1}^{N} I_t(i) I_t(j)}{\sum_{t=1}^{N} I_t(j)}$$

where the indicator function $I_t(i)$ takes the value 1 if a regional deficit occurs in year t at site i and zero otherwise.

The contingency table is reported in Table VIII. Obviously diagonal values take the value of 1. The table is not symmetric, which is indicative of the slightly different marginal probabilities of observing a regional deficit at the different sites. It can be inferred from the table that the highest conditional probabilities are those involving the two Jordan areas, namely Wadi Kaferin and Wadi Hisban catchments as expected due to their close locations. For example, if in a given year a regional deficit is observed at Wadi Hisban, there is a 78% probability of observing a regional deficit also at Wadi Kaferin. The table also indicates that Alentejo region exhibits higher contemporaneity with the two Italian regions than with other areas. On the other hand regional deficit occurrences in Siliana region appears to be linked more to regional deficit occurrences in the Jordan areas and less to those ones in the Italian and Portuguese areas.

Table VIII. Conditional probabilities of observing a regional deficit in an area (row) given a regional deficit occurs in another area (column)

	Alentej o region	Pozzillo catchment	Catania Plain district	Siliana catchment	Kaferin catchment	Hisban catchment
Alentejo region	1.00	0.54	0.55	0.38	0.30	0.44
Pozzillo catchment	0.64	1.00	0.73	0.50	0.50	0.44
Catania Plain	0.55	0.62	1.00	0.50	0.40	0.22
Siliana catchment	0.27	0.31	0.36	1.00	0.40	0.44
Kaferin catchment	0.27	0.38	0.36	0.50	1.00	0.78
Hisban catchment	0.36	0.31	0.18	0.50	0.70	1.00

4. Conclusions

An analysis of drought characteristics in six Mediterranean areas (one in Portugal, two in Sicily, one in Tunisia and two in Jordan) has been carried out, by means of run method. A preliminary analysis of stationarity on precipitation series indicates a generally negative trend affecting precipitation in all investigated regions, which partly explains the increasing frequency, severity and duration of droughts recently observed in the all areas.

Analysis of regional deficits in the investigated areas has shown that there is in general higher contemporaneity of deficit occurrences between eastern and central

Mediterranean regions than between central and western regions, with the exception of Siliana region whose deficit occurrences are more similar with those in the Jordan case.

Acknowledgements

The research has been carried out within the European Community INCO-DC project "A DSS for drought mitigation in Mediterranean regions" (DSS-DROUGHT, IC18-CT97-0169). The comparison with drought characteristics over the whole Sicily was carried out by using the results of an investigation made within the EU INTERREG IIC Project under the supervision of the Italian Department of National Technical Service.

References

Alecci, S., Arcidiacono, F., Bonaccorso, B. and Rossi, G. (2001) Identificazione delle siccità regionali e sistemi di monitoraggio. In: *Atti del Convegno su Siccità: Monitoraggio, mitigazione, effetti.* Cagliari, Villasimius, 21-22 settembre 2000.

Bergaoui, M. (2002) Local and regional droughts in Siliana catchment, Tunisia (This issue).

DICA (2000) *Regional Drought Identification Module REDIM.* Department of Civil and Environmental Engineering, University of Catania, Catania.

ISA (2000) Third annual report of project *"A DSS for mitigation of drought impacts"*, Technical University of Lisbon.

Paulo, A.A., Pereira, L.S. and Matias, P.G. (2002) Analysis of local and regional droughts in southern Portugal using the theory of runs and the standardised precipitation index (This issue).

Rossi, G. and Cancelliere, A. (2002) At site and regional drought identification by REDIM model (This issue).

Salihi, A. (2002) Drought identification and characterization in Jordan (This issue).

REGIONALISATION AND IDENTIFICATION OF DROUGHTS IN MEDITERRANEAN COUNTRIES OF EUROPE

J. ÁLVAREZ and T. ESTRELA
Centro de Estudios Hidrográficos del CEDEX, Madrid, Spain

Abstract: A methodology for the regionalisation and the identification of droughts at a pan European scale is described in this paper. It includes part of the activities carried out by CEDEX in the framework of the EC funded project entitled *Assessment of the Regional Impact of Droughts in Europe ARIDE* (Demuth and Stahl, Editors, 2001). The proposed methodology has been applied to identify and analyse regional drought events from the beginning of the XXth century at Europe and especially at Mediterranean European countries. Regionalisation and drought event analysis can be useful tools for decision makers in water resources.

Key words: drought events, regional droughts, Mediterranean Europe.

1. Introduction

A drought is characterised mainly by a lack of rainfall for a long period of time and the effects are usually stressed by a temperature increase. Both meteorological factors have a direct impact upon the hydrological cycle and can give rise to important reductions in water resources. If this decrease affects zones with a considerable water demand, this will have an effect upon the uses of the water as well as on the ecological status of water bodies.

The increasing demand on water in Europe during last decades has produced, especially in Mediterranean countries, stress situations on water resources exploitation systems. The strategic planning to be carried out in these systems must ensure that the impacts of droughts do no not adversely affect water resources availability and in-stream ecology. Regarding the length of the time period of droughts, it should be pointed out, as stated by Batini and Benedini (1998), that in the Mediterranean countries it is not a question of low rainfall values in one single year that cause a drought situation, but rather a sequence of years with low rainfall.

Precipitation, temperature, river flow and aquifer water level data are needed to study regional patterns of droughts. However, there are not many databases with long enough hydrological records for the whole of Europe, which makes it difficult the research in large-scale drought identification and regionalisation. This work has used data coming from the Climate Research Unit (CRU) database, which includes a consistent meteorological information to be used in scientific research. It covers the whole of Europe with data from the beginning of the XXth century.

G. Rossi et al. (eds.), Tools for Drought Mitigation in Mediterranean Regions, 123-146.

Rainfall series have been used since they are the most important input on hydrologic systems and this decision focuses the scope of the work to what it is commonly referred as meteorological droughts.

Gridded layers of monthly precipitation of Europe provided by CRU from January 1901 to December 1996 were used. Firstly, some basic comparisons were performed in order to validate the use of CRU database (Alvarez and Estrela, 1999) mainly comparing it with data collected in ARIDE project. Figure 1 shows the mean annual rainfall (1901-1996, calendar years) in Europe estimated from rainfall data of this database.

Prior to drought identification, several regions were established in Europe using the rainfall time series of CRU database. Regions were obtained starting from the definition of elemental boundary river basins. Figure 2 shows 81 elemental boundaries considered for this task. They were defined taking into account an existing GIS coverage developed by EUROSTAT, with approximately 920 basins. The selected basins were defined considering the main river basins and the orographic patterns existing in Europe. Monthly rainfall series were obtained in each one of these areas averaging gridded data inside each elemental boundary.

A cluster methodology has been applied to group basins with a common climatic pattern. The Pearson correlation coefficient has been used as a measure of similarity among series of anomalies derived from CRU data. Subsequently, information about seasonal patterns was also considered. The final result is assumed to be a suitable definition of a meteorological regionalisation that can be used to study drought events.

Afterwards, a model to identify drought events was applied and the main variables used in the so-called *theory of runs* were estimated for each region defined, and especially in this paper, for Mediterranean areas. This task was completed with a compilation of the available information about historical droughts which were compared to the series of events derived. Main droughts that recently occurred in Mediterranean countries were also described.

2. Application of a clustering technique

A preliminary step before identifying regional droughts consists on searching for common patterns of drought event occurrences at European scale. This task has been carried out using directly CRU monthly maps of rainfall from 1901 to 1996. The result of this analysis shows that the drought events seem to occur in a spatially and temporal random way with different intensities and extensions (Figure 3) depending on large-scale circulation anomalies.

The next step consisted on establishing regions where temporal rainfall series can be considered well correlated. The 81 elemental basin boundaries selected are supposed to have homogeneous temporal precipitation series and a procedure of aggregation by clustering was applied. The monthly distributed rainfall maps from January 1901 to

December 1996 were averaged into the elemental basin boundaries with GIS tools to obtain temporal rainfall series. As far as the similarities between time series allow grouping elemental areas in larger areas, some regions were defined and characteristics of droughts were found.

Some considerations related to the different components of the time series were made before of establishing comparisons. Indeed, if a regionalisation based on time series is searched, some of their characteristics may affect the evaluation of similarities. Long-term trends or cycles enhance serial correlation because they introduce order and regularity among data. Their effects have to be taken into account in order to improve the comparison between time series. Therefore, the time series have been decomposed into the aforementioned components and some of these components have been filtered out in order to facilitate comparisons.

It can be accepted that similar meteorological patterns for regionalisation may be found in the temporal series if components such as long-term trends and seasonal cycles are filtered out. It has been considered that those diminishing fluctuations of time lag greater than seasonal cycles and shorter than long-term trends are related to droughts. Once the correlation procedures have been applied, long term trends and seasonal cycles may be considered again to improve clustering.

A decomposition based on identifying long term trends in CRU data with a least square model was applied:

$$y_{i,j} = x_{i,j} - (a \cdot date_{i,j} + b) \tag{1}$$

Parameters a and b were estimated by means of a least squared model. $date_{i,j}$ is the order of $x_{i,j}$ in time series. Sub-indexes i and j represent months and years in time series.

Afterwards, seasonal cycles have to be identified and removed from data. It is usual to characterise seasonal cycles in monthly rainfall data by monthly means and monthly standard deviations (Salas *et al.*, 1985). Being μ_i and σ_i mean and standard deviation for month i. $z_{i,j}$ are final time series once long-term trend and seasonal cycles have been filtered out from monthly series:

$$z_{i,j} = \frac{y_{i,j} - \mu_i}{\sigma_i} \tag{2}$$

The elemental basins can be clustered using the Pearson correlation coefficient of monthly series of anomalies, which represent a similarity measure defining the relationships existing between anomalies. The Pearson correlation coefficient highlight relationships between nearby basins, which are mostly due to common atmospheric circulation patterns in Europe. In general terms, it may be stated that the correlation structure is lost as distance became larger between basins. As an example, it is remarkable that basins located in the Mediterranean Sea in Spain, Portugal or Greece show a meaningful number of negative correlation coefficients with others, especially with those located in the centre or northern Europe. This result highlights the differences existing between basins and specifically between southern and northern countries.

Different factors are related with correlation structure, such as the relief barriers, the proximity to the sea and continental effects. As an example of the former, it may be remarked the Scandinavian Mountain Chain closed to the coast. This factor makes a distinction with those other basins leeward as reflected by correlation coefficients.

Relationships pointed out by correlation coefficients in the Mediterranean area are of a shorter extent than those occurring in continental regions. This may be due to the particular features of the Mediterranean Sea where effects of cyclogenesis are meaningful, namely in the Gulf of Genoa and around Cyprus. Mediterranean Sea is located under the influence of global circulation patterns, mainly reflected by Atlantic westerlies winds and the Asiatic pressure systems, but the mentioned effect may change the zonal circulation into another cellular or meridional one (Wallen, 1970). Plenty of factors could explain differences in correlation coefficients between parts of Turkey to those others of Spanish Mediterranean coast.

Northern Atlantic coastal basins show larger similarities than southern basins do. As it was expected, England, Scotland, Wales and Ireland are joined into a consistent cluster.

Correlation coefficients estimated between monthly series of anomalies can be used for grouping basins in certain regions with the same climatic patterns. Hierarchical clustering is an exploratory method, suitable to the objectives of this work. It is based on selecting the strongest relationships between series and, step by step, reducing the number of series by gathering them into new clusters. A decision has to be done in selecting the final number of groups. The mean Pearson correlation coefficient between basins of every group may help to make this decision, as it is shown in Figure 4, where the evolution of the similarity measure is shown in blue colour and the selected final step in red one. Also the number of clusters are represented in this figure for the whole Europe. The occurrence of significant changes in the mean Pearson correlation coefficient of groups clustered and the qualitative knowledge of meteorological patterns was taken into consideration to select the point where clustering has been stopped.

Applying this technique, 10 groups were formed from the 81 original basins. These groups are constituted by: Iceland, Major part of the Scandinavian Peninsula, Northern Europe, Central Eastern Europe, Central Atlantic Europe, Southern Caspian Sea, Eastern Mediterranean Sea, Central Mediterranean Sea, Iberian Peninsula and Western Atlantic Islands. In Figures 5, 6 and 7 annual rainfalls and cumulative deviations for 1901 to 1996 period are shown for those groups that include Mediterranean areas.

3. Improving the regionalisation considering seasonality

The sequence of seasonal cycles influences droughts because it intensifies some main characteristics such as deficits, recovery time, etc. This factor has not traditionally been considered because the attention has been focussed on the coupled occurrence of anomalies in atmosphere circulation. Therefore, it is the aim of this work to distinguish groups of basins with similar seasonal cycles in order to improve the regionalisation derived from cluster analysis. This analysis will lead to split some basins belonging to the same cluster but with a lack of seasonal correspondence.

Most of clusters grouped show good fits if mean seasonal meteorological cycles are represented. Icelandic peaks are reached in October, while minimum rainfall values occur in May. Similar cycles are found in major part of Scandinavian Peninsula. Maximum values occur between August and October depending on the location of the basin. Minimum values are reached in May for Atlantic basins and from February to March if basins are leeward the Scandinavian Mountain Chain. The displacement shown suggests an improvement by distinguishing between windward and leeward basins. Northern and Central Eastern basins of Europe show good agreements in meteorological cycles. The peaks occur in June and July and minima in February and March.

A special case occurs when attention is devoted to upper Rhone that was initially clustered with Central Atlantic basins. These groups show low serial correlation and differences appear between seasonal cycles. That is the reason why a new cluster is searched and comparisons are checked in nearby basins. The best option found is to create a new group with upper Rhone basins, Po basins and others nearby lower Rhone in southern France.

Mediterranean basins show similar behaviour. Minima and maximum values occur mainly during summer and winter respectively. Even, peaks of rainfall reach spring and autumn for certain Mediterranean basins. A common problem appears in Central and Eastern Mediterranean Sea clusters because of the effect of mid and lower Danube basins with which are joined (Figure 8). Mid Danube basins have been clustered with Central Mediterranean basins, because of a high accordance with Adriatic coast anomalies. But when relationships with the rest of Apennine Peninsula basins are considered, some inconsistencies appear. In the Eastern Mediterranean Sea occurs the same when lower Danube basins are taken into account. So, Danube basins are disconnected from the Mediterranean clusters and a solution may be found if a new region formed by mid and lower Danube basins is considered. The scatter diagram of time series anomalies between basins produced a high accordance, much better than others existing in the original cluster. This reason makes acceptable to create a new cluster with these two basins.

Two major groups may be distinguished in the Iberian Peninsula. Mediterranean Spanish basins would constitute the first one. The rest of Atlantic basins and others in the south of Spain would form the second one.

Figure 9 shows a map of the final regionalisation reached. It combines the basis of clustering reached considering the time series of anomalies and the seasonal cycles occurring in those basins.

Table I shows the matrix with the Pearson serial correlation coefficients for the groups formed, where it is observed that in general terms the similarities are still maintained at a low degree after taking into account the seasonality.

Table I. Correlation coefficients between regions clustered

Iceland, g001	1,00	0,04	-0,27	-0,23	-0,18	-0,33	-0,26	-0,13	-0,02	0,04	-0,19	0,03	-0,14	0,03	0,17
Norwegian coastal basins, g002	0,04	1,00	0,45	0,15	-0,18	-0,01	-0,24	-0,31	-0,06	-0,28	-0,42	-0,38	-0,41	-0,19	0,01
Leeward Scandinavian basins, g003	-0,27	0,45	1,00	0,35	0,02	0,48	0,12	-0,20	-0,11	-0,26	-0,07	-0,28	-0,08	-0,23	-0,20
Northern basins of Europe, g004	-0,23	0,15	0,35	1,00	0,28	0,15	-0,04	-0,02	-0,10	-0,12	-0,10	-0,23	-0,11	-0,23	-0,26
Eastern basins of Europe, g005	-0,18	-0,18	0,02	0,28	1,00	0,26	-0,05	0,02	-0,06	-0,07	-0,02	-0,05	0,34	0,14	-0,08
Central European basins, g006	-0,33	-0,01	0,48	0,15	0,26	1,00	0,51	-0,18	-0,09	-0,26	0,25	-0,17	0,31	-0,11	-0,21
French and UK basins, g007	-0,26	-0,24	0,12	-0,04	-0,05	0,51	1,00	0,20	-0,08	0,13	0,61	0,06	0,11	-0,16	-0,06
Western Iberian basins, g008	-0,13	-0,31	-0,20	-0,02	0,02	-0,18	0,20	1,00	0,18	0,61	0,35	0,18	0,03	0,00	0,04
Western Atlantic Islands, g009	-0,02	-0,06	-0,11	-0,10	-0,06	-0,09	-0,08	0,18	1,00	0,17	0,01	0,06	0,02	0,11	0,09
Mediterranean basins in Iberia, g0010	0,04	-0,28	-0,26	-0,12	-0,07	-0,26	0,13	0,61	0,17	1,00	0,39	0,31	0,06	0,03	0,10
Po and Rhone basins, g011	-0,19	-0,42	-0,07	-0,10	-0,02	0,25	0,61	0,35	0,01	0,39	1,00	0,49	0,35	-0,04	-0,07
Central Mediterranean basins, g012	0,03	-0,38	-0,28	-0,23	-0,05	-0,17	0,06	0,18	0,06	0,31	0,49	1,00	0,49	0,27	0,02
Mid and lower Danube basins, g0013	-0,14	-0,41	-0,08	-0,11	0,34	0,31	0,11	0,03	0,02	0,06	0,35	0,49	1,00	0,38	-0,15
Eastern Mediterranean basins, g014	0,03	-0,19	-0,23	-0,23	0,14	-0,11	-0,16	0,00	0,11	0,03	-0,04	0,27	0,38	1,00	0,19
Southern Caspian basins, g015	0,17	0,01	-0,20	-0,26	-0,08	-0,21	-0,06	0,04	0,09	0,10	-0,07	0,02	-0,15	0,19	1,00

According to the regionalisation carried out, some statistics of averaged rainfall series has been obtained in order to highlight the differences between the regions (Table II).

4. Exploring droughts in Europe

Different methodologies to explore droughts in the time series have been proposed and described in many publications (Hisdal and Tallaksen, 2000; Menéndez, 1995; Cancelliere *et al.*, 1995). A varying threshold approach is selected in this work to identify the onset of a drought event by means of monthly percentiles defined from monthly duration curves (12 values of monthly percentiles were chosen). Also, an inter-event volume criterion has been used to take into account the partial recovery needed to consider that a drought is finished. This criterion is also useful for pooling minor droughts.

The partial recovery is evaluated using a critical parameter that expresses the percentage necessary of accumulated deficit of a drought to be surpassed by a monthly surplus with regard to the mean monthly values. Finally, those identified droughts that last for only 1 month have been filtered out.

Table II. Main figures of the climatic regions clustered

	Mean	Median	Maximum	Minimum	p25 %	p75%	Coefficient of variation
Iceland	985	983	1206	750	928	1035	0,10
Norwegian coastal basins	1271	1248	1562	1036	1192	1326	0,09
Leeward Scandinavian basins	572	570	695	442	543	607	0,08
Northern basins of Europe	562	567	659	458	532	587	0,07
Eastern basins of Europe	445	446	544	356	411	473	0,09
Central European basins	690	695	813	570	640	728	0,08
French and UK basins	889	886	1067	669	838	944	0,09
Western Iberian basins	646	636	917	495	571	695	0,14
Western Atlantic Islands	209	207	375	138	181	228	0,19
Mediterranean basins in Iberia	502	499	680	380	456	541	0,13
Po and Rhone basins	1093	1098	1329	784	1041	1152	0,08
Central Mediterranean basins	724	717	919	529	678	771	0,09
Mid and lower Danube basins	767	761	964	649	718	814	0,09
Eastern Mediterranean basins	601	597	719	461	572	632	0,08
Southern Caspian basins	394	396	526	293	369	420	0,09

Once the drought has been defined as described in precedent paragraph, some variables can be used to parameterise the model and characterise drought events (Tallaksen *et al.*, 1997; Estrada, 1996; Yevjevich *et al.*, 1983).

- Number of drought events identified, nd. This variable has a special relevance in parameterisation because it allows controlling the pooling effect among drought events.
- Deficit, maximum and mean deficit. Deficits are defined as accumulated monthly deficit values for each drought, si. The maximum value is also a reference to characterise the phenomena in a region. The mean deficit is obtained by dividing the sum of deficits by the total number of drought events identified.

$$\bar{s} = \frac{\sum_{i=1}^{nd} s_i}{nd} \qquad (3)$$

- Duration, maximum and mean duration. Duration express the length of the drought event from its onset to its recovery, di, which depends strongly on critical parameter value, calibrated considering the maximum duration identified. Therefore, this value has to be consistent with the needed pooling effect of minor droughts and the undesirable identification of too large droughts. The mean value of duration is estimated by dividing the sum of duration by the total number of drought events identified.

$$\bar{d} = \frac{\sum_{i=1}^{nd} d_i}{nd} \quad (4)$$

In general terms, low critical parameters allow to distinguish different events without pooling them so much. As increasing the critical parameter, mean duration increases and the number of events is reduced because of pooling effect. Mean deficit values show different behaviours. The most common one shows increments with low values of the critical parameter. As it reaches a certain value, a slight downward or a constant trend begins. Because of pooling surplus periods in drought events, minor deficits are balanced by a less number of drought events which makes the maintained trend in mean deficits reasonable. From certain critical values, other cases show important increments in deficit and duration values because of the absence of such balance in between (Figure 10).

Calibration has also taken into account data about the number of drought events identified, their maximum deficit and maximum duration values for each region. Results show similar patterns as critical parameter increases for all regions analysed. Number of droughts also shows a general downward trend. The opposite trend is shown when paying attention to maximum drought deficit and maximum duration until a sill is reached. As clearly shown (Figure 11), figures are useful to control the pooling effect of minor droughts and to obtain reasonable results adjusting parameter values. The slope change given at values of 0.1, 0.2 or 0.3 for most of cases guides the final selection of the critical parameter. Deficits are drawn in dashed lines and duration in continuous one. Different curves depend on different threshold levels (60% in red; 50% in magenta; 80% in green and 90% in blue).

Figure 12 shows the evolution of deficits for the identified droughts in two selected regions (Mediterranean basins in Iberia and Eastern Mediterranean basins) and significant differences are observed. Results for all regions have been obtained using common calibration parameters. A threshold value of 90% was selected and a desirable value for critical parameter would be in the surroundings of 0.1. For certain cases, values of 0.2 or even 0.3 for this parameter could be selected though the curves do not show a meaningful variation in deficits and duration, being the number of droughts identified the most sensitive variable. It is inferred that the identification of droughts and the conclusions do not differ much if 0.1 is selected as a critical parameter for all of them.

Some controversies could be arisen because the results (Table III) have been obtained using temporal rainfall series where the large trends have been removed. It is not the aim of this work to explain the reasons why these trends occur. They can be due to natural effects or to problems related with the availability of data in the first stages of XX th century. For a consistent application of the so-called *theory of runs*, it was decided to identify drought events once trends had been removed.

There are difficulties to compare the obtained results about drought identification with others compiled by other sources (IH, 1999; Vermes, 1998). The identification of droughts depends not only on climatic variables and models or indexes used but also on socio-economic factors and sensibility awaken to droughts. This study is constrained to identify droughts using rainfall series considering rainfall as the main force to derive a meaningful scarcity of water (see Hisdal and Tallaksen, 2000, related with drought event definition). No other hydrological factors as soil retention, aquifer, reservoir or ice storage that can lessen or increase drought effects have been considered. In spite of all these difficulties, drought event identification carried out shows a reasonable agreement with historic droughts collected from different sources (Table III).

Table III. Drought events identified with the proposed methodology in ARIDE project

Date	Drought events identified with proposed methodology in ARIDE project
1904-1905	Longest dry period in Central European basins
1916-1917	Worst drought in Island (previous important intermittent droughts from 1911 are also registered)
1917-1920	Worst drought in Southern Caspian basins
1921	Severe dry period in Central European basins Severe drought in Po and Rhone basins
1924-1926	Severest and longest drought in Mediterranean basins of Iberia (preceded by another severe drought initiated in former months of 1922 and 1923)
1929-1930	Longest drought in western Atlantic Islands
1933-1934	Worst drought in French and UK basins Severe drought period in Eastern Mediterranean basins
1937-1938	Frequent minor droughts in Northern European basins
1938-1939	Driest period in eastern basins of Europe
1939-1940	Dry period in Leeward Scandinavian basins
1942-43	Longest drought in Po and Rhone basins (followed by other minor droughts in latest 40's)
1945-1946	Driest and longest drought in Central Mediterranean basins Severe drought in Mediterranean basins of Iberia (preceded by minor droughts from 1942) Severest and longest dry period in mid and lower Danube basins followed by other minor droughts
1948	Driest drought in western Atlantic Islands (preceded and followed by other minor droughts) Dry period in eastern basins of Europe (preceded and followed by minor droughts)
1949	Driest drought in western Atlantic Islands (preceded and followed by other minor droughts)
1961	Dry period in Southern Caspian basins
1965	Dry period in Iceland
1968-1969	Severe drought in Norwegian coastal basins (followed by other drought in 1970)
1972	Severe dry period in Central European basins (preceded by minor droughts in 1971)
1973-1975	Worst drought in Eastern Mediterranean basins
1975-1976	Severest drought period in northern basins of Europe
1976	Severest dry period in Central European basins Severe drought in French and UK basins
1976-1980	Severest and longest drought in Norwegian coastal basins
1984	Longest dry period in eastern basins of Europe (continued in precedent and next year)
1985	Dry period in Iceland
1989-1990	Dry period in Eastern Mediterranean basins Minor droughts in French and UK basins
1989-1993	Dry period in mid and lower Danube basins
1991-1995	Worst drought in Western Iberian basins (preceded by other minor droughts) in late 80's

Droughts affecting Europe in recent years, and particularly those that have affected Mediterranean Europe, have brought to light the vulnerability of the water resource management systems to variations in the hydrological cycle. Table IV shows the main droughts occurred in Mediterranean countries in most recent years and makes reference to their main characteristics (Estrela *et al.*, 2000).

It can be seen from the above table that the most recent period of extensive drought in Mediterranean Europe took place between 1988 and 1992 (Figure 13).

Rainfall and runoff values in those years were much lower than the long-term average in many countries, especially those of Central and Southern Europe. This was not the case in the North, where the runoffs were much higher than in previous periods. The greatest impact of the drought was felt in zones where the pressure on the resources was greatest, especially the pressures caused by the demand for water for irrigation. The sharp drop in the runoff values for many rivers, together with the steady drop in water reserves in the reservoirs gave rise to problems in water supply for irrigation, so the effects on agricultural production were severe. As the distribution of the resources in the different regions was uneven, this caused political conflicts when it came to finding solutions involving a more even distribution of the resources in that situation.

Table IV. Summary of some recent droughts that affected Mediterranean countries

Date	Region	Characteristics
1971	Most of Europe	Exceptionally dry year The lowest winter rainfall in Spain for 30 years
1974	Scandinavia, France, Holland, Austria	Very slight rainfall from April to August in France
1988-1992	Most of Europe	Prolonged atmospheric circulation with rainfall deficiency. High temperatures in the summer and above average temperatures in winter. 3,000 rivers dried up in France at the end of 1990.
1990-1995	Spain, Portugal	Prolonged drought throughout Spain, except the North Coast. More intense between September 1994 and August 1995
1997	France, UK, Ireland, Portugal and Germany	Very low rainfall

The 6 years of drought in Greece, which affected the country in 1987-88 and 1992-93, included two extremely dry periods, 1989-90 and 1991-91, during which time the yields were the lowest recorded in the XXth Century. The severe drought, together with the prior over-exploitation of the Mornos Reservoir, brought about a critical situation that had to be rectified with measures affecting both the supply and the demand. The following are among the more noteworthy measures: reduction of up to 30% in domestic consumption, due to an increase in the price of water, imposing penalties and using awareness campaigns for the citizens; cutting down upon other uses, such as for farming on the Kopais Plain, and watering the parks and gardens of Athens with poor quality water; pumping dead water from the reservoirs, by means of pumping stations installed in the Mornos and Iliki Reservoirs; the use of groundwater for urban supply and other uses, allocating 110 hm^3 to supply Athens; and applying the main supply systems using new surface water sources, through the construction of a new reservoir and its water distribution channels (Koutsoyiannis and Mimikou, 1995).

In the Iberian Peninsula an extended drought occurred between 1990 and 1995. One of the places where this drought had its worst consequences was the South of Spain, with a very critical period between 1992 and 1995. In that zone, and specifically in the city of Seville, the rainfall was 70% below the average and the runoff was less than 30% of the average. The duration of the drought severely affected the management of the city's water, which was supplied using reserves from the long-term reservoirs.

Another type of drought was the one that occurred in France in 1976, for example. The French Electricity Company was aware of the fact that the amount of snow that had fallen in the winter of 1975 was not enough for the meltwaters to provide the large reservoirs used for hydro-electric power between March and September. In February 1976, measures were taken to use thermal power stations to generate energy and thus maintain the water levels in those reservoirs (Chastan, 1996).

5. Conclusions

An analysis of drought events in Europe and especially in Mediterranean countries from the beginning of the century has been presented. Results related to the evolution of deficits for the regions identified in Europe are shown. The use of the Climate Research Unit (CRU) database and the regionalisation of Europe using information of this database contributed to this aim.

The study of droughts is intrinsically related to regionalisation studies. A feasible regionalisation has been developed to assist the analysis of large-scale droughts in Europe. It has been carried out grouping monthly series of rainfall anomalies related to the main European river basins once their seasonal cycles and trends were removed. The algorithm for clustering is based on the Pearson correlation coefficient, which expresses a measure of similarity between series of anomalies. Afterwards, an improvement of regionalisation was searched by means of the knowledge of seasonal cycles.

A methodology to identify drought events has been implemented using averaged series of rainfall for the defined region. It is based on a selection of a 90% percentile in a varying threshold level approach for identifying the onset of drought events. This methodology was combined with an inter-event volume criterion to take into account a partial recovery to settle the end of a drought. A critical parameter of 0.1 was chosen for all regions.

Then, an identification of drought events and of their deficits and duration associated were estimated for each region in a monthly scale. It seems that droughts are independent from one region to another. A similar comment can be said about their intensities or variables related. Also some comparisons were done with historical droughts.

Regionalisation and drought event analysis constitutes a major issue for drought policies and implementation of drought awareness strategies. This work can constitute a useful contribution to these aims. Once regions with a homogeneous climatic behaviour have been delimited and drought events have been characterised, it is feasible to define tools and indicators to monitor drought events. Work developed may be also useful for water and socio-economic technicians, institutions or politicians because it gives the basis for studying other effects/impacts of droughts.

References

Alvarez, J. and Estrela, T. (Eds.) (1999) *Validation of Climate Research Unit GIS Database*. ARIDE Tech. Note Cedex, Madrid.

Batini, G. and Benedini, M. (1998) Drought management and prevention: Italy. *World Meteorological Organization Bulletin.* 47(4).

Cancelliere, A., Ancarani, A. and Rossi, G. (1995) Identification of drought periods on streamflow series at different time scales. In: N.X. Tsiourtis (Ed.) *Water Resources Management under Drought or Water Shortage Conditions*, Balkema, Rotterdam, pp. 49-57.

Chastan, B. (1996) Country paper of France. *Proceedings of Third Technical Review: Management and prevention of crisis situations: floods, droughts and institutional aspects.* EurAqua, Rome, October 23-25, pp. 31-42.

Demuth, S. and Stahl, K. (Eds.) (2001) *Assessment of the Regional Impact of Droughts in Europe*. Final Report to the European Union ENV-CT97-0553, Institute of Hydrology, University of Freiburg, Germany.

Estrada, F. (1996) *Caracterización y análisis de sequías. International Master for General and Applied Hydrology*. Centro de Estudios Hidrográficos del CEDEX. Madrid Spain.

Estrela, T., Marcuello, C. and Dimas, M. (2000) *Las aguas continentales en los países mediterráneos de la Unión Europea*. Ministerio de Fomento y Ministerio de Medio Ambiente. Octubre de 2000.

Hisdal, H. and Tallaksen, L.M. (Eds.) (2000) *Drought event definition.* ARIDE Tech. Report no. 6, University of Oslo, Oslo, Norway.

Institute of Hydrology (IH) (1999) *Drought Events in Europe*. Wallingford & Space Applications Institute (SAI), Ispra.

Koutsoyiannis, D. and Mimikou, M. (1996) Country paper of Greece. *Proceedings of Third Technical Review: Management and prevention of crisis situations: floods, droughts and institutional aspects.* EurAqua, Rome, October 23-25, pp. 63-77.

Menéndez, M. (1995) *Aspectos Hidrológicos de las Sequías. Sessions for drought studies. Droughts in Spain.* Centro de Estudios Hidrográficos del CEDEX, Madrid, Spain.

Salas, J. D., Delleur, J. W., Yevjevich, V. and Lane, W. L. (1985) *Applied Modelling of Hydrologic Time Series.* Water Resources Publications, Littleton, Colorado.

Stahl, K. and Demuth, S. (1999) *Method for Regional Classification of Streamflow Drought Series: Cluster Analysis.* ARIDE Technical Report no. 1, University of Oslo, Oslo, Norway.

Tallaksen, L.M., Madsen, H. and Clausen, B. (1997) On the definition and modelling of streamflow drought duration and deficit volume. *Hydrol. Sci. J.* 42(1): 15-33.

Vermes, L. (1998) *How to work out a drought mitigation strategy.* ICID guide. Guidelines for water management.

Wallen, C.C. (1970) *Climates of Northern and Western Europe*. Elsevier Publishing Company.

Yevjevich, V., Da Cunha L. and Vlachos, E. (Eds.) (1983) *Coping with Droughts.* Water Resources Publications, Littleton, Colorado.

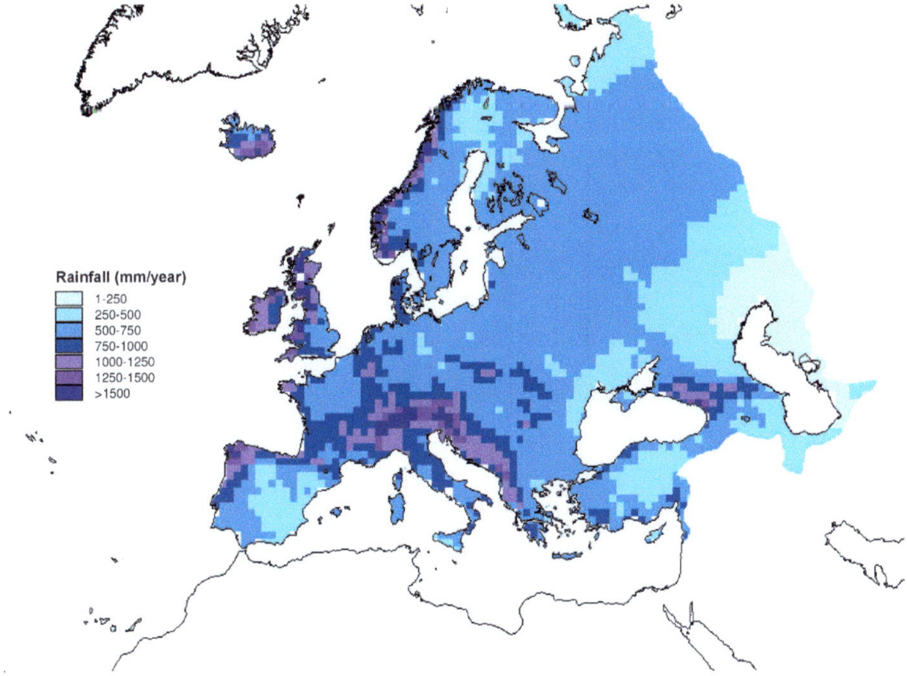

Fig. 1. Mean annual rainfall in Europe (1901-1996) estimated by CRU

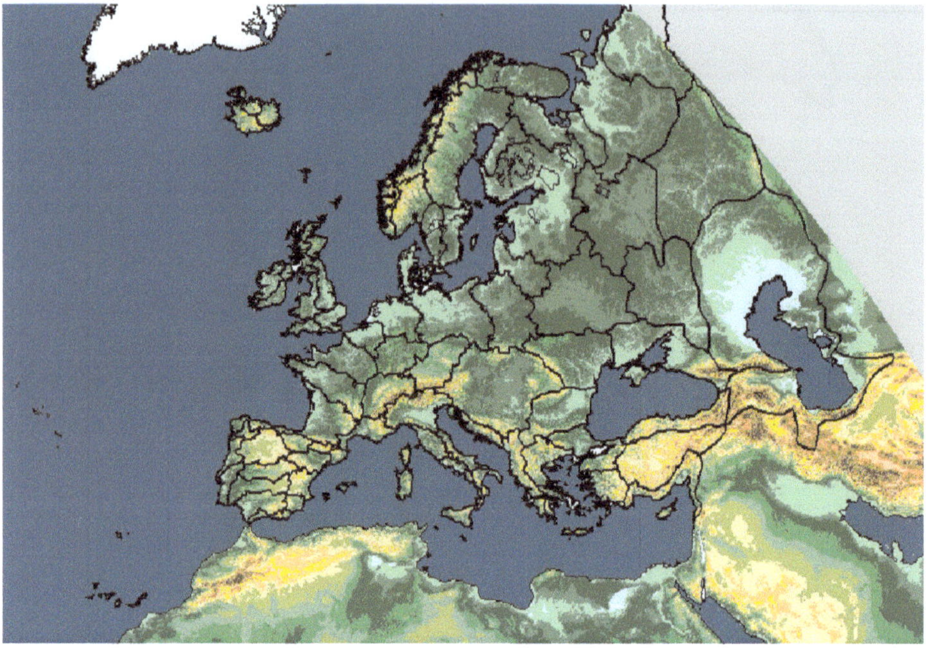

Fig. 2. Elemental boundary river basins considered across Europe for clustering over the digital elevation model of USGS of USA.

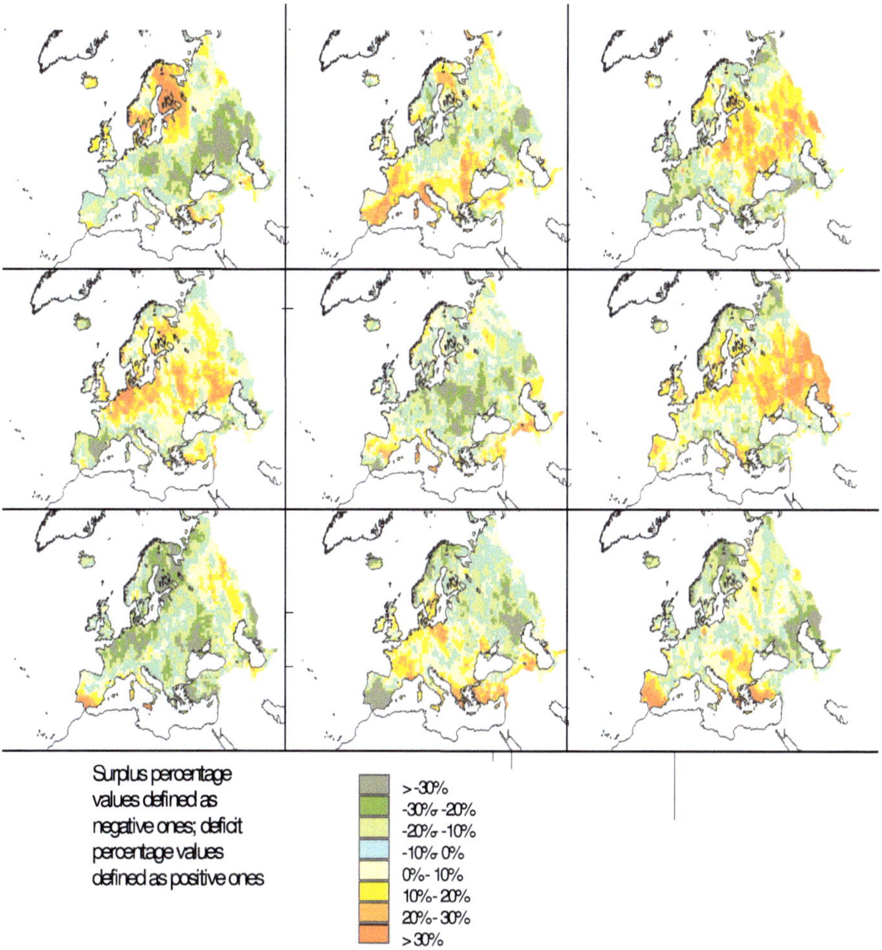

Fig. 3. Precipitation deficits and surpluses over Europe

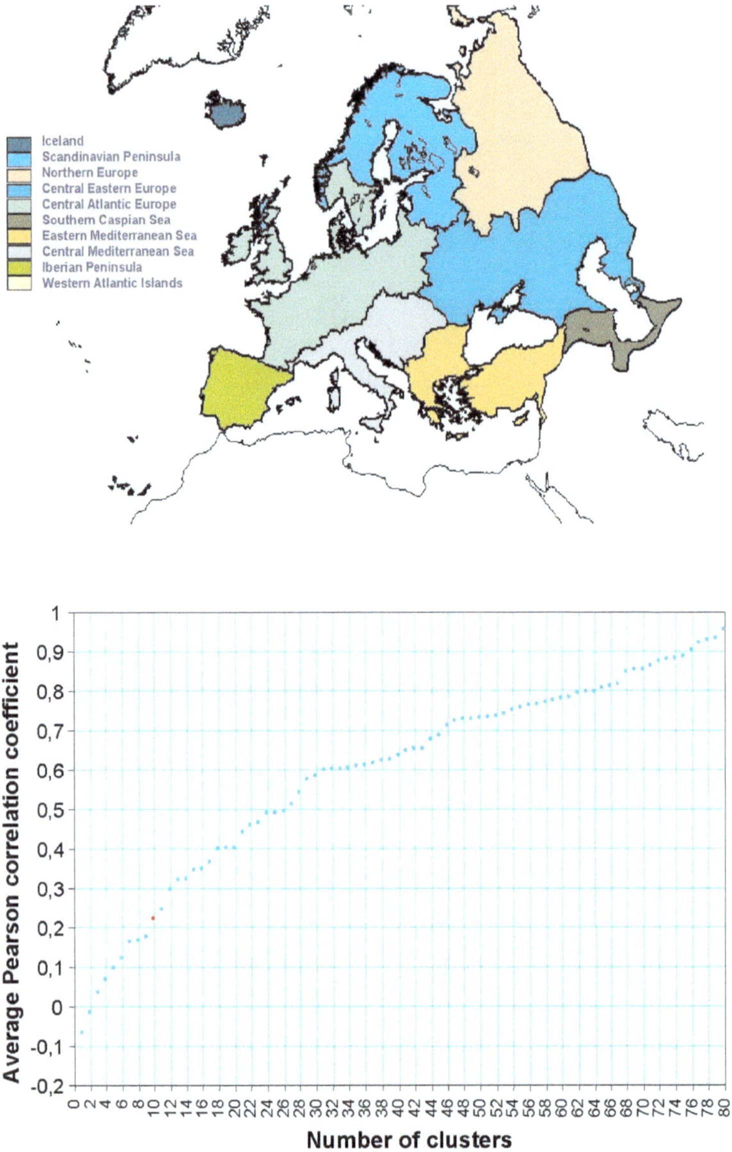

Fig. 4. Clustering and similarity measure for the whole Europe

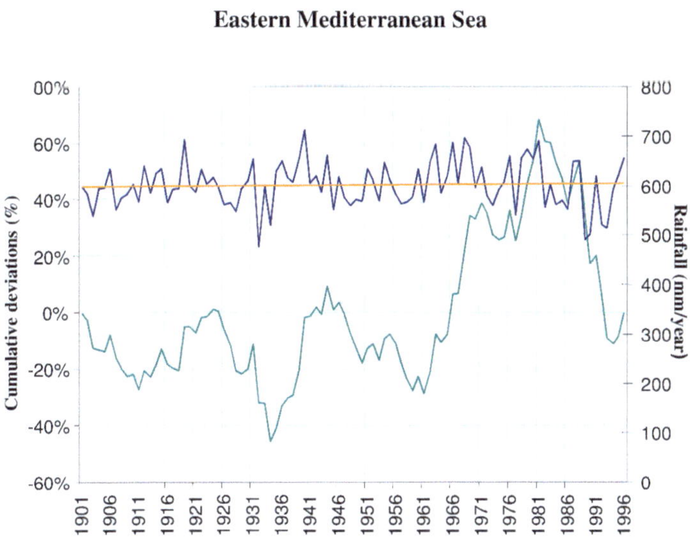

Fig. 5. Rainfall (blue) and cumulative deviation (green) for Eastern Mediterranean

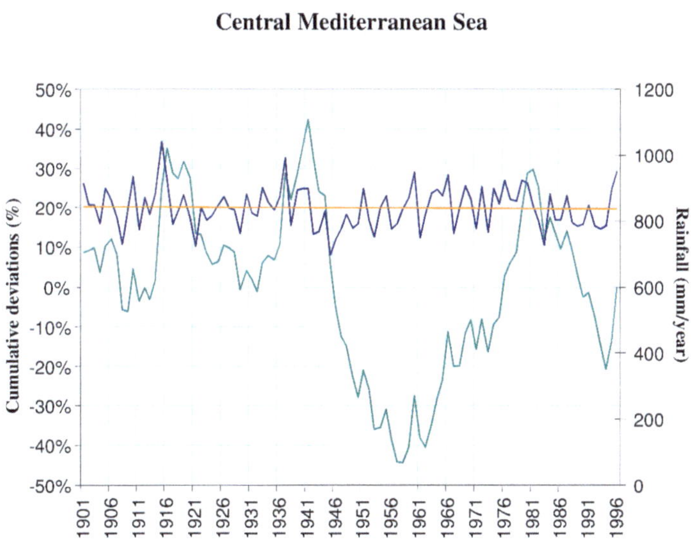

Fig. 6. Rainfall (blue) and cumulative deviation (green) for Central Mediterranean

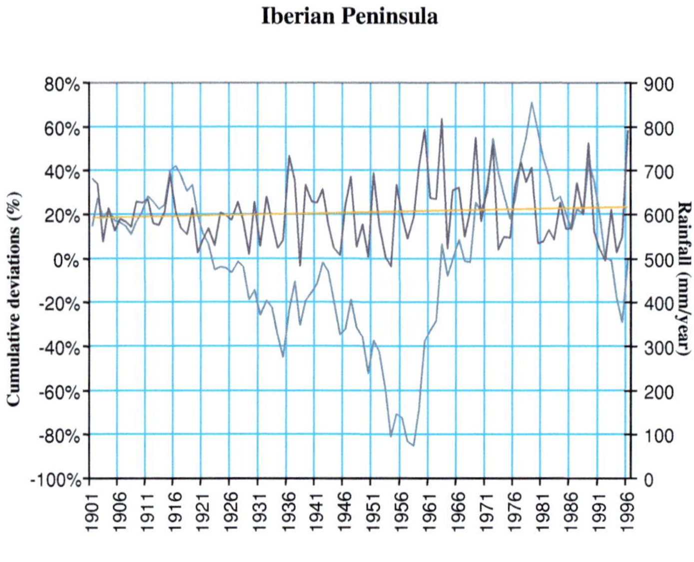

Fig. 7. Rainfall (blue) and cumulative deviation (green) for Iberian Peninsula

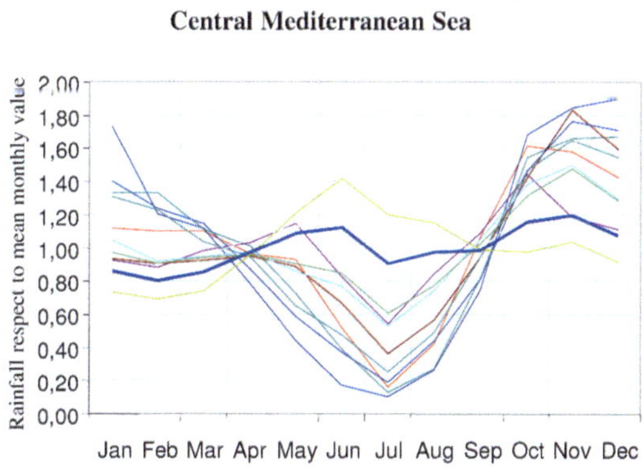

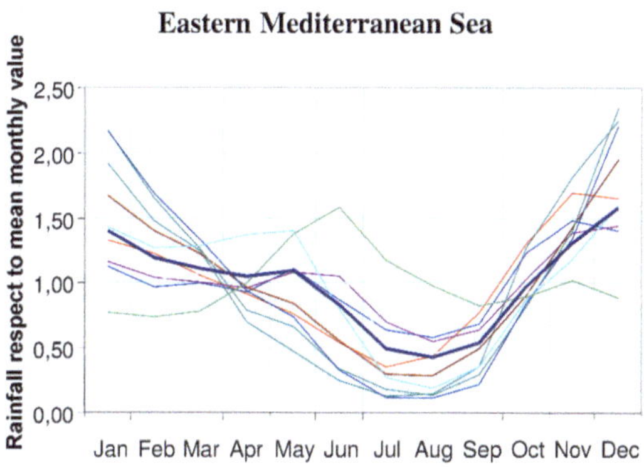

Fig. 8. Meteorological cycles in some regions clustered

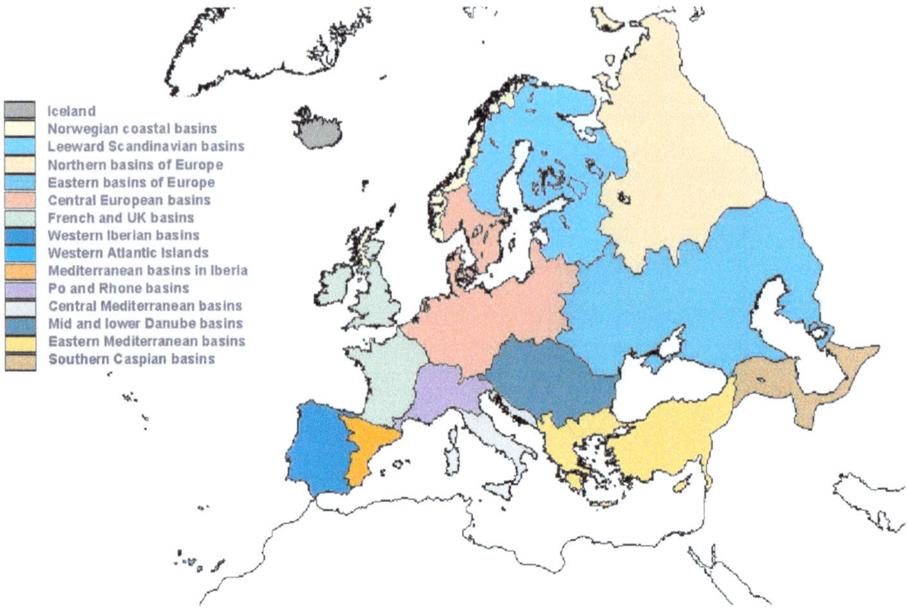

Fig. 9. Proposed meteorological regionalisation for Europe

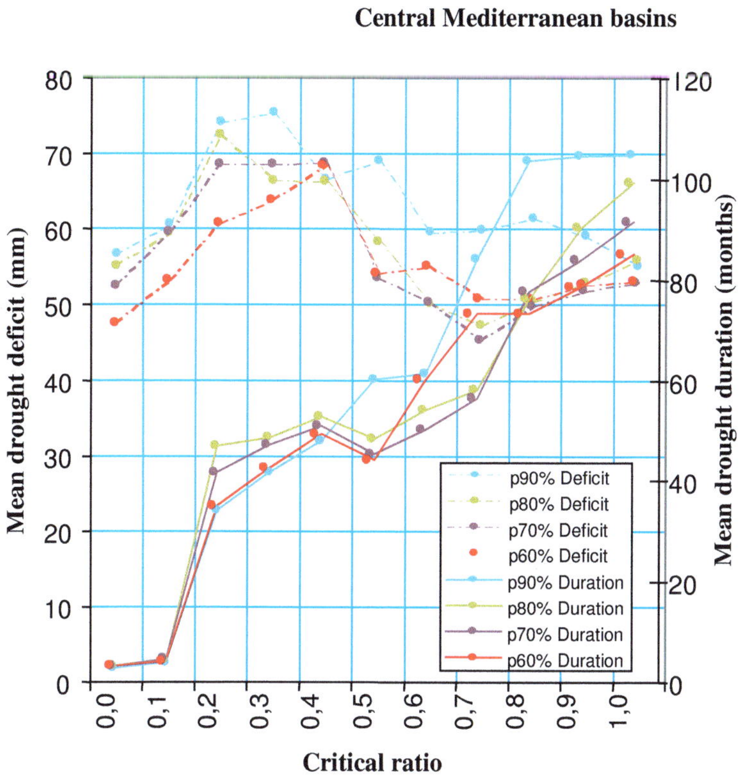

Fig. 10. Mean deficit and mean duration of droughts for Central Mediterranean basins

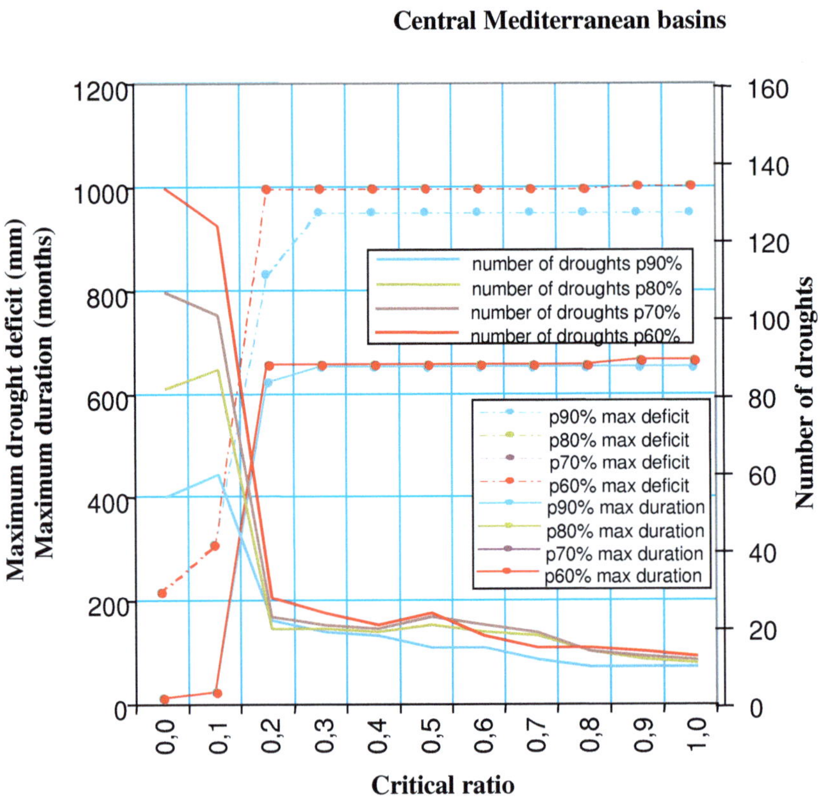

Fig. 11. Maximum deficit, maximum duration and number of droughts for Central Mediterranean basins

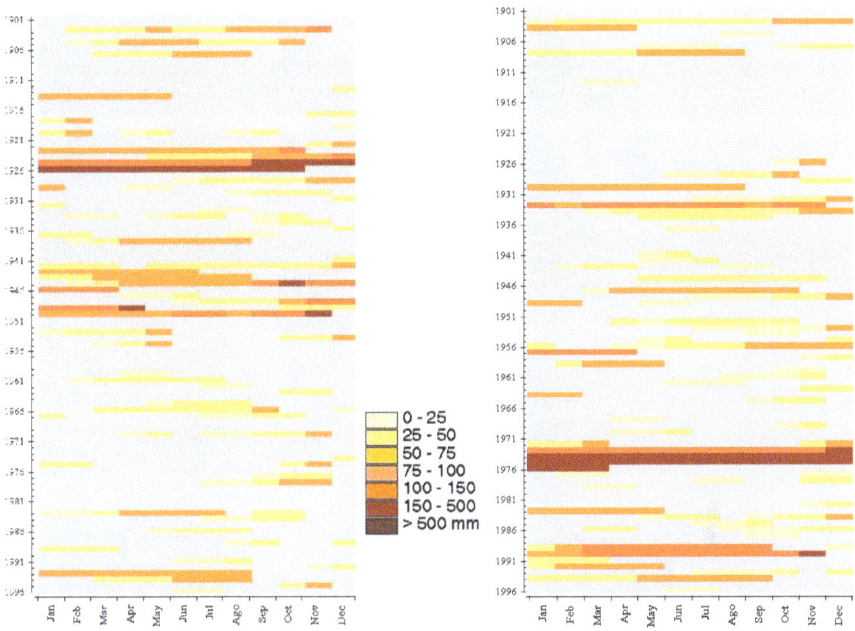

Fig. 12. Evolution of deficits in and Mediterranean basins in Iberia (left) and Eastern Mediterranean basins (right)

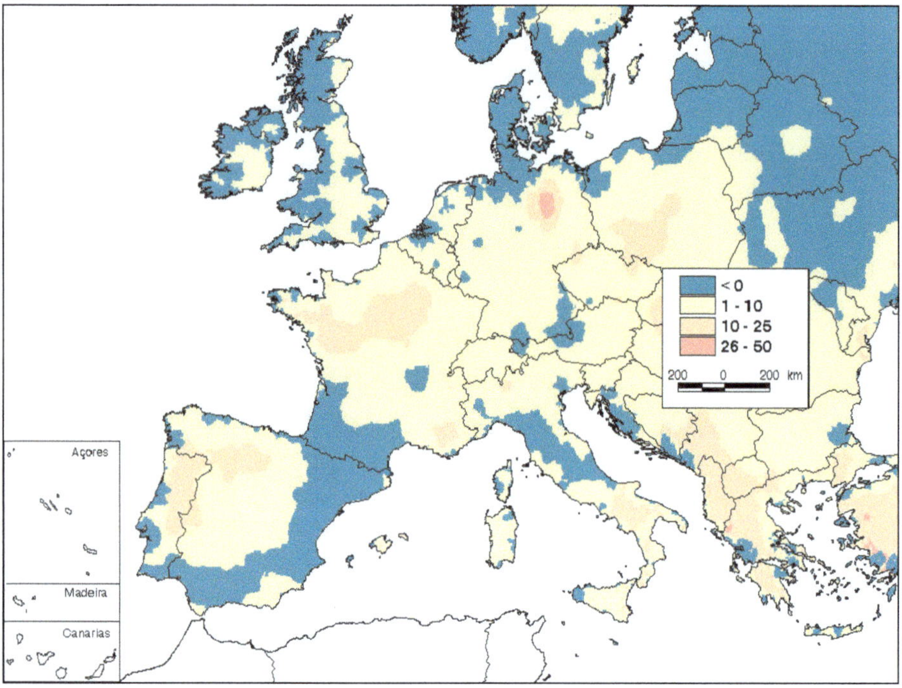

Fig. 13. Percentage reduction in rainfall values between 1988 and 1992 with respect to the mean values for the period 1900-1996. Taken from Estrela et al (2000)

REQUISITES FOR A DROUGHT WATCH SYSTEM

G. ROSSI
Department of Civil and Environmental Engineering, University of Catania, Italy

Abstract: An effective response to drought hazard depends on an early warning system able to provide timely and reliable information about the hydrometeorological events and water supply conditions. A review of some of the existing drought monitoring systems is presented. Also, an analysis of the institutional and technical aspects of the hydrometeorological networks and of the agencies responsible for drought management in selected Mediterranean countries is carried out. Finally, the role and the basic requisites of a drought watch system network for Mediterranean regions are discussed.

Key words: Drought monitoring, hydrometeorological networks, drought indices

1. Introduction

The severe droughts that recently occurred in several countries have pointed out that emergency measures are not sufficient to reduce effectively negative drought impacts, but drought management planning instruments should be developed in advance in order to implement successful actions both for reducing the vulnerability of water supply systems to the most severe water shortages and for mitigating the impacts of an ongoing drought.

An effective response to drought conditions depends strongly on the capability of an early perception of the incoming droughts and on the possibility of an appropriate monitoring of the evolution of climatic and water resources conditions (Rossi, 1995). Therefore, the need arises for a Drought Watch System (DWS) as an important tool of the decision making process oriented to face drought hazard by providing adequate information for drought declaration as well as to avoid severe water shortages through an effective water resources management under drought conditions.

Aim of this paper is to identify the role and the requisites of a Drought Watch System in order to assess its feasibility within the Mediterranean region context.

2. Role of a Drought Watch System

The main objective of a Drought Watch System is to help decision-makers to identify the drought warning conditions and to indicate the best drought mitigation measures on the basis of a continuous monitoring of the drought evolution in terms of meteorological and hydrological variables and water resources availability.

G. Rossi et al. (eds.), Tools for Drought Mitigation in Mediterranean Regions, 147-157.

Although the role of a permanent Drought Watch System as a very useful support for the implementation of effective drought management has been frequently discussed in several conferences and workshops, a very small number of such systems is actually in operation. This gap seems to depend basically on the following reasons:

1. the insufficient density, in many countries, of meteorological and hydrological gauging networks to provide adequate coverage for drought monitoring;
2. the fact that the permanent monitoring process of the hydro-meteorological conditions and of the water resources reserve status is generally shared among different agencies with different objectives, thus requiring a very difficult coordination effort;
3. the lack of universal standards in computing drought indices, due to the very different drought definitions in relation to the considered hydro-meteorological variables and to the affected economic sectors;
4. the necessity of a timely data acquisition for a near real-time data diffusion which resulted to be difficult in the past due to the limited number of remote monitored networks and to the lack of standard in procedures of acquisition and treatment of relevant drought information.

As a consequence of the recent severe droughts that occurred in many countries, and of the feared climatic changes, the awareness on the necessity of an effective preparedness to drought risk is increasing. Within such framework the importance of the implementation of national or international Drought Watch Systems is being generally recognised in drought-prone countries.

3. Experiences of operating Drought Watch Systems

Among the few drought monitoring systems in operation the following can be citated as examples:

- DROUGHT WATCH SYSTEM - National Drought Mitigation Center (NDMC) University of Nebraska-Lincoln, USA;
- DROUGHT WATCH SYSTEM – Bureau of Meteorology and Queensland Department of Natural Resources and Mines, AUSTRALIA;
- DROUGHT WATER WATCH – Centre for Ecology and Hydrology, formerly Institute of Hydrology, Wallingford, UK;
- DROUGHT MONITORING CENTRE FOR EASTERN AND SOUTHERN AFRICA - Nairobi (Kenya) and Harare (Zimbabwe).

Common feature of these DWS is the use of several drought monitoring indices related to different components of the hydrological cycle analysed at different time scales. This stems from the need to provide information to different user categories interested to several aspects of water deficit and from the difficulty of characterising drought through an unique index.

Particular emphasis is given to the graphical representation of the monitored variables in order to favour an immediate and easy assessment of the drought severity and its evolution. Access to information is ensured through public web sites with the objective of its dissemination to as many users as possible, besides the public institutions.

Relevant characteristics of the main drought monitoring systems are summarised in Table I in which are indicated, for each system, the responsible agency, the web site address for users access and the main information provided.

Table I. Characteristics of the principal existing Drought Watch Systems

Agency and WEB site	**Available information**	**Time-scale**
DROUGHT WATCH SYSTEM Bureau of Meteorology and Queensland Dept. of Natural Resources and Mines, Australia www.bom.gov.au/silo/	Precipitation: anomalies, deciles, percentage, deficiency maps Temperature: mean, anomalies, highest, lowest	day, week, 1, 3, 6, 9, 12, 18, 24, 36 months day, week, 1, 3, 6, 9, 12, 18, 24, 36 months
DROUGHT WATCH SYSTEM National Drought Mitigation Center (NDMC) – University of Nebraska-Lincoln, USA www.ndmc.unl.edu Department of Commerce NOAA's Drought Information Center ww w.drought.noaa.gov Department of Health and Human Services*Drought/Heat Information* www.hhs.gov/hottopics/drought.html	Percent of normal rainfall SPI Palmer Drought Index or Crop Moisture Index Daily streamflow Soil moisture anomaly Vegetation Condition Index	month, season, year 3-6-12-24-48-72 months month, week day day week
DROUGHT WATER WATCH Centre for Ecology and Hydrology Wallingford UK www.nerc-wallingford.ac.uk	Precipitation maps River flow maps Ground water level maps	3, 12 months month month
DROUGHT MONITORING CENTRE FOR EASTERN AND SOUTHERN AFRICA, Nairobi (Kenya) and Harare (Zimbabwe) Africa News www.africanews.org/environ Environment and Development in the Third World, *Dakar, Senegal* www.enda.sn/	Precipitation amount by categories (<40 mm; 40-100;100-200; >200mm) Drought severity index Temperature anomalies Precipitation anomalies Normalised difference vegetation index Expected weather impacts on agriculture	10-days, month 10-days, month 10-days, month 10-days, month 10-days, month 10-days, month

The most complete Drought Watch System is probably the one operating at the National Drought Mitigation Center (NDMC) of the University of Nebraska (where is also operating the International Drought Information Center which provides a significant contribution in coping with drought since 12 years). Actually the NDMC drought watch system provides the spatial and temporal distribution of 6 drought indices for the entire U.S. territory (Le Comte, 2000). Further, a drought classification, that attempts to synthesise all six indices is periodically reported. Such drought severity classification and the ranges for the 6 drought indices are presented in Table II.

Table II. Drought severity classification and ranges for drought indices, adopted by NDMC

Category	Percent of normal rainfall	Standardised Precipitation Index	Palmer Drought Index or Crop Moisture Index	Daily streamflow (percentiles)	Soil moisture anomaly (percentiles)	Vegetation Condition Index
D0 - Abnormally Dry	<75% for 3 months	0.99 to –0.99	-1.0 to –1.9	21-30	21-30	36-45
D1 - First Stage Drought	<70% for 3 months	-1.00 to –1.49	-2.0 to –2.9	11-20	11-20	26-35
D2 - Severe Drought	<65% for 6 months	-1.50 to –1.99	-3.0 to –3.9	6-10	6-10	16-25
D3 - Extreme Drought	<60% for 6 months	-2 or less	-4.0 to –5.4	2-5	2-5	6-15
D4 - Exceptional Drought	<60% for 12 months		-5.5 or less	0-1	0-1	1-5

As an example of graphical representation, in Figure 1 the spatial distribution of the Palmer Drought Index as downloaded from the NDMC web site is reported.

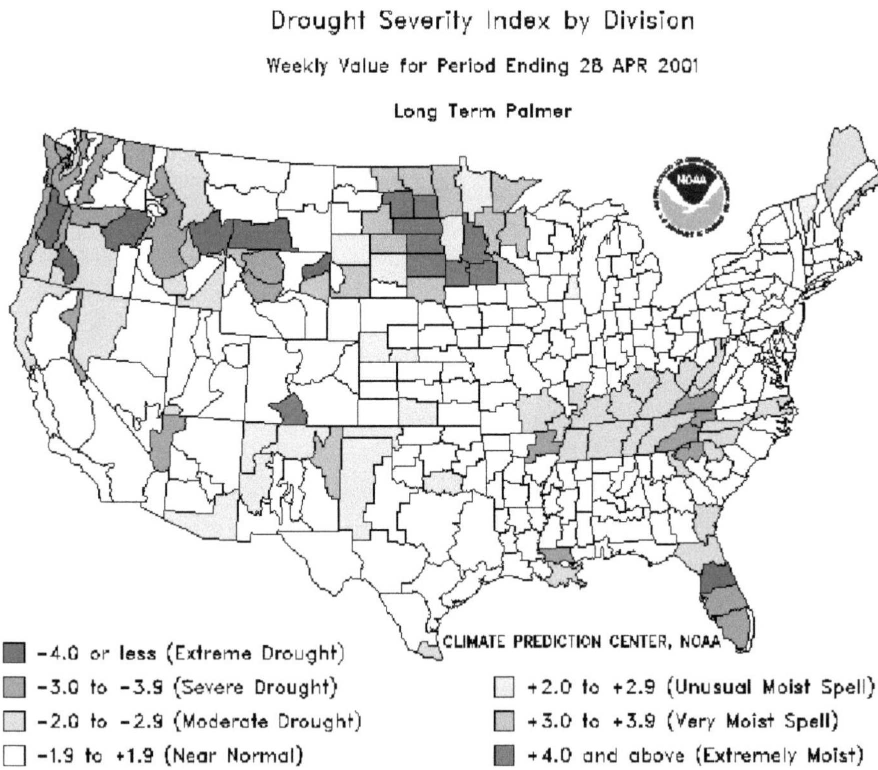

Fig. 1. Output of the NDMC drought watch system relative to a weekly map of the Palmer Drought Index

4. Requisites for the development of a Drought Watch System

Assessment of drought conditions requires the collection, transmission, storage, treatment and dissemination of a very large amount of information. Therefore, the choice of the appropriate data and the coordination of the different sources of information are critical issues for developing an effective system. In particular the development of a DWS requires:

1. the choice of appropriate drought indicators able to assess the drought conditions (e.g. alert, alarm and emergency status) as well as the level of risk of severe water shortages. These indicators should allow for synthesising the spatial and temporal variation of droughts into a few parameters, thus simplifying the communications with decision-makers;
2. the identification of the different gauging station networks which have to be connected in order to monitor adequately the meteorological and hydrological variables especially when operated by different agencies;
3. the definition of gauging station networks for monitoring the water resources conditions (surface reservoirs, aquifers);
4. the definition of the minimum information on the water consumption and expected demands (to be acquired periodically from the water supply agencies) and the definition of the basic network for collecting and elaborating these information;
5. the realisation of an organisational structure able to support the real time acquisition of the meteorological, hydrological and water resources reserve data, and the timely transfer of the drought information to decision-makers responsible for emergency actions and prevention plans for coping with drought;
6. the definition of models for the evaluation of drought conditions and for the assessment of the level of the available water resources reserve below which alert conditions should be declared;
7. the definition of the information system architecture and procedures for the quick transmission of the collected drought information to decision-makers.

An efficient DWS needs to select appropriate indicators to recognise the drought situations. As well known, the simplest indicators of water shortage are based on the evaluation of rainfall deficiency, like the one used by the Australian Bureau of Meteorology in the nation-wide drought monitoring system for the past decades (Power, 2000). Other indicators describe the water deficit on the basis of hydrological and meteorological variables or by analysing the available water resources in surface reservoirs, aquifers or snow pack. The choice of these indicators depends on the availability of hydrological data and on the characteristics of water supply systems.

Among these indicators, the SPI index (McKee et al., 1993) is a relatively new drought index based only on precipitation, that is receiving widespread acceptance in many countries and is actually tested in more drought areas. The SPI can be calculated for a variety of time scales suitable for both short-term agricultural and long-term hydrological applications.

A check should also be made on the convenience of using in the Mediterranean context some indices currently used in regions with different climatic and institutional characteristics (Cancelliere et al., 1996). The Palmer Drought Index, for example, based on long-term records of temperature and precipitation and the Surface Water Supply Index (SWSI) based on stream-flow and reservoir storage data should be investigated for possible adoption.

The drought indices should take into account the water resources stored in surface reservoirs or aquifers, because these resources represent a very significant part of municipal water supply which should have priority in case of restriction needs.

In order to obtain a comprehensive assessment of the shortage risk, an integrated approach to climate and water supply monitoring has to be used by combining meteorological, hydrological and water resources information into a single product for the purpose of detecting drought conditions and monitoring their evolution.

5. Perspectives for the development of DWS in the Mediterranean Region

The development of a DWS network in the Mediterranean Region represents an important step for the prevention of drought damages and improvement of drought preparedness and, more in general, for a sustainable water resources management in countries which are characterised by highly drought-vulnerable water systems. The basic information on the institutional and technical features of the organisations responsible for hydrometeorological monitoring and drought mitigation actions in the countries involved in the INCO-DC Project "DSS-DROUGHTS" are summarised in Table III. It is confirmed that, with the exception of the Portugal, where the Water Institute operates a limited monitoring system, these Mediterranean countries do not have any drought early warning system, although they have several reliable meteorological and hydrologic networks. However the existence of specific tasks of the Government for drought declaration and of the water management agencies for the implementation of drought mitigation measures points out the necessity of the development of a certain drought monitoring instrument. Also promising signals of an increasing attention to the drought phenomenon (e.g. computation of drought indices, preparation of periodic reports on water deficits, development of prototypal drought bulletins), show that a certain awareness, often not very clearly expressed, exists about the appropriateness of a drought watch system.

A few experiences of such systems have already been developed in Mediterranean region (e.g. MED-HYCOS system funded by World Bank to WMO) and different studies have been carried out for European countries in connection with European research projects (e.g. the cited ARIDE).

The Recommendations of the Expert Group on Early Warning Systems for Drought Preparedness and Drought Management, issued during the meeting held in September 5-7, 2000 in Lisbon, Portugal, and recently published by the World Meteorological Organisation (Wilhite et al., 2001) underline the need of incorporating

Table III. Preliminary information for the development of a Drought Watch System

	Italy	Jordan	Portugal	Tunisia
Existing hydro-meteorological networks	• Meteorological and hydrologic network operated by Hydrographic and Marine Survey • Meteorological network operated by Meteorological Service • Climatic network operated by Ministry of Agriculture	• Meteorological network operated by Department of Meteorology • Streamflow gauges network operated by Ministry of Water and Irrigation	• Meteorological network operated by Portuguese Meteorological Institute • Network operated by Instituto da Água	• Meteorological network operated by National Institute of Meteorology Streamflow and groundwater gauges network operated by Department of Water Resources • Reservoirs gauges network operated by Department of Dams
Governmental authority for drought declaration	Prime Minister or Minister of Interior on request of Regional Departments or Provinces	Cabinet of Minister and the Prime Minister Office	Ministry of Environment and Land Use Planning	Ministry of the Agriculture
Agency responsible for drought emergency actions	Department of Civil Protection (at national and/or regional level)	Ministry of Agriculture, Ministry of Interior, Ministry of Water and Irrigation, Department of Civil Defense	Civil Protection Service	The National Commission
Existing Drought Monitoring Service (DMS)	No specific DMS exists but information on a small sample of hydrometeorological stations is currently available at National Hydrographic and Marine Survey.	No specific DMS exists	Water Institute (Instituto da Água) through a network including 42 meteorological stations, 41 reservoirs and 15 wells	No specific DMS exists, although periodic reports on the climatic conditions are prepared for the Government.
Drought indices used	A prototypal drought bulletin using SPI (k=3,6,12,24 months) has been developed within European INTERREG program	Rainfall amount and recently SPI are used by the Department of Meteorology to characterize drought condition	Cumulated rainfall amount, monthly reservoir volume, groundwater levels.	Agrometeorological water balance, rainfall deficit, reservoir storage, groundwater levels

a drought early warning system in a more general national drought policy which addresses the complexity of the natural hazard through a set of linked elements including:

- institutional arrangement (e.g. drought task force or commission with principal responsibility for drought preparedness, monitoring and assessment, mitigation, composed of governmental agencies and key stakeholders, NGO, etc.;
- assessment of social, economic and environmental drought impacts;
- development of a drought preparedness plan, including the identification of drought prone areas and vulnerable economic sectors and the definition of actions to be taken to address vulnerabilities and reduce drought impacts.

In particular it is emphasised that the development of the drought monitoring system should give priority to the improvement of the existing observation networks, to the establishment of new networks and to the development of analytical and predictive tools and models.

Having in mind also these Recommendations, the efforts for realising a DWS in Mediterranean countries must integrate the following actions:

- analysis of the institutional framework responsible for the water resources planning and for the emergency actions during drought;
- development of methodology for drought risk assessment (on the basis of the hydro-meteorological variables);
- development of methodology for water shortage risk assessment of water supply systems (on the basis of a balance resources-water uses);
- development of a methodology for the assessment of the economical, environmental and social impacts of drought;
- development of operating guidelines for the definition of long-term measures aiming to the reduction of the systems vulnerability to drought;
- definition of guidelines for the development of an emergency plan to cope with drought including the definition of the tools of co-ordination among the responsible organisations at different levels (local, regional, national);
- development of training programs for the personnel working at the Drought Monitoring Agency and of a public campaign for making people aware of the importance of a drought monitoring service.

A proposed scheme of a DWS for European and Mediterranean region is depicted in Figure 2, where the data flows are represented by arrows, databases by ellipses and involved organisations by boxes. In particular the DWS collects data from several national and international data sources and provided information to Governmental bodies and water management agencies. Dissemination to general users can be effectively implemented through INTERNET by using different levels of access to the information.

The choice of the national scale for the proposed DWS stems from the significant differences in existing institutional framework of drought monitoring, water resources management organisations and governmental authorities. However in many cases the information exchange with other national Drought Watch Systems can effectively improve the quality of the service. Also a relevant support to a fast development of these systems in various countries could derive from an intense exchange of the prototypal analyses and feasibility studies carried out in each country in order to achieve common standards on acquisition data, computing indices, and drought information disseminating.

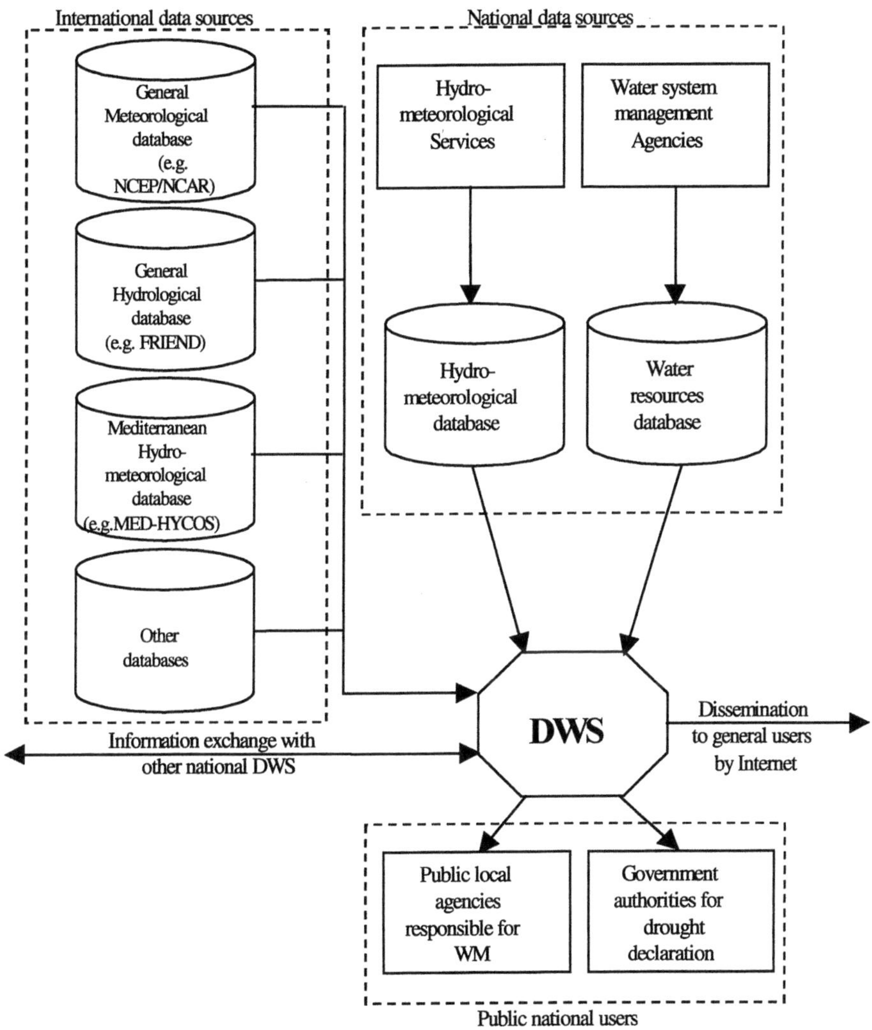

Fig. 2. Scheme of a Drought Watch System for Mediterranean countries

The feasibility analysis for the development of a drought watch system should include specific investigations oriented to:

- the improvement of the hydro-meteorological monitoring networks;
- the development of a water resources status monitoring network (surface reservoir and aquifers);
- the development and/or acquisition of meteorological radar;
- the realisation of a system for acquisition, managing and storage of satellite images (NOAA, METEOSAT);
- the realisation of a center for collecting, managing and storing the data acquired by the different networks;
- the definition of the drought indices and the development of mathematical models for the drought spatial distribution analysis;
- the development of the necessary software for the preparation of a periodical drought bullettin.

6. Conclusions

A DWS is recognised as a key instrument for an effective drought management effort in the Mediterranean region.

In spite of the fairly good quality of hydrometeorological monitoring networks, some delay is observed in European, North-Africa and Near-East countries in the development of drought early warning systems similar to those already in operation in Australia, in USA and in few other countries.

In order to obtain an effective decision support to the measures to be taken for reducing drought vulnerability and mitigating drought impacts, it is necessary that the monitoring system includes meteorological and hydrologic conditions information also through remote sensing, as well as information on the water supply systems conditions.

The introduction of the concept of a timely drought information into the general national policy for drought mitigation is a necessary condition for a successfull implementation of a Drought Watch System. A large exchange of experiences between the different Mediterranean countries oriented, at middle-term, to realise a network among national drought watch system seems, an important issue to be discussed at scientific and political level.

Acknowledgements

The paper is based upon the results of the EC-INCO DC Project "A DSS for mitigation of drought impacts in Mediterranean regions" (Contract IC18-CT97-0169). Preliminary analyses were supported by the Italian National Research Council, Group for Prevention from Hydrogeological Disasters and by the Italian Department of National Technical Services within the INTERREG IIc Project on Drought.

References

Cancelliere, A., Rossi, G., Ancarani, A., 1996. Use of Palmer Index as a drought indicator in Mediterranean Regions. *Proceedings of the Conference "From flood to drought"*, SunCity, SouthAfrica, 5-7 august, 1996.

Demuth, S., K. Stahl (eds.) 2001, *ARIDE-Assessment of the Regional Impact of Drought in Europe, Final Report.* Institute of Hydrology, University of Freiburg, Freiburg, Germany, pp.125-137.

Le Comte, D., 2000. Drought Monitoring and Forecasting in the United States. *Proceedings of the Conference Drought: Monitoring, Mitigation, Effects.* Cagliari-Villasimius, 21-23 september, 2000.

McKee, T.B., Doesken, N.J. and Kleist, J. (1993) The relationship of drought frequency and duration to time scales. 8th Conference on Applied Climatology, Anaheim, California, pp. 179-184.

Power, S., 2000. Drought in Australia : Impacts, Mitigation, Monitoring, Policy and Prediction. *Proceedings of the Conference Drought: Monitoring, Mitigation, Effects.* Cagliari-Villasimius, 21-23 september, 2000.

Rossi, G., 1995. Role and requisites of a drought watch system. In: *Climate Change and Hydrogeological Hazards in the Mediterranean Area*, , edited by L. Ubertini, F. Castelli and R. Bras, Grafica Salvi, Perugia, Italy, pp. 95-102.

Wilhite, D., Sivakumar, M.V.K., Wood, D. (eds.), 2000. *Proceedings of the Meeting on "Early Warning Systems for Drought Preparedness and Drought Management"*, 5-7 september 2000, Lisbon, Portugal.

Part 3

Modelling irrigation management under drought conditions

IRRIGATION SCHEDULING SIMULATION: THE MODEL ISAREG

L. S. PEREIRA, P. R. TEODORO, P. N. RODRIGUES and J. L. TEIXEIRA
Agricultural Engineering Research Center, Institute of Agronomy, Technical University of Lisbon, Portugal

Abstract: This paper presents the new Windows version of the irrigation scheduling simulation model ISAREG. It describes the main features of the model, data requirements, simulation options, basic calculation procedures, and approaches used to scheduling irrigations for drought water scarcity. The main windows for handling the model are shown. including those for presentation of results.

Key words: modelling, irrigation scheduling, soil water balance.

1. Introduction

The irrigation scheduling simulation models are essentially of two types: water fluxes simulation models, when the water balance is computed from the water fluxes entering and leaving the root zone, and soil water balance simulation models, when the water balance computations are performed by estimating the input and output quantities of water to the soil reservoir with a predetermined time step (Pereira *et al.*, 1995).

Water flux models are used when the soil water fluxes have to be known or when the quantities of solutes transported by the water have to be estimated. They require detailed information on the soil hydraulic properties, crop canopy characteristics and other parameters influencing water (and nutrients) extraction by the crop roots. The complexity of processes simulated and of model parametrisation favour that a crop growth and yield model is associated with water flux models. On the contrary, the soil water balance models are easier to parametrise and require only essential soil water characterisation and basic crop data, and use simplified water-yield functions to evaluate the effects of water deficits in terms of yield reductions. The water flux models need careful calibration of several parameters and the adoption of auto-calibration for many others that are difficult to obtain in the field. The second need only the calibration of few observable parameters. Nevertheless, they also need field validation of assumptions before generalised use. When both type of models are applied to the same crop, soil and environment they produce similar results, *e.g.* ISAREG and WAVE models as described by Liu and Fernando (1998).

The use of soil water balance simulation models for irrigation scheduling is definitely recommended when to be applied in the irrigation practice by planners, farmers, managers or farm advisers. This is the case for the ISAREG model, which has been successfully applied in several regions (Liu *et al.*, 1997; Fernando *et al.*, 1998), and more recently to generate and evaluate deficit irrigation schedules for cereal and horticultural crops in the Mediterranean region (Rodrigues *et al.*, 2002; Zairi *et al.*, 2002; Oweis *et al.*, 2002), as well as for citrus orchards in Sicily (Alba *et al.*, 2002) and

G. Rossi et al. (eds.), Tools for Drought Mitigation in Mediterranean Regions, 161-180.

in Jordan (Shatanawi *et al.*, 2002). More recently, the model is also operational in a GIS framework (Fortes *et al.*, 2001). The main characteristics of the model are described below and refer to the new Windows version of the model.

2. Model features

The ISAREG is a soil water balance simulation model aimed at establishing irrigation scheduling programs for a given soil-climate-crop combination, or at the evaluation of selected irrigation schedules (Teixeira and Pereira, 1992; Liu *et al.* 1997). In the present version, the model ISAREG is combined with two other computer programs, EVAP56 that computes the reference evapotranspiration (ET_o) using the Penman-Monteith method, and KCISA for computing the required crop parameters using the updated FAO methodology (Allen *et al.*, 1998). Computations can be performed for daily, 10-day or monthly time steps. The model can be used under different *alternative options*:

1. To schedule irrigations aiming at maximum yields. Simulations are performed using a soil water threshold selected to avoid water stress of the crop;
2. To simulate an irrigation schedule adopting user selected irrigation thresholds, including for controlled water deficits of the crop, and various forms to compute the application depths;
3. To evaluate an irrigation schedule when water is applied at given dates, *i.e.* when irrigation frequency and depths are pre-determined or known from the field. This option is used when validating the model against available experimental data (Oweis *et al.*, 2002) or using field trials installed and explored for that purpose (Alba *et al.*, 2002).
4. To search for an optimal irrigation scheduling under conditions of limited water supply, with constant or variable irrigation depths (this option is only available in the MS-DOS version of the model);
5. To execute the water balance without irrigation, and
6. To compute the net irrigation water requirements.

The model is able to consider restrictions in water availability for irrigation in several manners when the options 2 and 3 are considered: (a) periods when water is not available; (b) total water depth available for the season; and (c) water depths available during selected time periods.

To support users decisions for selecting the most appropriate schedule, the model uses a simplified linear relationship between the relative yield loss and the relative seasonal evapotranspiration deficit as proposed by Stewart *et al.* (1976) and Doorenbos and Kassam (1979) using a crop specific yield response factor K_y (see section 7). This parameter may be selected by the user.

The model applies to the field scale. Simulations can be performed for a single crop or for a succession of up to three crops. Up-scaling can be performed with GIS (Fortes *et al.*, 2001). The operation of the model is commanded from a main menu

(Figure 2). This menu handles all data files required for computations and accedes to the operational and computational menus further presented.

3. Soil water balance

The soil water balance is performed for the complete effective rooting depth as:

$$\theta_i = \theta_{i-1} + (P_i - Q_{r\,i} + I_{n\,i} - ET_{a\,i} - DP_i + GW_i) \,/\, (1000\, z_{r\,i}) \quad (1)$$

where

- θ_i average soil water content in the root zone [mm mm^{-1}] on day i,
- θ_{i-1} soil water content on the previous day[mm mm^{-1}],
- P_i precipitation on day i [mm],
- $Q_{r\,i}$ runoff from the soil surface on day i [mm],
- $I_{n\,i}$ net irrigation depth on day i [mm] (irrigation water infiltrating the soil),
- $ET_{a\,i}$ actual crop evapotranspiration on day i [mm],
- DP_i deep percolation on day i [mm],
- GW_i upward flux of water from a shallow water table on day i [mm], and
- $z_{r\,i}$ root depth on day i [m].

The model computes θ_i and then calculates the required irrigation depths $I_{n\,i}$ when the appropriate thresholds for the soil water are reached as described below. The ET_a computations base on the product between the reference evapotranspiration and a crop coefficient (Allen *et al.*, 1998) as described below. The model considers two alternatives for the estimation of the groundwater contribution GW_i: the simplified procedure proposed by Doorenbos and Pruitt (1977) or a recently developed algorithm (Liu and Fernando, 1998; Fernando *et al.*, 2001) as indicated in Section 4.4. Deep percolation is estimated either empirically (Teixeira and Pereira, 1992) or using an algorithm describing the soil water depletion from saturation to field capacity (Liu and Fernando, 1998; Fernando *et al.*, 2001). The runoff depth $Q_{r\,i}$ can be predicted using the SCS curve number method (SCS, 1982) but, for most applications, runoff can be neglected.

The water balance computations are performed considering that the actual available soil water, R [mm], may be referred to three storage zones (Figure 1):

i. the excess water zone above field capacity (θ_{FC}), corresponding to gravitational water, which is not immediately available for the plants,;
ii. the optimal yield zone between θ_{FC} and the optimal soil water threshold θ_p, and
iii. the water stress zone between θ_p and the wilting point θ_{WP}, where water still is available to the plants but requires increased energy for extraction by the plant roots.

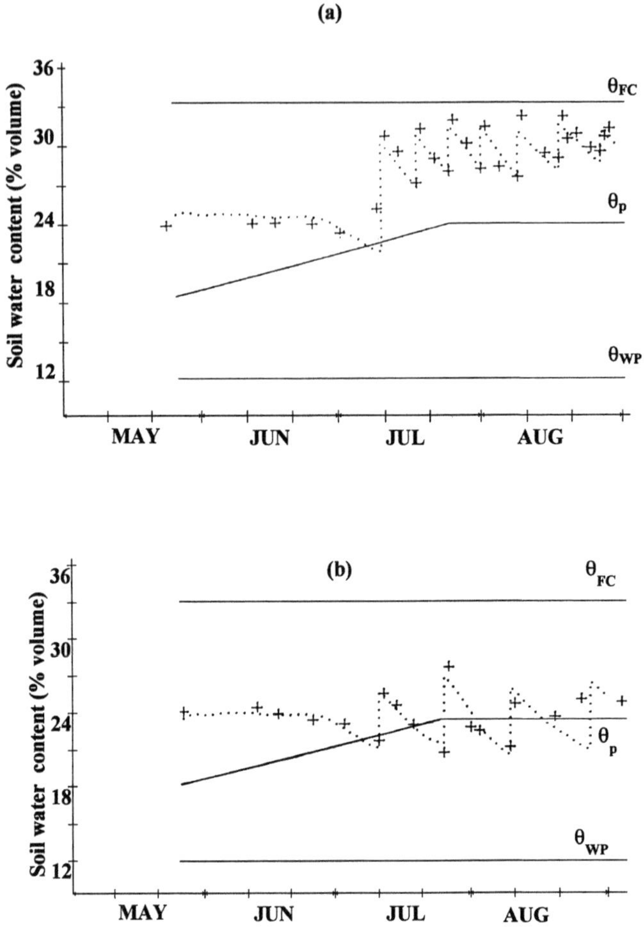

Fig. 1. Observed (+ +) and simulated (. . . .) soil water content for a maize crop in a loam soil at Coruche using irrigation schedules for (a) no stress, and (b) for deficit irrigation. (source: Teixeira and Pereira, 1992) The curves θ_{FC} *and* θ_{WP} *represent soil moisture at field capacity and at wilting point, respectively, and the curve* θ_p *represents the threshold for no stress*

The size of these zones of the soil reservoir varies from crop to crop with the root depth and the crop sensitivity to water stress, expressed by the soil water depletion fraction for no stress p (adimensional). However, the size of these reservoir zones mainly depends from the soil hydraulic characteristics. The upper boundary for the *excess water zone* is the soil water at saturation (θ_s) when the root depth z_r is maximal, thus corresponding to the full size of the soil reservoir. The upper limit of the *optimal yield zone* is R_{max} [mm], which corresponds to the total available soil water (*TAW*), thus when the soil is at field capacity, θ_{FC}:

$$R_{max}=1000\, z_r\, (\theta_{FC} - \theta_{WP}) \qquad (2)$$

where

θ_{FC} soil water content at field capacity [mm mm^{-1}]

θ_{WP} soil water content at wilting point [mm mm^{-1}]

z_r root depth [m].

The lower limit of the *optimal yield zone* corresponds to R_{min} [mm], which is defined by:

$$R_{min} = (1 - p)\, R_{max} \tag{3}$$

where p is the soil water depletion fraction for non stress [ ].

This lower limit of the optimal yield zone defines the optimal yield threshold (line θ_p in Figure 1) *i.e.* the minimum soil water content to avoid crop water stress. The difference between R_{max} and R_{min} consists of the readily available soil water ($RAW = p\ TAW$).

The *water stress zone* is limited by R_{min} and R=0, *i.e.* the soil water reserve is fully depleted when the soil water is at the wilting point.

When the available soil water is in the optimal yield zone, the actual evapotranspiration equals the crop potential or maximum evapotranspiration ET_m [mm], *i.e.* $ET_a = ET_m$ with

$$ET_m = K_c\, ET_o \tag{4}$$

where (Allen *et al.*, 1998)

K_c crop coefficient [ ]

ET_o reference evapotranspiration [mm].

In the water stress zone, the actual evapotranspiration is constrained to $ET_a < ET_m$ because $R < R_{min}$, *i.e.* the available water is below the soil water threshold for no stress. Thus:

$$ET_a = (R / R_{min})\, ET_m \tag{5}$$

The variation of the available water R with time is not linear when $R < R_{min}$ but follows a parabolic decay curve which exponent is a function of R and ET_m (Teixeira and Pereira, 1992).

4. Data

4.1 INPUT DATA

Data required to perform the soil water balance calculations are stored in several files that are commanded from the main menu (Figure 2). They refer to:

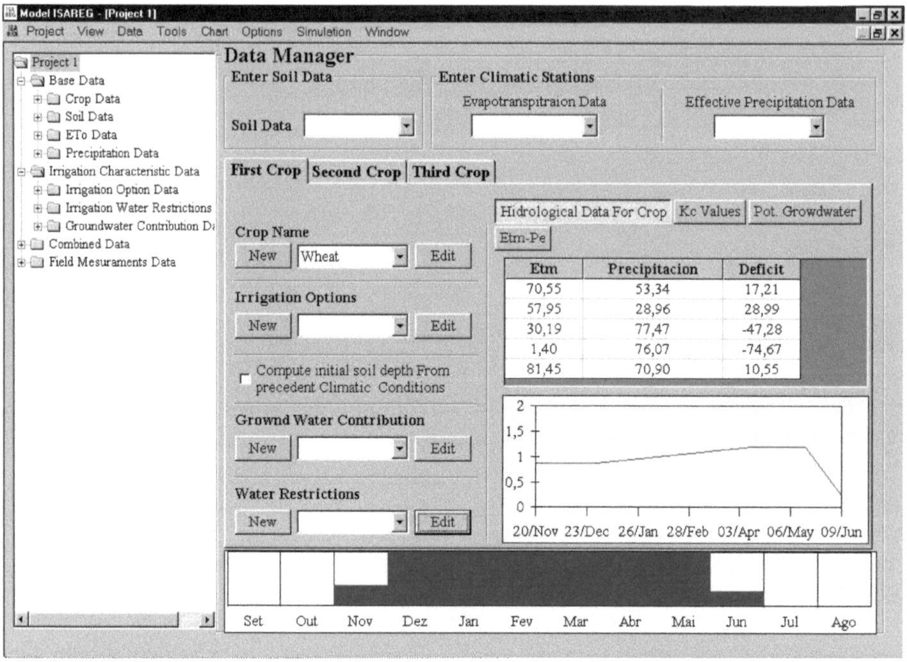

Fig. 2. Main menu

a) *Meteorological data*: effective precipitation, P_e [mm], and reference evapotranspiration, ET_o [mm] to compute the soil water balance; maximum and minimum temperature, T_{max} and T_{min} [°C], maximum and minimum relative humidity, RH_{max} and RH_{min} [%], or dry and wet bulb temperatures [°C], solar radiation
[MJ m^2 day^{-1}] or sunshine duration [h] and wind speed [m s^{-1}] to compute ET_o with EVAP56; and wind speed and minimum relative humidity to adjust the crop coefficients to the local climate with KCISA. Default procedures are used when not all climatic variables but precipitation and maximum and minimum temperatures are available.

b) *Crop data*: dates of the crop development stages, the correspondent crop coefficients, K_c; root depths, z_r [mm]; crop height h [m], the soil water depletion fraction for non stress, p, and the yield response factor, K_y. The user may input these data or use the computation facilities provided by the KCISA program (Rodrigues *et al.*, 2000).

c) *Soil data* referring to a multi-layered soil: for each layer the respective depth, d [m], and the total available soil water, TAW [mm m^{-1}] or the soil water contents at field capacity and wilting point, θ_{FC} and θ_{WP} [mm mm^{-1}]. Additional information on the depth, textural and water holding characteristics of the soil surface evaporative layer are also required to compute K_c for the initial crop growth period with KCISA as described below.

d) *Groundwater data*, and complementary parameters to compute the groundwater contribution *GW* [mm day^{-1}] and deep percolation *DP* [mm day^{-1}].

e) *Irrigation options data*, which refer to a data file created by the user to store alternative simulation options and further compare the corresponding results without having to repeat operations.

f) *Water restrictions data*, referring to limitations imposed to the simulated schedules relative to the net amounts of water available for irrigation and the periods when they may be supplied.

The operation with data files, *i.e.* creating, modifying or just reading is performed through appropriate windows such as that for crop data in Figure 3.

4.2 CROP AND SOIL DATA

Crop data is handled differently according to the type of crop: (a) field, horticultural and tree crops (as in the example in Figure 3), (b) winter field crops in environments where the soil is frozen during a few months, (c) forage crops having several cuts during the growing season, and (d) perennial crops having a constant crop coefficient.

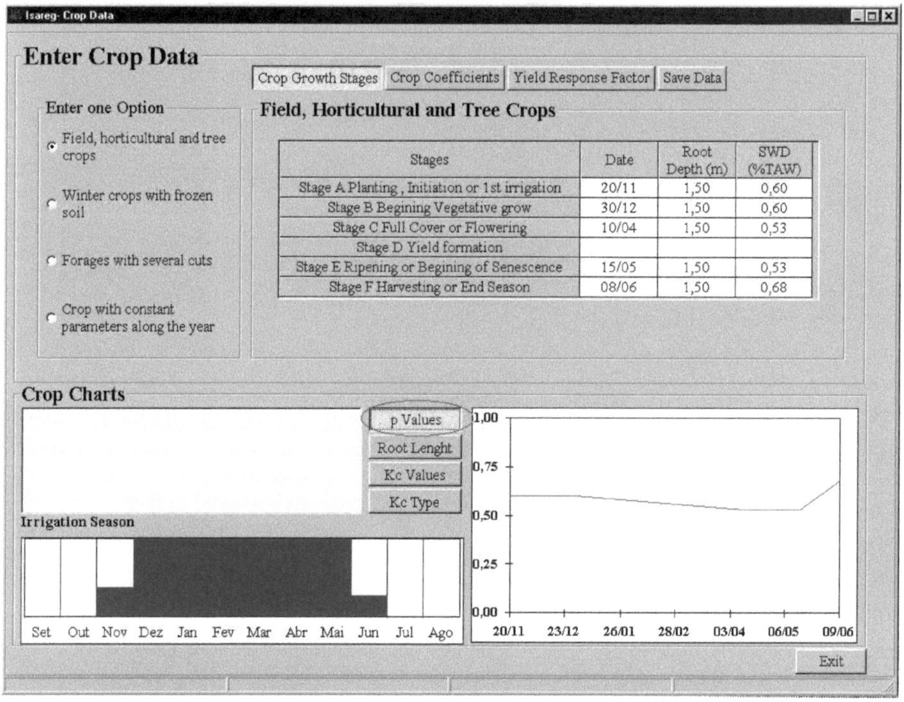

Fig. 3. Window for entering crop data, case of crop coefficients for field, horticultural and tree crops

Crop data (Figure 3) refer to dates of crop development stages, crop coefficients (K_c) for each stage, soil water depletion fraction for no stress (p), crop height (h), root depths (z_r) from planting to maximum development, and yield response factor (K_y). Two options exist: these data are directly introduced by the user or data in existing files are selected. This is often the case for the crop coefficients K_c, the depletion fractions p and the root zone depths. K_c and p are adjusted to local and season climate using the methodology proposed by FAO (Allen *et al.*, 1998). Thus, K_c and p may change from one year to the other. The program KCISA (Rodrigues *et al.* 2000) is used for this purpose and is incorporated in the new ISAREG software package (see Figure 4 with example of output calculation by KCISA). Several crop data files may be acceded to help the user selecting the duration of crop growth stages, crop coefficients, crop height, root depths, the soil water depletion fraction for no stress and the crop yield response factor K_y.

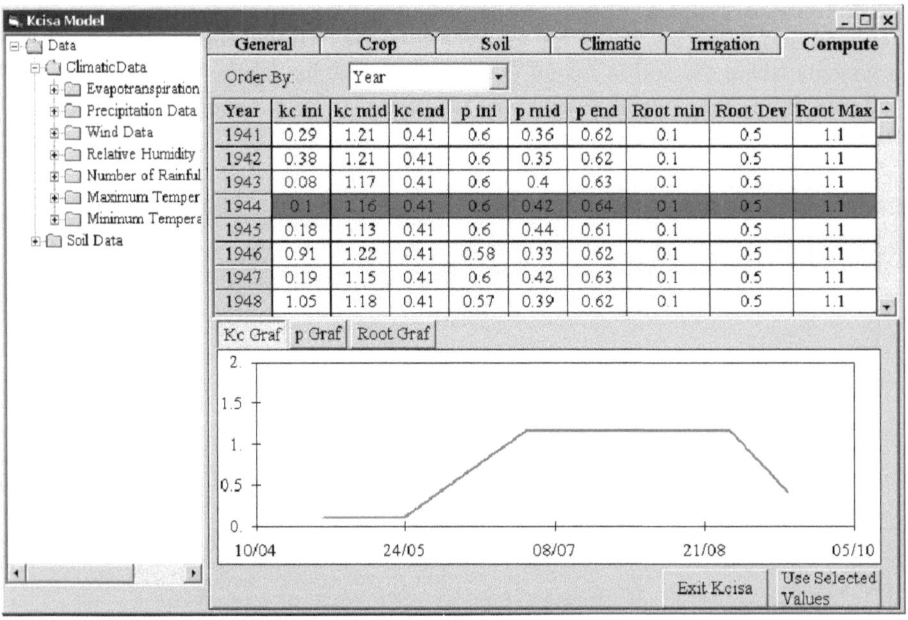

Fig. 4. Window with results of KCISA referring to wheat crop parameters to be used in a sequential water balance relative to a long weather data set

A salinity sub-model is now added to the model. Related crop data concerns the electrical conductivity threshold for crop salt stress and the rate of yield decrease when the soil salinity is larger than that threshold (Allen *et al.*, 1998). Data files containing tabled values relative to these variables can be acceded through a specific window (Figure 5). This information is used to adjust the crop coefficients and the soil water depletion fraction for no stress to the prevailing salinity conditions when it is above the crop threshold and, therefore, salt stress is likely to occur (Campos *et al.*, 2001).

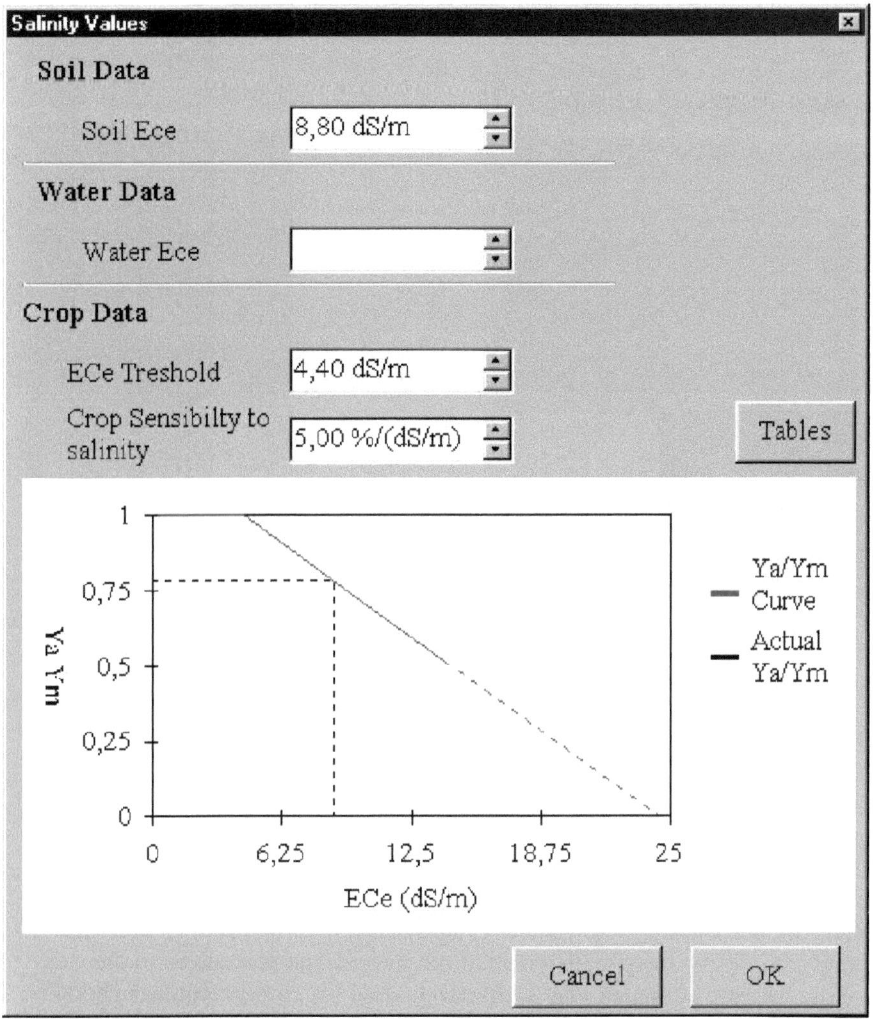

Fig. 5. Window for input of salinity data for consideration of crop salt stress effects on evapotranspiration and yield

Soil data refers to a multi-layered soil. The user may either utilise an existing soil file or input data on depth and thickness of the soil layers and respective water holding capacity. The latter can be inputted as (Figure 6):

- total available soil water, *TAW* [mm],
- the soil water content at the field capacity and the wilting point, θ_{FC} and θ_{WP} [mm mm^{-1}],
- the soil water content at the field capacity and the wilting point, W_{FC} and W_{WP} [kg kg^{-1}] and the bulk density of the soil [kg dm^{-3}].

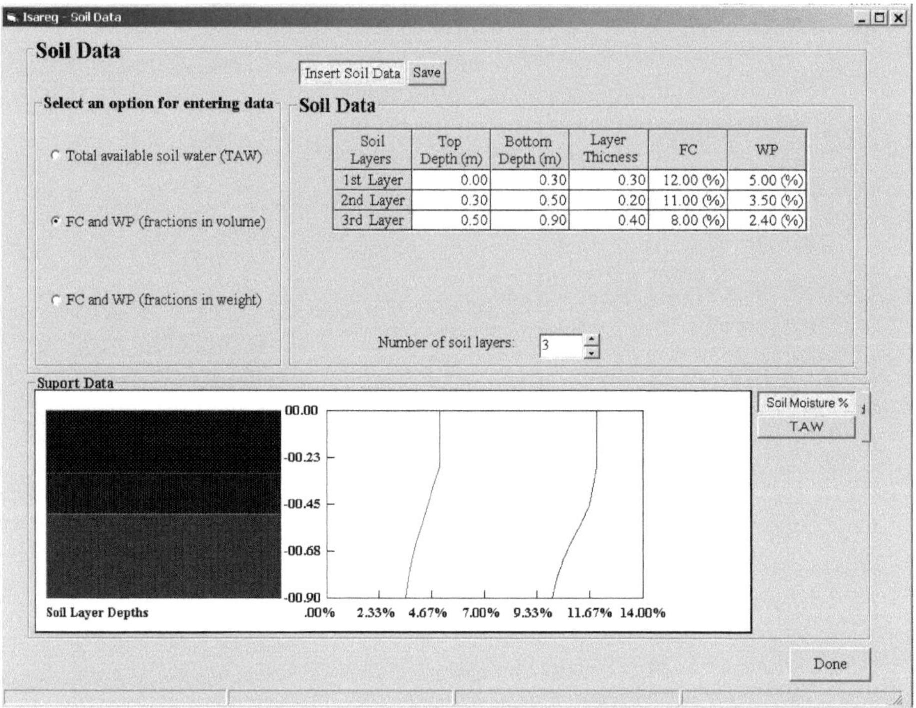

Fig. 6. Window to input soil data, case of a three-layer sandy soil

4.3 WEATHER DATA

The meteorological data for performing the water balance consists of rainfall and reference evapotranspiration and concerns monthly, 10-day, or daily data. When daily rainfall data are not available, the number of rainy days must be given. The model do not compute specifically the effective rainfall but through the procedures of the water balance only, the part of rainfall that is effectively used for crop evapotranspiration is calculated. Information on the depth of non-used rainfall is produced in the table of results. The input of rainfall data is performed through a specific window (not shown) where the user identifies the precipitation file or introduces data through the keyboard.

The reference evapotranspiration data has three input alternatives (Figure 7): using the keyboard, reading an existing file or using the program EVAP56. When the operation is completed, the ET_o data may be displayed as in the graphic of Figure 7. When the option is to compute with EVAP56, calculations are performed with the FAO-Penman-Monteith method (Allen *et al.*, 1998) using the menu in Figure 8. Computations are performed for daily, 10-day or monthly weather data on maximum and minimum temperature, dry and wet bulb temperature or maximum and minimum relative humidity, solar radiation or sunshine duration, and wind speed.

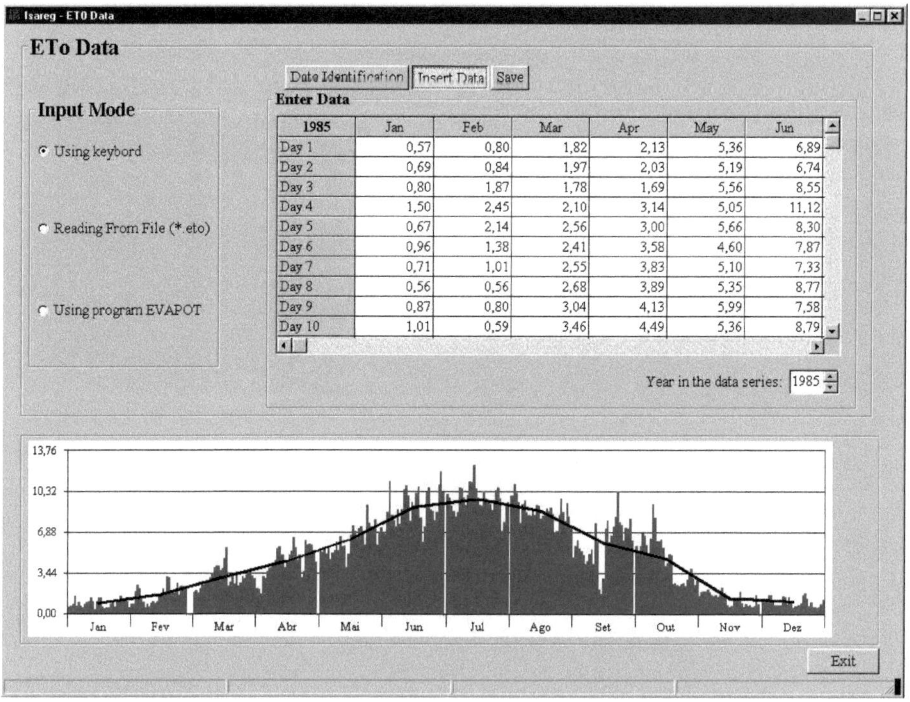

1985	Jan	Feb	Mar	Apr	May	Jun
Day 1	0,57	0,80	1,82	2,13	5,36	6,89
Day 2	0,69	0,84	1,97	2,03	5,19	6,74
Day 3	0,80	1,87	1,78	1,69	5,56	8,55
Day 4	1,50	2,45	2,10	3,14	5,05	11,12
Day 5	0,67	2,14	2,56	3,00	5,66	8,30
Day 6	0,96	1,38	2,41	3,58	4,60	7,87
Day 7	0,71	1,01	2,55	3,83	5,10	7,33
Day 8	0,56	0,56	2,68	3,89	5,35	8,77
Day 9	0,87	0,80	3,04	4,13	5,99	7,58
Day 10	1,01	0,59	3,46	4,49	5,36	8,79

Fig. 7. Window to input ET_0 *data, case for daily data, and to display inputted data.*

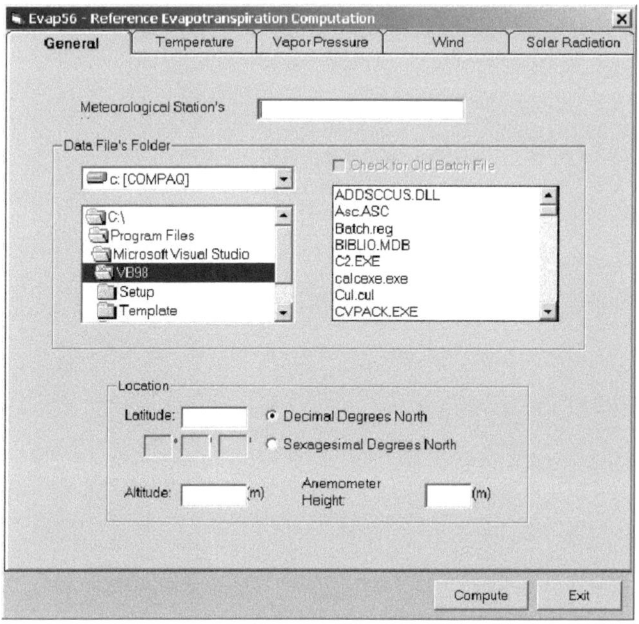

Fig. 8. Main menu of EVAP56 where the weather station and the data files are identified and computations are commanded by the user

When not all these climatic variables are observed, computations are performed with alternative methodologies proposed by Allen *et al.* (1998). If local data for computing the actual vapour pressure is not available, it is computed from T_{min}. Solar radiation may be estimated from data of a neighbouring station or calculated from T_{max} and T_{min}. Wind speed can be estimated using data from a neighbouring station or regional data.

4.4 GROUNDWATER CONTRIBUTION AND DEEP PERCOLATION DATA

The estimation of the groundwater contribution GW_i may be performed through using one of the following alternatives:

- Adopting a constant *GW* selected by the user in agreement with his own data.
- Using the simplified procedure proposed by Doorenbos and Pruitt (1977), where the user has to input the maximal *GW* (GW_{max}) at selected dates. The model then uses a linear interpolation and computes *GW* from GW_{max} as a function of the available soil water *R* as described by Teixeira and Pereira (1992).
- Using a recently developed algorithm (Fernando *et al.*, 2001) which requires that a set of parameters be inputted (Figure 9). Some refer to the water table depths and the crop leaf area index (*LAI*) at several dates during the crop season, others characterise the soil water storage at field capacity and the wilting point, the capillarity in the root zone and an ET_m threshold. The accuracy of computations with this algorithm depends upon the quality of these parameters, which should be validated/calibrated for any specific soil and environmental condition of application.

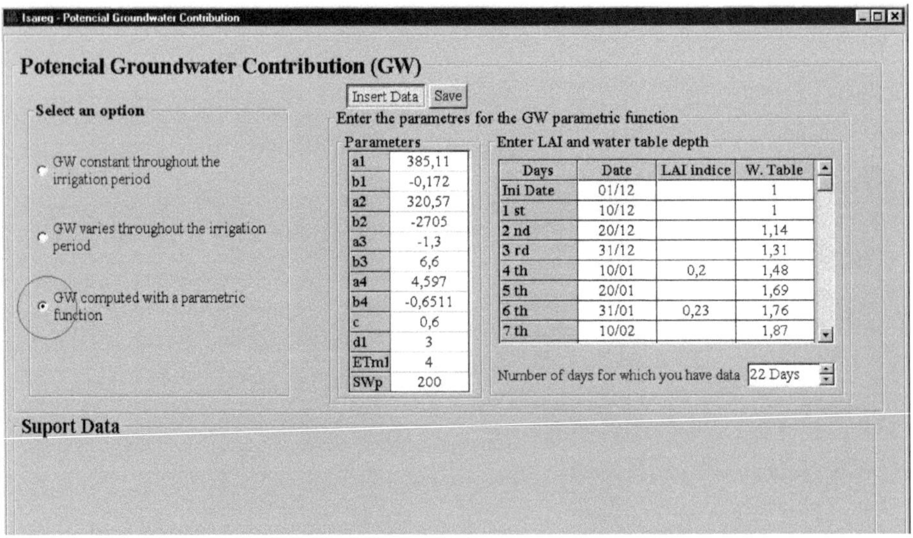

Fig. 9. Window for input of data for computing groundwater contribution and percolation

Deep percolation is generally estimated as $DP_i = 0$ when $\theta_i \leq \theta_{FC}$ and $DP_i = 1000\ (\theta_i - \theta_{FC})\ z_{ri}$ otherwise. However, when the above mentioned GW algorithm is selected, DP_i is estimated through an algorithm describing the soil water depletion from saturation to field capacity (Fernando *et al.*, 2001). The required parameters are inputted together with those for *GW* calculation.

5. Irrigation scheduling

The latest date for scheduling an irrigation to avoid water stress is when the actual soil water content θ_i equals the threshold soil water content θ_p (Figure 1), thus when $\theta_i = \theta_p$ given by:

$$\theta_p = (1 - p)\ (\theta_{FC} - \theta_{WP}) + \theta_{WP} \tag{6}$$

where

- p soil water depletion fraction for no stress (fraction of *TAW* that can be depleted before water stress occurs),
- θ_{FC} soil water content at field capacity
- θ_{WP} soil water content at the wilting point.

Irrigations may be scheduled when the soil water depletion threshold designated as "management allowed depletion", *MAD*, is attained (Martin *et al.*, 1990; Pereira and Allen, 1999). Generally, $MAD < p$ when there is risk aversion or uncertainty, and $MAD > p$ when plant water stress is an intentional part of soil water management, as applied when deficit irrigation is adopted. Then, irrigation is scheduled when $\theta_i = \theta_{MAD}$ given by:

$$\theta_{MAD} = (1 - MAD)\ (\theta_{FC} - \theta_{WP}) + \theta_{WP} \tag{7}$$

The model ISAREG accepts several approaches to define the irrigation timings and depths (Figure 10) depending upon the objectives of the user as indicated in the various options listed in Section 2. They include:

- variable irrigation frequency when the selected threshold is p (Equation 6),
- variable frequency when the threshold is defined by
  - a *MAD* fraction as for Equation 7
  - a user selected ET_a/ET_m ratio,
  - a percentage of *TAW*, or
  - a percentage of the soil water content at field capacity θ_{FC}, which is useful when soil water observations are performed;
- fixed frequency, *i.e.* constant time intervals between successive irrigations, as it is usual when a rotation delivery schedule is adopted by irrigation system managers;
- irrigation at fixed dates.

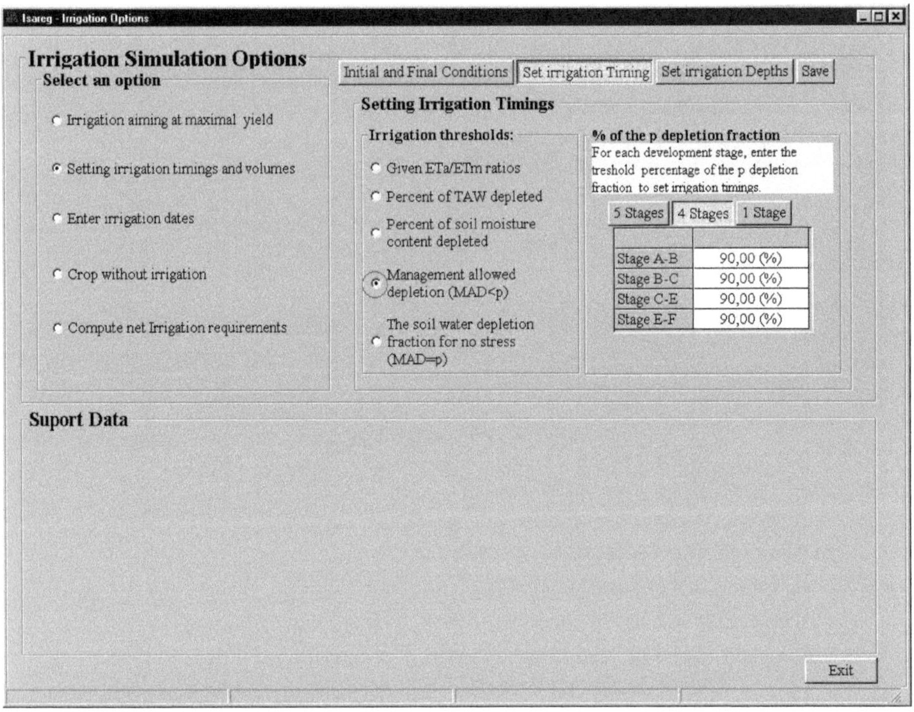

Fig. 10. Window for selection of simulation options, irrigations timings or thresholds and application depths

The net irrigation depth $I_{n\,i}$ to be applied may be computed as

$$I_{n\,i} = 1000\, z_{r\,i}\,(\theta_{FC} - \theta_i) \tag{8}$$

where $z_{r\,i}$ is the root zone depth at the day *i*. However, the model may compute the application depths in different ways such as:

- variable water depths defined by the quantities required to
  - refill the soil water to the field capacity θ_{FC} as in Eq. (8)
  - refill the soil water to a selected percentage of *TAW*, or
  - refill the soil water to a given percentage of θ_{FC},.
- user selected application depths, in agreement with the irrigation methods being considered, which is the approach more frequent in the irrigation practice.

An example of an optimal schedule for sprinkler irrigation with variable irrigation depths is reproduced in Figure 11.

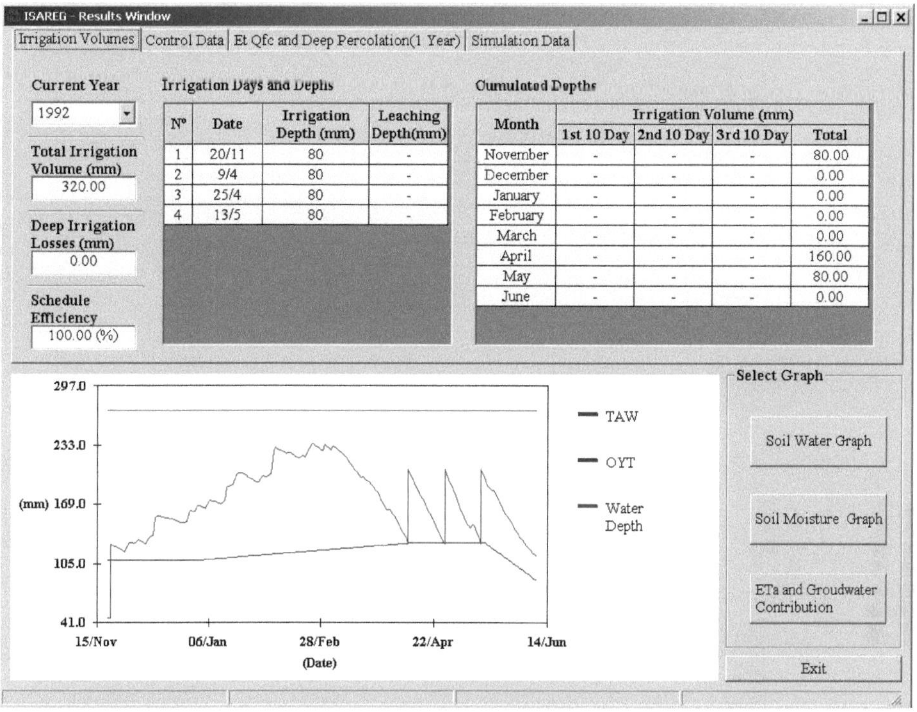

Fig. 11. Window for presentation of results referring to an example of irrigation scheduling with variable frequency and variable application depths

6. Seasonal irrigation requirements

The irrigation water requirement for a complete growing season is defined as

$$IWR = (ET_c - P_e - GW - \Delta S) \,/\, (1 - LR) \tag{9}$$

where

- P_e effective precipitation [mm], defined as gross precipitation less all runoff and deep percolation "losses",
- GW upward flow of groundwater into the root zone during the growing season [mm],
- ΔS change in soil water storage in the root zone during the growing season [mm], from planting to harvesting, and
- LR leaching requirement [ ].

All terms in the numerator are automatically computed by the model, thus the Eq. (9) reduces to

$$IWR = \Sigma I_{n\,i} \tag{10}$$

The leaching requirement, *LR*, that is the percentage of irrigation water that must pass through the root zone to keep the salinity of the soil below a specified value, can be estimated, for non-cracking soils, as

$$LR = EC_{iw} / (5\, EC_e - EC_{Iw}) \tag{11}$$

where EC_{iw} is the electrical conductivity of the irrigation water and EC_e is the desired (non stress) average root zone salinity expressed as the electrical conductivity of the saturated soil extract. Values for EC_e tabled in Allen *et al.* (1998) are available through purposeful windows in the model (Figure 5). The leaching fraction may be considered for every irrigation (Figure 12) or computed as a total.

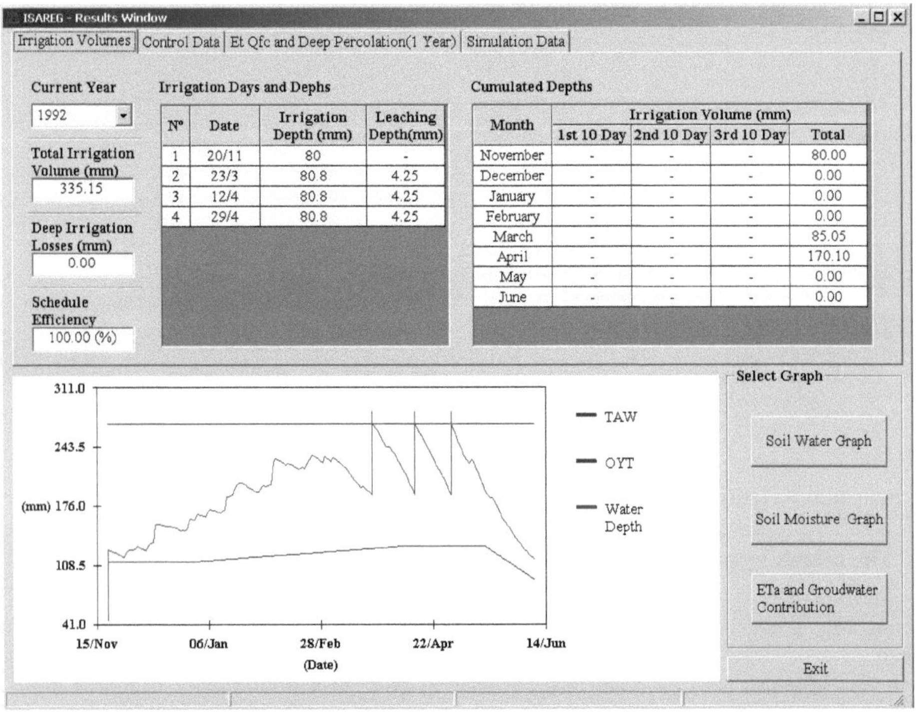

Fig. 12. Window showing the output of an irrigation schedule for wheat considering the application of the leaching fraction in every irrigation

ISAREG is able to perform the sequential water balance referring to any crop using a long weather data series and therefore to perform a simplified frequency analysis of the respective irrigation requirements. This is useful to perform a more in-depth study for selected crop demand conditions, such as those for wet, average or dry years. An example for presentation of results on seasonal irrigation requirements and respective statistical analysis is given in Figure 13.

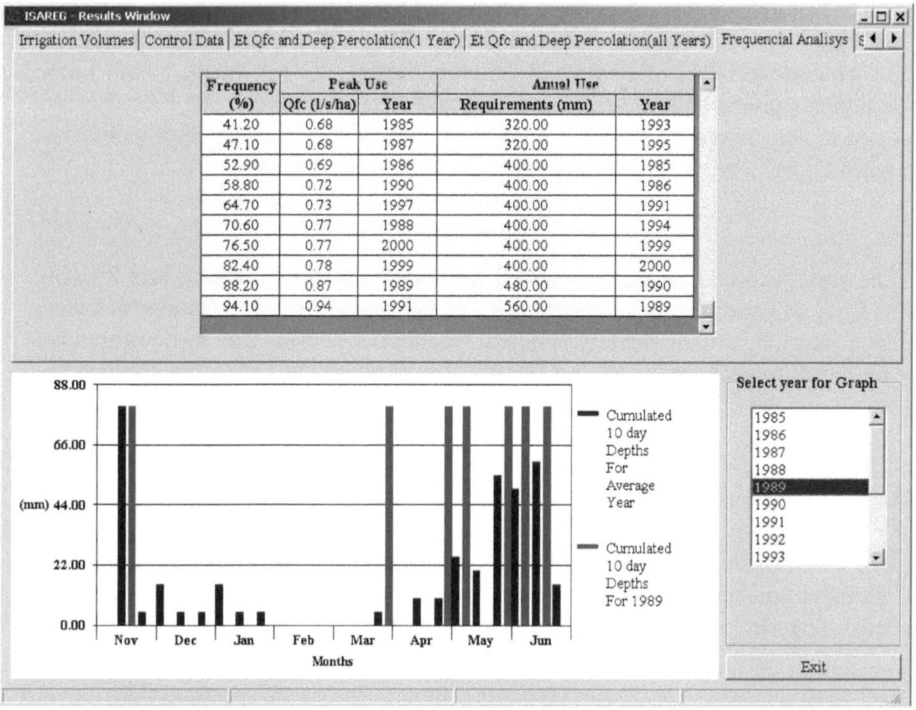

Frequency (%)	Peak Use Qfc (l/s/ha)	Peak Use Year	Anual Use Requirements (mm)	Anual Use Year
41.20	0.68	1985	320.00	1993
47.10	0.68	1987	320.00	1995
52.90	0.69	1986	400.00	1985
58.80	0.72	1990	400.00	1986
64.70	0.73	1997	400.00	1991
70.60	0.77	1988	400.00	1994
76.50	0.77	2000	400.00	1999
82.40	0.78	1999	400.00	2000
88.20	0.87	1989	480.00	1990
94.10	0.94	1991	560.00	1989

Fig. 13. Window for presentation of results of seasonal irrigation requirements for a large time series with application of an empirical statistical distribution of irrigation requirements

The gross irrigation water requirements are computed as

$$GIWR = IWR / Eff \tag{12}$$

where *Eff* is the efficiency of the irrigation system. The value for *Eff* depends on the spatial scale for *GIWR*. If for a single field, the *Eff* is the application efficiency of the field irrigation system and *GIWR* will be the amount of water that the farm system must be designed to apply. If *GIWR* is for a farm, *Eff* is the product of the application and distribution efficiencies within the farm. For an irrigation project or scheme, *GIWR* will represent the total water necessary to take at the source and *Eff* is the global efficiency, product of the efficiencies of the transport, the distribution and the farm irrigation system. Indicative farm irrigation performances are presented by Pereira and Trout (1999). *GIWR* are computed in base of the outputs of ISAREG when its results are exported to another Windows application such as Excel.

7. Yield impacts of irrigation scheduling

To evaluate alternative irrigation schedules concerning the respective impacts on yields, a simplified water-yield function proposed by Stewart *et al.* (1976) is adopted.

To each deficit irrigation corresponds a relative evapotranspiration deficit $(1 - ET_d/ET_m)$ defined from the ratio between the deficit crop ET (ET_d) and the potential crop ET (ET_m). The corresponding relative yield deficit is then $(1 - Y_d/Y_m)$, where Y_d and Y_m are the crop yields produced when ET equals ET_d and ET_m respectively. All ET values are expressed in mm, cumulated to the crop season period. It has been empirically observed that, when $ET_d/ET_m > 0.5$, these relative deficits are linearly related:

$$(1 - Y_d/Y_m) = K_y\,(1 - ET_d/ET_m) \tag{13}$$

The yield response factor K_y is tabled for several crops (Doorenbos and Kassam, 1979). K_y is an input parameter of ISAREG as mentioned before. Adopting this yield reduction function, several alternative deficit irrigation schedules may be compared not only by the differences in water use (*GIWR*) or consumption (*NIR*), but through the relative ET and the relative yield deficits (losses) due to the irrigation deficits.

8. Model validation

Irrigation scheduling simulation models should preferably be validated to be used in a different soil and climate, or with a crop very different from those previously simulated. The window in Figure 14 is available for this purpose.

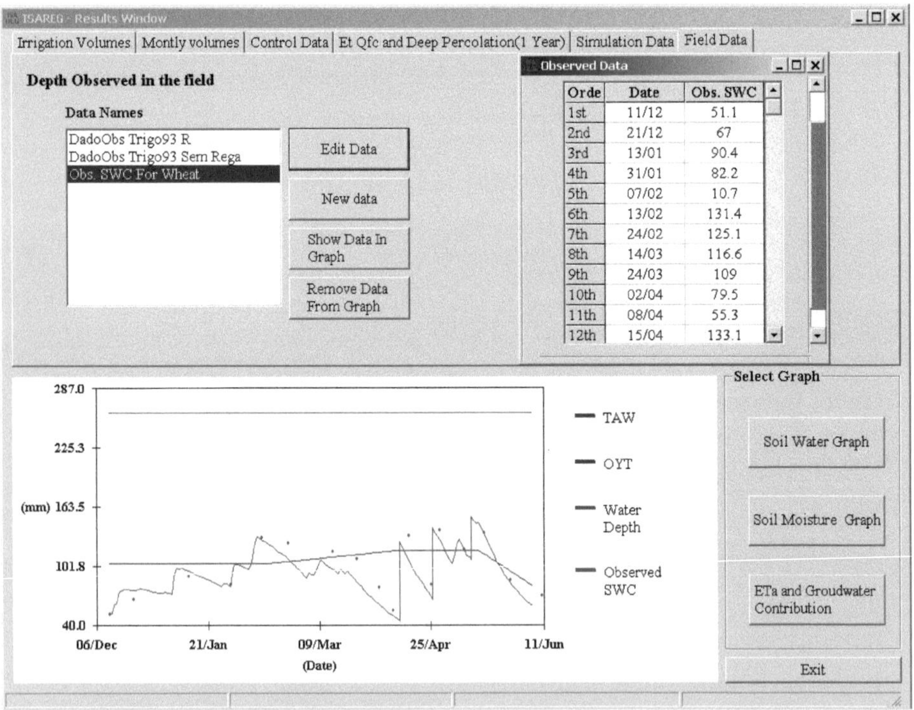

Fig. 14. Window to input observed soil water content data for validation of the model

Validation is performed with option 3 (see Section 2) using as input data those observed in field trials (weather, soil, crop and irrigation data) and comparing simulated and observed soil water data Examples of validation are given in Figure 1 and in other papers in this book (Alba *et al.*, 2002: Oweis *et al.*, 2002; Zairi *et al.*, 2002).

9. Conclusions

The model ISAREG is an appropriate tool for simulation of irrigation scheduling at field scale. Its former version for MS-DOS has been used in various regions for different crops, as illustrated in other papers in this publication. The new Windows version herein presented is due to the need to respond to the recent evolution of the PC's and aims at easier users' utilization. The new version adopts several new computational facilities but still requires improvements in the DLL and following users validation.

Acknowledgements

This research is funded by the research contract IC-18CT-970169 with the European Union, the Contract Interreg IIC 99.09.5942.1 with the Ministry of Agriculture, and FCT, Foundation for Science and Technology, through funding CEER/ISA.

References

Alba, I., Rodrigues, P.N., Pereira, L.S. (2002) Irrigation scheduling of citrus in Sicily to cope with water scarcity. (This issue).

Allen, R.G., Pereira, L.S., Raes, D., Smith, M. (1998) *Crop Evapotranspiration. Guidelines for Computing Crop Water Requirements.* FAO Irrig. Drain. Pap. 56, FAO, Rome, 300p.

Campos, A., Fabião, M. S., Pereira, L S., Gonçalves, J.M., Fang, S. X., Mao, Z., Dong, B. (2001) Farm irrigation water savings in the Upper Yellow River Basin. Irrigation scheduling simulation to improve management. In: *International Conference on Agricultural Science and Technology, Vol. 6: Information Technology for Agriculture*, ICAST, Ministry of Science and Technology, Beijing, China, pp. 421-426.

Doorenbos, J. and Kassam, A.H. (1979) *Yield Response to Water.* FAO Irrig. and Drain. Pap. 33, FAO, Rome.

Doorenbos, J., Pruitt, W.O. (1977) *Guidelines for Predicting Crop Water Requirements.* FAO Irrig. and Drain. Pap. 24, 2nd ed., FAO, Rome.

Fernando, R.M., Pereira, L.S., Liu, Y., Li, Y., Cai, L. (1998) Reduced demand irrigation scheduling under constraint of the irrigation method. In: L.S. Pereira and J.W. Gowing (Eds.) *Water and the Environment: Innovation Issues in Irrigation and Drainage*, E &FN Spon, London, pp. 407-414.

Fernando, R. M., Pereira, L. S., Liu, Y. (2001) Simulation of capillary rise and deep percolation with ISAREG. In: *International Conference on Agricultural Science and Technology, Vol. 6: Information Technology for Agriculture*, ICAST, Ministry of Science and Technology, Beijing, China, pp. 421-426.

Fortes, P.F., Pereira, L.S., Rodrigues, P.N., Calejo, M.J., Teixeira, J.L., Platonov, A.E. (2001) Integrating an irrigation scheduling model in a GIS for improving water use and environmental issues. In: A.A. Soares and H.M. Saturnino (Eds.) *Environment and the Water: Competitive Use and Conservation Strategies for Water and Natural Resources.* ABID, Brazilian Assoc. Irrig. Drain., Brasilia, pp. 11-19.

Liu, Y., Fernando, R.M. (1998) Irrigation scheduling. In: L.S. Pereira, A. Musy, R.J. Liang, M. Hann, (Eds.) *Water and Soil Management for Sustainable Agriculture in the North China Plain.* Instituto Superior de Agronomia, Lisbon, pp. 167-235.

Liu, Y., Teixeira, J.L., Zhang, H.J., Pereira, L.S. (1997) Model validation and crop coefficients for irrigation scheduling in the North China Plain. *Agric. Water Manag.* 36: 233-246.

Martin, D.L., Stegman, E.C,. Fereres, E. (1990) Irrigation scheduling principles. In: G.J. Hoffman, T.A. Howell and K.H. Solomon (Eds.) *Management of Farm Irrigation Systems.* ASAE, St. Joseph, MI, pp. 155-203,

Oweis, T., Rodrigues, P.N., Pereira, L.S. (2002) Simulation of supplemental irrigation strategies for wheat in Near East to cope with water scarcity. (This issue).

Pereira, L.S., Allen, R.G.. (1999) Crop water requirements. In: H.N. van Lier, L.S. Pereira, and F.R. Steiner (Eds.) *CIGR Handbook of Agricultural Engineering, vol. I: Land and Water Engineering,* ASAE, St. Joseph, MI, pp. 213-262.

Pereira, L.S., Trout, T.J. (1999) Irrigation methods. In: H.N. van Lier, L.S. Pereira, and F.R. Steiner (Eds.) *CIGR Handbook of Agricultural Engineering, vol. I: Land and Water Engineering,* ASAE, St. Joseph, MI, pp. 297-379.

Pereira, L.S., van den Broek, B.J., Kabat, P., Allen, R.G. (Eds.) (1995) *Crop-Water Simulation Models in Practice.* Wageningen Pers, Wageningen.

Rodrigues, P.N., Pereira, L. S., Machado, T. G. (2000) KCISA, a program to compute time averaged crop coefficients. Application to field grown vegetable crops. In: M.I. Ferreira and H.G. Jones (Eds.) *Proceedings of the 3rd International Symposium on Irrigation of Horticultural Crops.* ISHS, Acta Horticulturae Nr. 537, Leuven, vol. I, pp. 535-542.

Rodrigues, P.N, Machado, T., Pereira, L.S., Teixeira, J.L, El Amami, H., Zairi, A. (2002) Feasibility of deficit irrigation with center-pivot to cope with limited water supply and drought in Alentejo, Portugal. (This issue).

SCS (1982) *National Engineering Handbook.* USDA Soil Conservation Service U.S. Printing Office, Washington D.C.

Shatanawi, M., Al-Zu'bi, Y., Al-Jayoussi, O. (2002). Irrigation management dynamics in the Jordan valley under drought conditions. (This issue).

Stewart, J.L., Hagen, R. M., Pruit, W.O. (1976) *Water production functions and predicted irrigation programs for principal crops as required for water resources planning and increased water use efficiency.* University of California, Dep. Land, Air and Water Resources, Davis, and USDI/BR, Denver, CO.

Teixeira, J.L., Pereira, L.S. (1992) ISAREG, an irrigation scheduling model. *ICID Bulletin*, 41(2): 29-48.

Zairi, A. El Amami H., Slatni A., Pereira L.S., Rodrigues, P.N., Machado, T. (2002) Coping with drought: deficit irrigation strategies for cereals and field horticultural crops in Central Tunisia. (This issue).

COPING WITH DROUGHT: DEFICIT IRRIGATION STRATEGIES FOR CEREALS AND FIELD HORTICULTURAL CROPS IN CENTRAL TUNISIA

A. ZAIRI[(1)], H. EL AMAMI[(1)], A. SLATNI[(1)], L.S. PEREIRA[(2)], P.N. RODRIGUES[(2)] and T. MACHADO[(2)]

[(1)] *National Research Institute for Agricultural Engineering, Water and Forestry, Tunisia*

[(2)] *Agricultural Engineering Research Center, Institute of Agronomy, Technical University of Lisbon, Portugal*

Abstract: Preparedness measures to cope with droughts and water scarcity in semi-arid climates in irrigated agriculture include the identification of irrigation scheduling strategies that minimize the water demand with acceptable impacts on yields. Those strategies may be produced by simulation and focus on different levels of water demand. The irrigation scheduling simulation model ISAREG, validated for Tunisia, is used to simulate those strategies. The generated irrigation scheduling strategies are applied to deficit irrigation of winter wheat, tomato and potato crops under semi-arid conditions in Central Tunisia. The alternative schedules are evaluated through the combined use of indicators relative to the reduction in demand for irrigation water, the consequent yield reduction and the impacts on farmer's income. The economic evaluation of those strategies is performed with the help of a linear programming model. Results indicate that, when there is limited water availability, each crop reacts differently to the water restrictions applied in terms of yield decrease and gross margin per unit surface and unit of water applied. For average demand, the adoption of deficit irrigation is generally feasible for all crops considered. However, for the tomato crop, there is a trend to restrict the cultivated surface when water scarcity increases. Under high and very high demand conditions, the gross margin per unit of water applied decreases for the potato and the tomato crops but increases for the wheat crop. Then, adopting deficit irrigation is not feasible for the tomato and potato crops, the option being to decrease the cultivated surface. On the contrary, the wheat crop responds well to deficit irrigation, with a good water valorisation, including under drought conditions.

Key words: water scarcity, irrigation scheduling, modelling, economic optimisation, wheat, potato, tomato.

1. Introduction

Irrigated agriculture represents only 7 % of the Tunisian cropped area, but is responsible for 33 % of the agricultural product. The increase in the irrigated area from 176 000 ha in 1980 to 334 000 ha in 1995, and the related agricultural intensification have contributed to the continuous rising of water demand. Thus irrigation represents near 80 % of the total water demand in Tunisia (DG/GR, 1996). Along with the diversity of water sources (*e.g.* dams, diversions from wadis, groundwater wells), water management is also diverse as it is strongly influenced by the regional climate and the socio-economic and cultural conditions.

G. Rossi et al. (eds.), Tools for Drought Mitigation in Mediterranean Regions, 181-201.

The frequent occurrence of droughts requires pro-active and rational water management approaches that could be helpful to both managers and farmers. Those measures include the reinforcement of the hydraulic infrastructures, increased waste water reuse, the enforcement of restrictions on water volumes supplied for irrigation, and changes in water prices. However, their application requires better knowledge of the climatic variability and respective impact on crop yields.

A recent study shows that a decrease of 10 % in water supply to cereals and fodder crops would result in a reduction of about 2 % in the net agricultural product, mainly as a consequence of a reduction in the irrigated area (DG/RE, 1997). Irrigated wheat surface (60 000 ha) is less than 5 % of the total cereal cropped area in Tunisia but, when drought occurs, the irrigated wheat area contributes with 20 % to the total cereal production, while also playing a main role in seed production (El Amami, 1995).

In order to prepare appropriate irrigation strategies for wheat and main field horticultural crops, tomato and potato, when water supply restrictions have to be imposed, the present study was performed to assess how deficit irrigation (English and Raja, 1996) could be practiced. Thus, the impacts of several alternatives of deficit irrigation for crops are evaluated in terms of water saving, yield reduction and farmers income reduction, the latter referring to the fraction of intended area to be cropped that is allocated to the irrigated crop. The study applies to an irrigation system in the semi-arid Central Tunisia (Siliana region).

2. Materials and methods

2.1. CASE STUDY SILIANA IRRIGATION PROJECT

The Siliana irrigation project has 1240 ha irrigated area supplied by the Lakhmès storage dam (7 Mm^3) and diversions from the Oued Ouzafa. During drought, supplies are supplemented by groundwater, which potential is evaluated in 3 Mm^3. The Siliana project is located in Central Tunisia (35° 58' N, 7° 02' E) at an average elevation of 470 m, near the Serj Mountains. The climate is classified as semi-arid superior, with a cold winter and a hot and dry summer. The annual rainfall varies from 200 to 550 mm, and is erratically distributed inter-annually and seasonally. The average monthly rainfall and reference evapotranspiration ET_o are shown in Figure 1.

Main soils are alluvial silty-clays and brown steppe silty-clays. In average, the textural classes are 10 % sand, 45 % silt and 45 % clay. The average soil water at field capacity and at wilting point are $\theta_{FC} = 0.40\ m^3m^{-3}$ and $\theta_{WP} = 0.20\ m^3m^{-3}$. The main crops in this irrigation project are cereals, horticultural crops including tomato and potato, fruit trees, mainly olive trees, and forage crops.

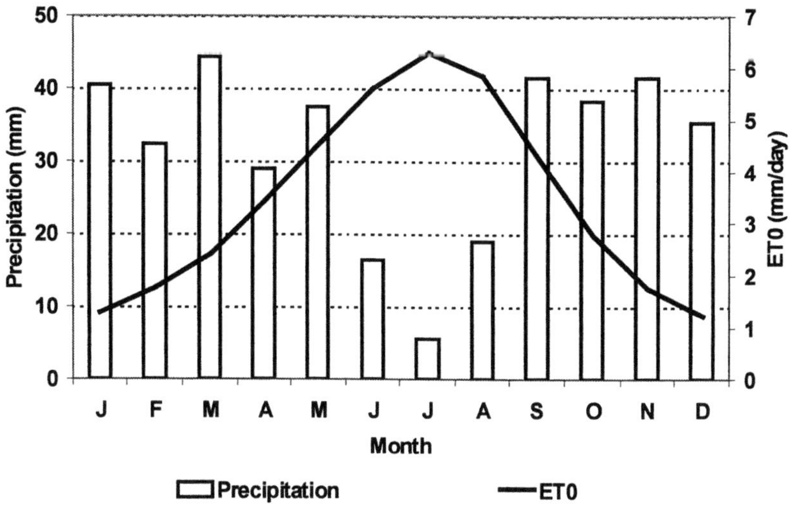

Fig. 1. Average monthly rainfall and reference evapotranspiration for Siliana (1981-2000)

2.2. MODELLING TOOLS

The ISAREG model (Teixeira and Pereira, 1992) has been used to simulate different irrigation water management strategies. The ISAREG model performs the soil water balance using different options to define and evaluate the irrigation schedules (Pereira *et al.*, 2002). The reference evapotranspiration was computed with the FAO-Penman-Monteith method (Allen *et al.*, 1998). The crop data parameters were determined with the KCISA program (Rodrigues *et al.*, 2000), following the updated methodology proposed by FAO (Allen *et al.*, 1998), where the crop coefficients (K_c) and the soil water depletion fraction for no stress (p) have been adjusted to the Siliana climate.

The calibration and validation of the ISAREG model for Tunisia is described in Teixeira *et al.* (1995) and Zairi *et al.* (1998) and is partly reproduced in section 3.1. Yield losses due to water stress are estimated utilising the yield response factors proposed by Doorenbos and Kassam (1979).

In order to evaluate the impacts of the deficit irrigation strategies on the farmer's income, a linear programming model was used. This technique aimed at determining the optimal allocation of water resources that would maximize farmer's income considering the availability of land, labour and water (considering stable water prices). The economic analysis of farm management and planning activities fit the assumptions of linear programming such as: limited factors of production and farm activities are subjected to constraints that influence the technical performance and, most importantly, all quantities of inputs used and outputs produced may be assumed proportional to the level of the activity. The mathematical optimisation model is built to reach the combination of alternative farm activities which will maximize the objective function proposed, and that does not violate any of the fixed constraints or involve any negative activity levels. Mathematically, it can be described by (Hazell and Norton, 1986):

$$max\, Z = \sum_{j=1}^{n} c_j X_j \tag{1}$$

subject to:

$$\sum_{j=1}^{n} a_{ij} X_j \le b_i \quad i = 1,\cdots,m \tag{2}$$

and to:

$$X_j \ge 0 \quad j = 1,\cdots,n \tag{3}$$

where: X_j = level of the jth farm activity, within n possible activities,
c_j = calculated gross margin of a unit of the jth activity,
a_{ij} = the quantity of the ith resource necessary to produce one unit of the jth activity,
with: m being the number of resources,
b_i = the amount of the ith resource available.

In the present study, the activities considered are every one of the irrigation strategies generated for the wheat, potato, and tomato crops, which are characterized by the respective irrigation depth and relative yield losses. The irrigation depths are obtained from the ratio between the net irrigation computed by ISAREG and the application efficiency for sprinkler systems (E_a = 75 %). The objective function is evaluated by the gross margin, obtained after subtracting the variable costs related with the specified activities from the farm gross product. Current prices for land, labour and crop products in Tunisia are utilized.

The economic results are expressed in terms of the gross margin per unit surface of land cropped and per unit volume of water applied for the different supply scenarios. So, the main problem is to determine which irrigation strategy is the best in terms of maximization of the farmer's income taking into consideration several levels of water availability. The output of the linear problem is shown in terms of the percentage of the land available for a given activity that is allocated to that activity after balancing the economic return from water and land.

2.3. WATER SUPPLY STRATEGIES

In the Siliana project, sprinkler irrigation systems are used (hand moved and semi-permanent set systems). The existing sprinkler systems have been designed to produce application rates close to 6 mm h^{-1} to overcome the low infiltration rates of the soils. Farmers adopt a set time of 8 hours. Thus, considering an application efficiency of 75 %, the net irrigation depths are approximately I_n = 40 mm (Zairi, *et al.*, 2000) for both wheat and horticultural crops.

When water availability is non-limiting, the frequency of irrigations is not restricted and varies along the season according to the crop demand. Under limited water availability, the supply is made with restrictions:

- for irrigation of the wheat and potato crops, water is applied in supplement to rainfall. Thus, the farmers are advised to reduce the number of irrigation events according to the available water supply;
- for the tomato crop, which is almost entirely dependent upon irrigation, when water is scarce farmers are told to adopt large time intervals between successive irrigations. The frequency of irrigations is generally constant along the crop season.

Observations in the practice of irrigation at Siliana allow defining several alternative supplemental irrigation schedules for wheat and potato crops, as summarized in Table I. The non-restricted schedule (SC) is only constrained by the system characteristics and corresponds to the maximal yield strategy. The other schedules are defined by decreasing the number of irrigation events and, consequently, the total depth applied, so corresponding to deficit irrigation. The deficit irrigation strategies for the tomato crop are also those observed in the Siliana project and correspond to increased time intervals between irrigations (Table II). The schedule SC also corresponds to achieving the maximal yields.

Table I. Deficit irrigation strategies and corresponding available water supplies for the wheat and potato crops at Siliana using set sprinkler systems (I_n = 40 mm)

Strategies	Symbol	Supply conditions
System constraints	SC	Variable frequency
Light deficit irrigation	LDI	Seasonal irrigation reduced by 40 mm
Deficit irrigation	DI	Seasonal irrigation reduced by 80mm
Large irrigation deficit	LID	Seasonal irrigation reduced by 120mm
Very large irrigation deficit	VLID	Seasonal irrigation reduced by 160 mm
Extreme irrigation deficit	EID	Seasonal irrigation reduced by 200 mm
Very extreme irrigation deficit	VEID	Seasonal irrigation reduced by 240 mm

Table II. Deficit irrigation strategies and corresponding available water supplies for the tomato crop at Siliana using set sprinkler irrigation (In = 40 mm)

Strategies	Symbol	Supply conditions
System constraints	SC	Variable frequency
Light deficit irrigation	LRF	Irrigation frequency > 7 days
Restricted frequency	RF	Irrigation frequency > 10 days
Highly restricted frequency	HRF	Irrigation frequency > 12 days
Extremely restricted frequency	ERF	Irrigation frequency > 15 days

The alternative irrigation schedules are evaluated by the reductions in season irrigation depths, the corresponding crop ET deficits, measured by the relative ET ratio between actual crop ET and potential crop ET, and the relative yield loss given by RYL = 1 – Y_a/Y_c [%], where Y_a and Y_c are the crop yields corresponding to actual and potential crop ET, respectively.

3. Model validation in Tunisia

The ISAREG model was validated for wheat at an experimental station located at Hendi Zitoun, Central Tunisia, which environmental conditions are close to those of Siliana, including soils and wheat varieties. Data of two irrigation seasons were used (Zairi *et al.*, 1998):

- 1992/93 – field trials with sprinkler set systems, using net irrigation depths $I_n = 40$ mm, for three treatments with irrigation thresholds $\theta_{MAD} = 0.3$, 0.6 and 0.9 θ_p, where θ_p is the soil water content when the depletion fraction equals p (soil water depletion fraction for no stress) as referred in the companion paper by Pereira *et al.* (2002);
- 1997/98 – field trials with surface irrigation ($I_n = 100$ mm) using two treatments with irrigation thresholds $\theta_{MAD} = 0.6$ and 0.9 θ_p.

The ISAREG model validation was performed by comparing the soil water content observed in the field experimental plots throughout the crop season with those simulated by the model. Figure 2 shows the results of the model simulation for wheat at Hendi Zitoun in 1992/93 for the treatment where $\theta_{MAD} = 0.9\ \theta_p$. It shows that the model appropriately describes the soil water content, expressed in % in volume, along the entire crop season.

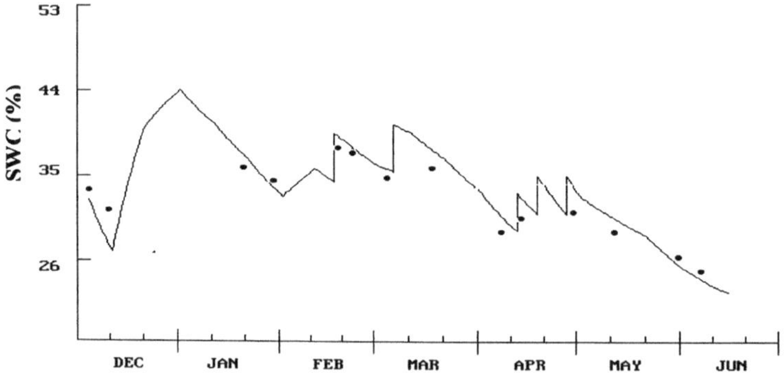

Fig. 2. Simulated (–) and observed (•) soil water content (expressed in % in volume) for a wheat treatment with $\theta_{MAD} = 0.9\ \theta_p$ at Hendi Zitoun in 1992/93 (from Zairi et al., 1998)

Results in Figure 3 concern the surface irrigation experiment for 1997-98. It shows that the regression line relative to the simulated vs. observed soil water content is close to 1:1 and that the dispersion of residuals of estimates is similar for high and low values of the soil water content. Figure 3 also includes the confidence interval relative to the 95 % probability. Results indicate that the model is able to predict with acceptable errors the soil water content for well irrigated and deficit irrigated crops. More recent observations with summer crops (results not published) confirm that assumption. The validation for yield loss predictions is presented by Teixeira *et al.* (1995).

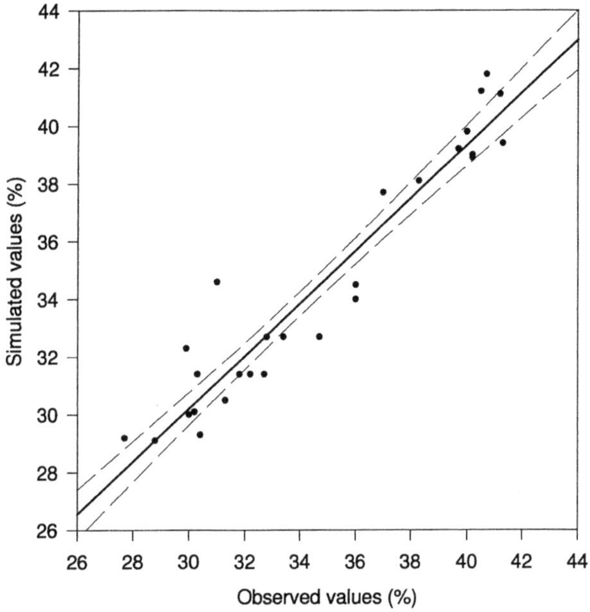

Fig. 3. Simulated vs. observed soil *water content (% in volume), for all field trials performed in 1997/98 at Hendi Zitoun (from Zairi et al., 1998)*

4. Crop water requirements and yield responses to water deficits

4.1. CROP WATER AND IRRIGATION REQUIREMENTS

The planting dates and duration (days) of crop development stages are given in Table III. The crop coefficients (K_c) and the soil water depletion fractions for no stress (p) computed with the KCISA program for the three crops are also presented in Table III, as well as the root depths along the crop season.

The annual net irrigation requirements (NIR) for the wheat, potato and tomato crops were computed with ISAREG model by performing a sequential water balance for the time series 1981-2000. Adopting a normal distribution for the computed NIR, the crop irrigation requirements for the average, high and very high demand conditions were computed, which correspond to the probabilities for being exceeded of 50, 20 and 5 %, respectively. Results are summarised in Table IV. For the wheat crop results for the 20 and 5 % probabilities are quite similar, so only those for the very high demand conditions are considered. Table IV also include the seasonal rainfall and crop evapotranspiration for the selected years. The available soil water at planting (ASWP) was computed with ISAREG in a sequential soil water balance starting at the end of the summer of the first year, when soil water depletion is maximal for the environmental conditions of Siliana.

Table III. Crop parameters for average demand conditions for the wheat, potato and tomato crops

Parameters	Crop development stages			
	Initial	Development	Mid season	End season
Wheat:				
(planting 15/11)				
Period length (days)	37	69	56	31
Crop coefficients	0.89	0.89-1.15	1.15	1.15-0.25
Root depth (m)	0.05-0.38	0.38-1	1	1
Depletion fraction p	0.60	0.60-0.60	0.60	0.60-0.70
Potato:				
(planting 01/02)				
Period length (days)	25	25	50	20
Crop coefficients	0.73	0.73-1.15	1.15	1.15-0.40
Root depth (m)	0.20-0.35	0.35-0.50	0.50	0.50
Depletion fraction p	0.50	0.50-0.38	0.38	0.38-0.47
Tomato :				
(planting 01/04)				
Period length (days)	30	45	50	25
Crop coefficients	0.42	0.42-1.16	1.16	1.16-0.90
Root depth (m)	0.05-0.52	0.52-1.00	1.00	1.00
Depletion fraction p	0.54	0.54-0.31	0.31	0.31-0.40

Table IV. Demand conditions for wheat, potato and tomato crops for Siliana project

Crop	Demand condition	Available soil water at planting (mm)	Season rainfall (mm)	Season ET_c (mm)	Season irrigation requirement (mm)
Wheat	Average	8	247	420	240
	Very high	22	178	468	320
Potato	Average	77	155	314	160
	High	86	106	377	280
	Very high	61	54	346	320
Tomato	Average	77	142	772	640
	High	5	100	726	720
	Very high	5	31	758	840

4.2. CROP YIELD RESPONSES TO DEFICIT IRRIGATION

The relative ET and relative yield loss referring to the simulated deficit irrigation schedules for the wheat crop are given in Figure 4. Results for the average year show that when irrigation demand is reduced by 50 %, from 240 to 120 mm, a relative yield loss close to 20 % (Figure 4) is expected. However, under severe drought conditions (very high demand), a reduction of 50 % of the seasonal irrigation, from 320 to 160 mm, induces much higher relative yield losses, close to 30 % (Figure 4). When only 120 mm would be applied, then the RYL would be close to 40 %.

Considering stronger restrictions in water supply under severe drought conditions, *i.e.* only one sprinkler irrigation applying 40 mm, the ET_c would reduce to 220 mm, resulting RYL = 47 %. This case may be out of the limit of the model applicability because validation experiments did not include so large water deficits. However, other studies using the same wheat variety concluded that grain production is only viable when ET_c is above 200 mm (Zairi *et al.*, 1996). Rainfed wheat is generally feasible under average climatic conditions but not under drought.

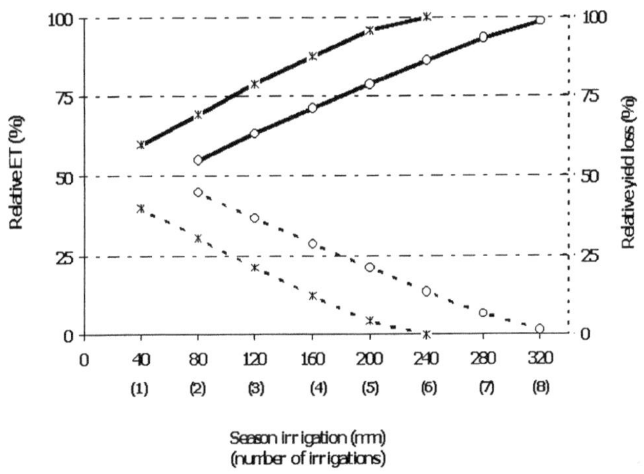

Fig. 4. Response of the wheat crop to different supplemental irrigation strategies in Siliana for average (—✱—) and very high (—○—) demand conditions

Differences in wheat responses to deficit irrigation between average climatic conditions and drought are due to both the seasonal ET, respectively 420 and 468 mm, and the season rainfall contributing to satisfy the crop demand, 247 and 178 mm respectively. Thus, a similar reduction in the applied depths produce different RYL, much higher under drought. Results confirm the strategies used in the farmers practice and indicate that the increased water productivity associated with deficit irrigation supports its feasibility as shown from the economic analysis hereafter (section 5.1). Results are similar to those obtained in North Syria (Oweis *et al.*, 2002) and Alentejo, Portugal (Rodrigues *et al.*, 2002).

It may be concluded from the analysis above that under drought it is not appropriate to apply less than 80 mm in two irrigation events, one at planting to provide for crop establishment, and the second by the beginning of February to avoid heavy stress at heading. Preferably, two other irrigations by April would be desirable to provide for grain formation and grain filling.

Results for the potato crop are given in Figure 5. For average demand conditions, reducing the net irrigation from 160 to 120 mm would produce a relative yield loss close to 9 %, thus easy to be practised. Reducing demand by 50 %, so to only two irrigation events totalling 80 mm, leads to RYL < 30 %. However, for high and very high demand, the application of only 120 mm would cause much higher relative yield losses, 35 % and 44 % respectively (Figure 5). This difference in responses is due to season ET but mainly to the season rainfall (Table IV), which make that net irrigation requirements are much higher under drought than for average demand conditions. Therefore, while mild deficits can be considered, the adoption of strong irrigation deficits is less effective when water availability is scarce.

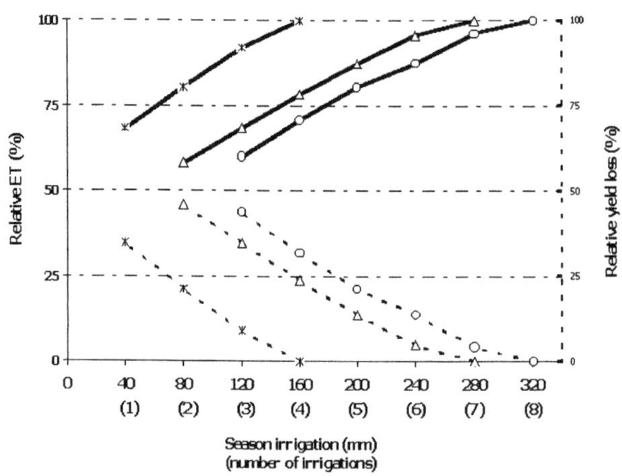

Fig. 5. Response of the potato crop to different supplemental irrigation strategies for average (—✳—), high (—△—) and very high (—○—) demand conditions

Results for the tomato crop are shown in Figure 6. It can be seen that reducing the season irrigation to 400 mm produces RYL near to 30 % for average demand, 40 % for high demand, and 50 % for very high demand conditions. Differently from the potato crop, results do not indicate that deficit irrigation could be viable for the average year because only a small fraction of the crop ET is satisfied by rainfall and soil water. Thus, since most of ET has to be provided by irrigation, including for the average year, any reduction in the applied water results in ET and yield decrease. Results are similar to those for maize in Alentejo (Rodrigues *et al.*, 2002). This indicates that the adoption of deficit irrigation for summer crops is difficult and does not produce the high demand reduction that is desired to cope with droughts. The economic feasibility analysis confirms these difficulties (see section 5.3.), which generally lead farmers to reduce the irrigation area and apply there a near to optimal irrigation scheduling when a drought occurs.

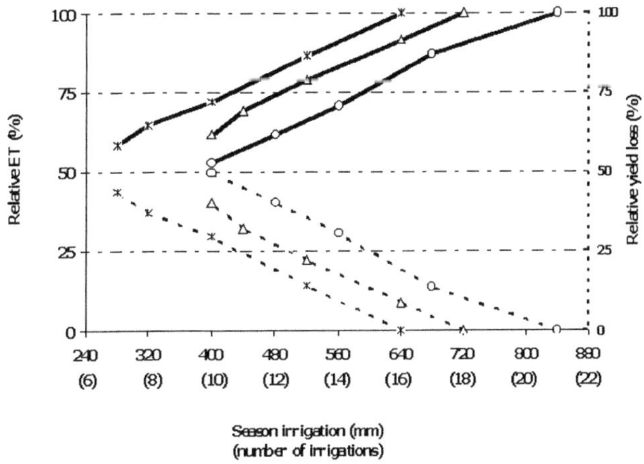

Fig. 6. Response of the tomato crop to different supplemental irrigation strategies for average (—✳—), high (—△—) and very high (—○—) demand conditions

5. Economic evaluation of deficit irrigation

5.1. WHEAT

The economic results for wheat when average climatic demand is considered (Table V and Figure 7), show that gross margins per unit volume of water applied (GM/m^3) increase from 0.417 USD/m^3 for the optimal irrigation schedule up to 1.586 USD/m^3 for the heavier deficit schedule. This indicates that the increased water productivity is highly valorised. Differently, under drought conditions the GM/m^3 increases in a much smaller range, from 0.275 to 0.639 USD/m^3. This positive trend for the gross margin per unit water applied indicates that it is generally feasible to adopt deficit irrigation, including if strong supply restrictions are imposed when severe droughts occur.

As expected, the gross margins per unit surface cropped (GM/ha) decrease when allowed water deficits increase (Table V and Figure 7 a) due to the consequent yield losses and the respective decrease in the gross product of the activities. However, the GM/ha are always positive, thus indicating favourable conditions to adopt of deficit irrigation. To be noted that the GM/ha for very high demand conditions are not very different from those for the average despite the significant difference between water volumes applied. This is explained by the fact that sub-products, mainly straw, have much higher prices when a drought occurs because they have then a high demand as fodder for sheep and cows.

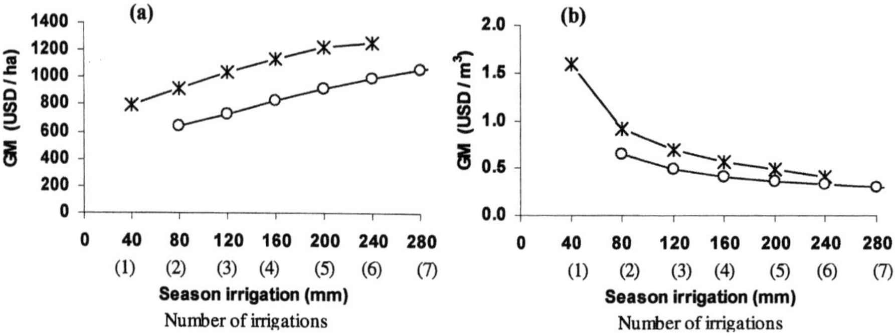

Fig. 7. Gross margins per unit surface (a) and per unit volume of water applied (b) for alternative deficit irrigation strategies for the wheat crop under average (—✳—), and very high (—○—) climatic demand conditions

The results for the optimization model are given as the percentage of area aimed to be cropped that is effectively allocate to each irrigation strategy (activity) considered (Table V). For average demand conditions, all irrigation strategies are selected by the model, *i.e.* for all cases it results 100 % of the surface allocated to the considered deficit strategy. For very high demand, the model selects some strategies at 100 % level while for others it selects an alternative combination of strategies to obtain the higher farm income. This indicates that the strategies not selected (LID and EID) are sub-optimal. However, it is feasible to apply the respective deficit irrigation schedules.

Table V. Economic analysis of the wheat crop responses to deficit irrigation

Deficit irrigation strategies	Season irrigation (mm)	Relative ET (%)	Relative yield loss (%)	Gross margin (USD/ha)	Gross margin (USD/m^3)	Percent surface allocated to each irrigation strategy
Average demand condition						
SC	240	100.0	0.0	1253	0.417	100/SC
LDI	200	95.7	4.3	1228	0.491	100/LDI
DI	160	87.5	12.5	1140	0.570	100/DI
LID	120	78.7	21.3	1033	0.688	100/LID
VLID	80	69.4	30.6	915	0.915	100/VLID
EID	40	59.9	40.1	793	1.586	100/EID
Very high demand condition						
SC	320	98.5	1.5	1102	0.275	100/SC
LDI	280	93.2	6.8	1062	0.303	100/LDI
DI	240	86.4	13.6	995	0.331	100/DI
LID	200	78.8	21.2	910	0.364	50/DI+50VLID
VLID	160	71.3	28.7	827	0.413	100/VLID
EID	120	63.2	36.8	731	0.487	50/VLID+50/VEID
VEID	80	55.1	44.9	639	0.639	100/VEID

5.2. POTATO

The results relative to the potato crop are shown in Table VI and Figure 8. Under average demand conditions, all deficit strategies are economically acceptable, with 100 % of the surface allocated to every optional strategy (Table VI). This means that yield products can remunerate both the land and the water under such conditions. This is evidenced by the fact that the GM/m^3 grows from 1.74 to 3.29 USD/m^3 for the well irrigated and the high deficit irrigation strategy, respectively. The GM/ha decreases from 3485 to 1648 USD/ha for the same irrigation schedules.

Under high demand conditions, every water supply restriction results in a decrease of the GM/ha, from 3278 to 922 USD/ha, while the GM/m^3 do not increase like for average demand conditions but only slowly increases from 0.936 to 1.043 USD/m^3 (LID strategy) and then decreases for the larger irrigation deficits, down to 0.922 USD/m^3 (EID strategy). This means that the increase in water productivity relative to deficit irrigation is not sufficient to compensate for the decrease in land productivity. Therefore the model selects the deficit strategies that produce an increase in the GM/m^3 and rejects the heavier deficit strategies (Table VI) proposing as alternative to crop just a percentage of the area available and to apply there the LID deficit irrigation scheduling corresponding to a total 160 mm irrigation depth.

Table VI. Economic results of the potato crop alternative irrigation strategies

Deficit irrigation strategies	Season irrigation (mm)	Relative ET (%)	Relative yield loss (%)	Gross margin (USD/ha)	Gross margin (USD/m^3)	Percent surface allocated to each irrigation strategy
Average demand condition						
SC	160	100.0	0.0	3485	1.742	100/SC
LDI	120	92.0	8.8	3037	2.024	100/LDI
DI	80	80.5	21.4	2366	2.366	100/DI
LID	40	68.3	34.8	1648	3.296	100/LID
High demand condition						
SC	280	100.0	0.0	3278	0.936	100/SC
LDI	240	95.7	4.8	3065	1.021	100/LDI
DI	200	87.5	13.8	2606	1.042	100/DI
LID	160	78.4	23.8	2087	1.043	100/LID
VLID	120	68.5	34.7	1516	1.010	75/LID
EID	80	58.2	46.0	922	0.922	50/LID
Very high demand condition						
SC	320	100.0	0.0	3209	0.802	100/SC
LDI	280	96.2	4.2	3032	0.866	100/LDI
DI	240	87.5	13.8	2537	0.845	85/LDI
LID	200	80.5	21.4	2159	0.863	71/LDI
VLID	160	71.0	31.9	1612	0.806	57/LDI
EID	120	60.1	43.9	976	0.650	42/LDI

For very high demand (severe drought) results are similar to those for high demand but less favourable since only a mild deficit is economically viable (Table VI). When available water would be more restricted, the option would be to reduce the cropped area and apply there a mild deficit irrigation scheduling.

The results in Figure 8 elucidate the differences among gross margins variation under average and high and very high demand conditions. Two different trends can be detected: for average demand, the GM/m^3 steadily increases when stronger deficits are considered, similarly to the behaviour of the wheat crop (Figure 7), indicating feasibility for adopting deficit irrigation.

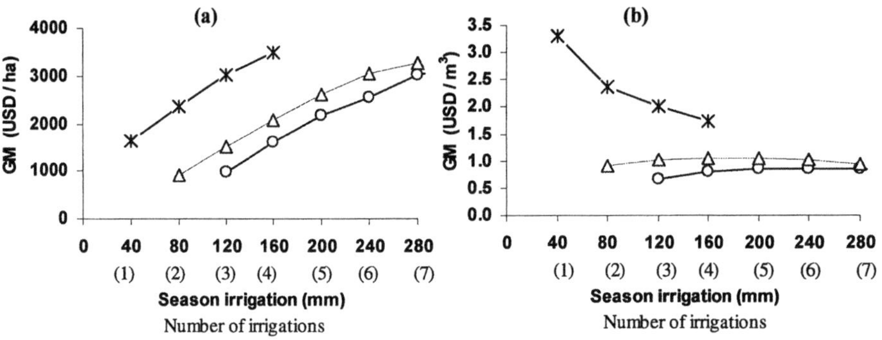

Fig. 8. Gross margins per unit surface (a) and per unit volume of water applied (b) for alternative deficit irrigation strategies of potato crop in Siliana for average (—✱—), high (—△—) and very high (—○—) demand conditions

For high and very high demand conditions, a threshold relative to GM/m^3 is apparent, after what the GM/m^3 decreases, indicating that deficit irrigation is not anymore feasible. Below these threshold values, the best return from the limited water available is obtained by reducing the cropped area and applying there moderate or mild deficit irrigation schedules.

5.3. TOMATO

For the tomato crop, the economic evaluation of alternative irrigation schedules produces results similar to those observed for the potato crop (Table VII).

For average demand conditions, a reduction in irrigation water depths produces a slight increase in GM/m^3, smaller than for the potato crop (Figures 8 and 9). However, the ERF strategy is refused and the alternative would be to reduce the surface cropped. Under high and very high demand conditions, adopting severe water supply constraints in water application induces a decrease in GM/m^3 (Figure 9), thus making it preferable to reduce the area cropped. Nevertheless, this reduction in GM/m^3 is not as large as the one observed in the potato crop (Figure 8), indicating that the tomato crop valorises the water applied better than potato.

Table VII. Economic results of the tomato crop alternative irrigation strategies

Deficit irrigation strategies	Season irrigation (mm)	Relative ET (%)	Relative yield loss (%)	Gross margin (USD/ha)	Gross margin (USD/m^3)	Percent surface allocated to each irrigation strategy
Average demand condition						
SC	640	100.0	0.0	4551	0.568	100/SC
LRF	520	86.5	14.2	3835	0.590	63/SC + 37/HRF
RF	400	71.9	29.5	3046	0.609	25/SC + 75/HRF
HRF	320	64.6	37.1	2689	0.672	100/HRF
ERF	280	58.4	43.7	2326	0.664	87/HRF
High demand condition						
SC	720	100.0	0.0	4413	0.490	100/SC
LRF	640	91.6	8.8	3979	0.497	100/LRF
RF	520	78.9	22.2	3315	0.510	100/RF
HRF	440	69.3	32.2	2804	0.509	85/RF
ERF	400	61.6	40.4	2338	0.467	77/RF
Very high demand condition						
SC	840	100.0	0.0	4206	0.400	100/SC
LRF	680	87.0	13.6	3600	0.423	100/LRF
RF	560	70.8	30.6	2700	0.385	82/LRF
RF	560	70.8	30.6	2700	0.385	82/LRF
HRF	480	61.6	40.3	2207	0.367	71/LRF
ERF	400	52.7	49.7	1734	0.346	59/LRF

The threshold values for the gross margin per unit water applied are 0.67 USD/m^3, 0.51 USD/m^3 and 0.42 USD/m^3 under average, high and very high demand conditions, respectively (Table VII). These thresholds are lower for the tomato than for the potato crop. The differences for the thresholds between the two horticultural crops relate to the costs of production and yield values, which are higher for tomato.

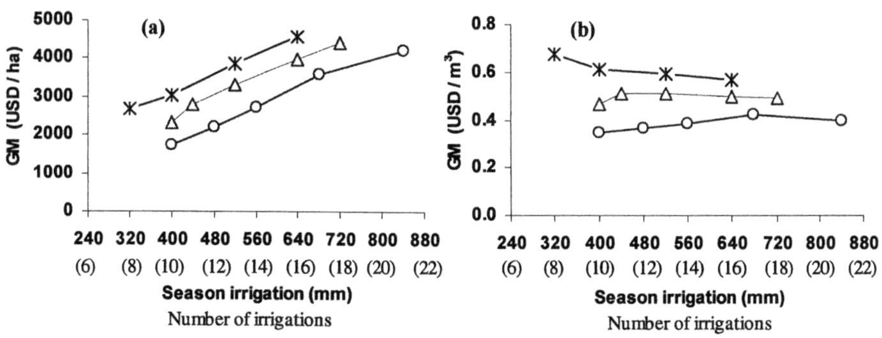

Fig. 9. *Gross margins per unit surface (a) and per unit volume of water applied (b) for alternative deficit irrigation strategies of tomato crop in Siliana for average (—✱—), high (—△—) and very high (—○—) demand conditions.*

6. Water management issues

An irrigation scheduling simulation model is able to generate information required to support water system operation and drought mitigation actions. The previous analysis is based on the global water demand for the whole crop season, *i.e.* the irrigation dates were not analysed. Nevertheless, in order to define the operation rules, the generated irrigation schedules should be used as water demand patterns at different time scales to provide information for operation of the water supply system, and to advise farmers. Particularly, for defining reservoir operation rules under drought, it is necessary to identify the most appropriate irrigation schedules based on sustainable socio-economical aspects.

The objectives for establishing reservoir operation rules are the optimisation of the water supplies to meet the demand requirements and to guarantee a minimal volume of water stored at the end of every irrigation season to allow for irrigation at the initial stage of the next crop season considering the inflows expected to occur in the next time steps (Lebdi, 1999). For the wheat crop, this first irrigation is often required to assure the setting up of the crop, due to the limited soil water at planting. Indeed, the available soil water at planting (ASWP) doesn't exceed 30 % of TAW for the probability of 50 % (Figure 10), which indicates the need for irrigation at planting except for the wet years.

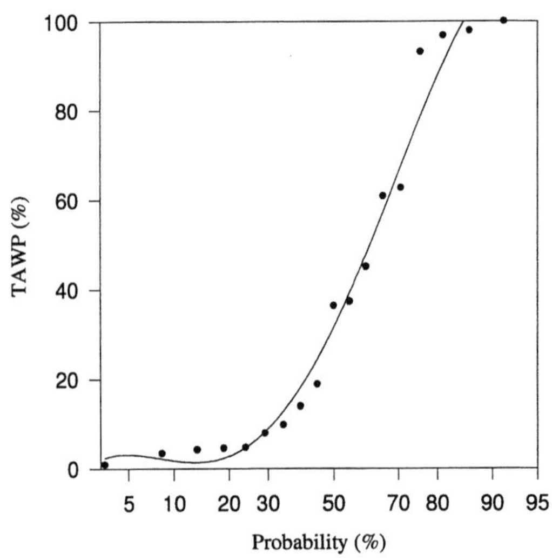

Fig. 10. Probabilities of available soil water at wheat crop seeding ASWP in Siliana

For the wheat crop under very high demand conditions (severe drought), the economic analysis shows that adopting the deficit strategy VEID reducing the demand from 320 to 80 mm and producing a relative yield loss close to 45 % is feasible (section 5.1.). However, this option is in contradiction with the objectives of the Cereal Supplementary Irrigation National Program of Tunisia that aims at achieving a yield threshold of 4.5 tons/ha for the whole of 60000 ha of cereals irrigated area. Therefore, all alternatives leading to a relative yield loss larger than 35 % (Table V) should not be considered. Instead, the irrigation schedule VLID, where the seasonal applied depth is limited to 160 mm, could be selected.

When the VLID schedule would be adopted for drought conditions as referred above, the minimum irrigation depth at planting is 40 mm but when the objective would be to avoid yield losses (SC strategy), an irrigation of 80 mm would be required (Figure 11). Further irrigations should then be planned by February, at the elongation phase, and by March and April, at heading and yield formation (Figure 11 b). When yield losses should be avoided, a last irrigation should be given later to assure grain filling (Figure 11 a).

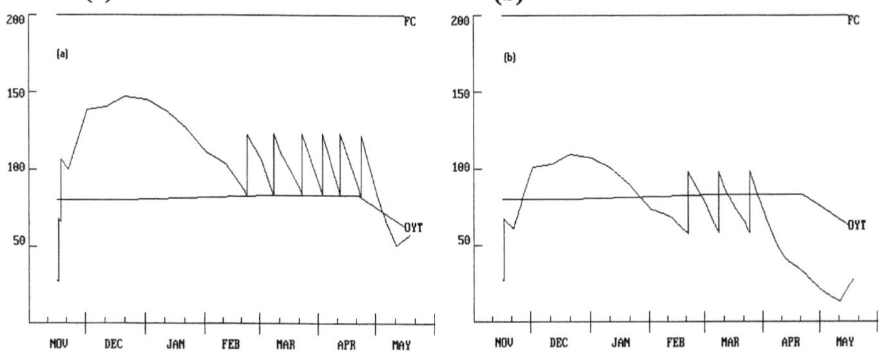

Fig. 11. Simulated available soil water ASW (mm) for two *irrigation schedules for wheat under severe drought: (a) without supply restrictions; (b) adopting the VLID strategy*

For the potato crop, the feasible irrigation schedules under severe drought are the SC and LDI alternatives including when the cropped area is reduced (Table VI). The simulated available soil water for both are given in Figure 12.

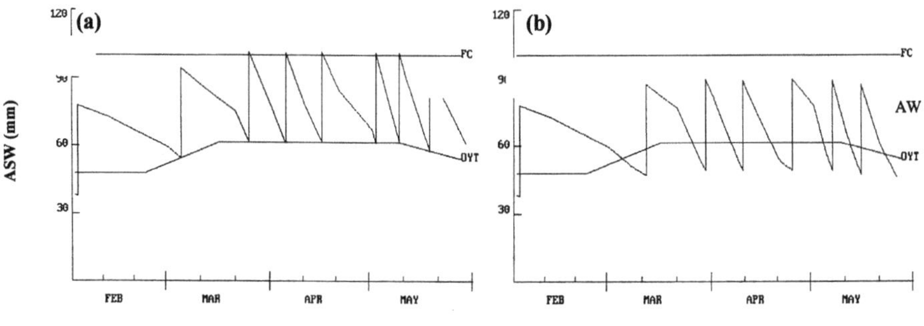

Fig. 12. Simulated available soil water ASW (mm) for the two feasible irrigation schedules for potato under severe drought: (a) SC strategy; (b) LDI strategy

For both, an irrigation at planting is required as well as several irrigations from March through May. Since the LDI strategy may lead to a very small relative yield loss, 4.2 %, with less irrigation, it probably would be easily adopted by farmers, even when the cropped area should be reduced due to water availability limitations.

In case of the tomato crop (Figure 13) the two feasible irrigation schedules for very high demand conditions are SC and LRF, the latter also when the tomato cropped area has to be reduced (Table VII). For both, a very large irrigation should be applied at planting (Figure 13) due to the very low available soil water at planting (5 mm). The peak demand for water occurs in July, 240 mm for the SC strategy, but it may be reduced by 1/3 when the LRF strategy is adopted. This reduction in the peak demand is a good reason for its implementation in the practice.

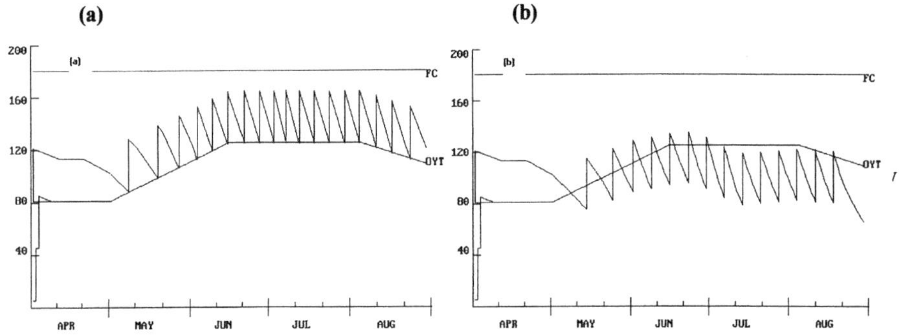

Fig. 13. Simulated available soil water ASW (mm) for the two feasible irrigation schedules for tomato under severe drought: (a) SC strategy; (b) LRF strategy

7. Impacts of water prices

An analysis was performed for understanding the impacts of water prices on the selection of the deficit schedules under very high demand conditions, *i.e.* under drought. The hypothesis tested is based on the assumption that a strong competition for water occurs under those conditions, and that rising the water price may contribute to a more rational water resource management favouring the water uses that assure the best economic return.

The methodology adopted to study the effect of water price variation in the optimal choice of the irrigation strategy is to consider a non-limiting water availability, *i.e.*, assuming that increased water prices have similar effects to those of water availability restrictions in the selection of deficit irrigation schedules under very high climatic demand conditions. Results are presented in Figure 14.

Results show:

- *Wheat crop*: when the water price is less than 0.18 USD/m^3, the full irrigation strategy SC is chosen; for higher prices, deficit strategies that use less water are progressively selected. The VLID schedule is not anymore feasible when the water price is higher than 0.36 USD/m^3.

- *Potato crop*: the full irrigation strategy SC is chosen for water prices up to 0.425 USD/m^3. Above this value, the strategy LDI is selected, up to the limit of 0.60 USD/m^3.
- *Tomato crop*: the full irrigation strategy SC is selected when the water price is not higher than 0.40 USD/m^3. When this price is exceeded, the strategy LRF is proposed but limited to 0.50 USD/m^3.

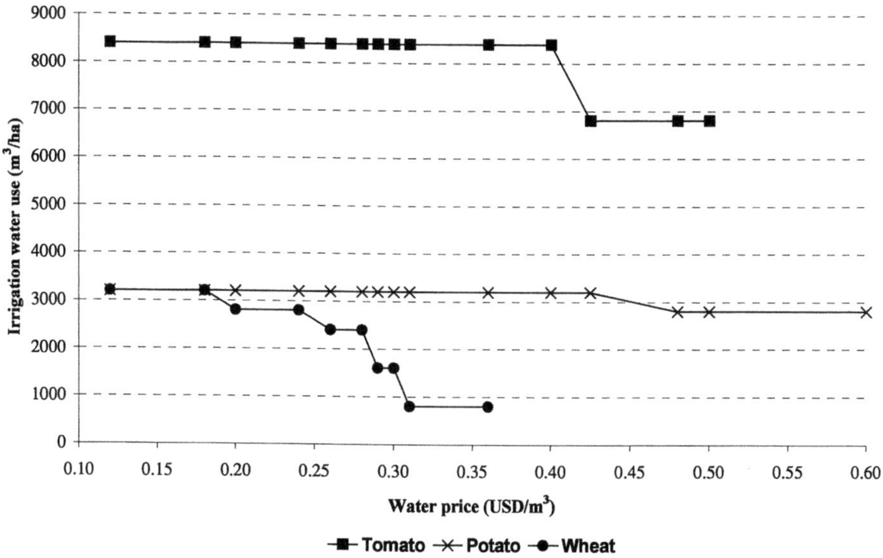

Fig. 14. Gross irrigation water use as a function of the water price

These results confirm the assumption that water prices may play a role similar to that of imposing water supply limitations. However the crop response is highly dependent from the prices of the commodities: wheat has low trade prices and low water demand, thus responds quickly to variations in water prices with the high deficit schedules becoming not feasible before the water price affects the cash crops. Tomato has the highest demand for water and the highest commodity price and responds much later than wheat to water prices variation. This indicates that when a water prices policy would be implemented to induce reductions in demand it should consider the specific responses of the crops where to be applied and the consequences on the respective impacts on the areas cropped.

8. Conclusions

The analysis above shows that the use of simulation models is useful to establish and evaluate strategies for supplemental irrigation of cereals or deficit irrigation of horticultural field crops to cope with drought and water scarcity. However, these models alone do not provide all the information required for decision-making, thus need to be coupled with economical analysis models or incorporated in a DSS where results of the generated simulation scenarios could be evaluated. The methodology utilised confirms that combining the use of an irrigation scheduling simulation model to generate alternative deficit irrigation strategies and an economic model applied to those strategies is helpful to support the selection of the best irrigation schedules to be applied when water availability is limited.

Results of this analysis show important differences in responses of crops to deficit irrigation as influenced by the climatic demand conditions. For the wheat crop, which growth season develops during the rainy season and irrigation is supplemental to rainfall, adopting deficit irrigation is generally feasible, including under drought demand conditions and considering heavy water supply restrictions.

Results relative to deficit irrigation of the potato crop depend upon the amount of the winter rains and the available soil water at planting. When these conditions are favourable as for average climatic demand, relatively important water demand reductions may be achievable without heavy yield losses, thus making deficit irrigation generally feasible. On the contrary, under severe drought conditions, when the climatic demand is very high, only mild deficit irrigation is feasible due to lack of rainfall and low soil moisture at planting. For tomato, cropped during the summer, out of the rainy season, every deficit strategy aiming at large reductions in the seasonal irrigation volumes produces large yield losses, so making it difficult to adopt such water saving strategies. Only mild deficits are then viable.

The analysis of selected irrigation schedules for the crops under study shows to be important to provide support for water systems operation, including reservoir operation rules. Indeed, the simulation results produce the demand pattern for the irrigated crops, which could be used for management practices.

An analysis on impacts of water prices on the crops irrigation demand has shown that increasing the water prices lead to adopt demand reduced irrigation schedules similarly to the imposition of water supply constraints. However crops react differently according to the values of the commodities produced. Relatively small increases in water prices lead to progressively severe deficit irrigation strategies for wheat. On the contrary, for the potato and tomato crops, deficit irrigation would be induced for much higher prices than for wheat despite water demand for these crops is larger, mainly in case of tomato. The use of the water price to induce water savings should therefore be crop specific, otherwise cash crops would be largely in advantage over other crops.

Acknowledgements

This research is being funded by the research contract ERBIC-18CT-970169 with the European Union, and the Contract Interreg IIC 99.09.5942.1 with the Portuguese Ministry of Agriculture.

References

Allen, R.G., Pereira, L.S., Raes, D. and Smith, M. (1998) *Crop Evapotranspiration. Guidelines for Computing Crop Water Requirements.* FAO Irrig. Drain. Pap. 56, FAO, Rome, 300p.

DG/GR (1996) *Plan stratégique pour le développement de l'économie de l'eau en irrigation.* Dir. Gen. Génie Rural, Tunis, 14 p.

DG/RE (1997) *Etude du Sector de l'Eau – Approches à un Modèle de la Prévision de la Demande de l'Eau.* DG/RE, Betchel and SCET – Tunisie, Tunis, 71 p.

Doorenbos, J. and Kassam, A.H. (1979) *Yield Response to Water.* FAO Irrig. Drain. Pap. 33, FAO, Rome.

El Amami, H. (1995) *Synthèse des Résultats de la Campagne de Céréales Irrigués 94/95.* Ministère de l'Agriculture, Tunis, 10 p.

English, M. and Raja, S.N. (1996) Perspectives on deficit irrigation. *Agric. Water Manag.* 32: 1-14.

Hazell, P.B.R. and Norton, R.D. (1986) *Mathematical Programming for Economic Analysis in Agriculture.* Macmillan Publishing Company, New York.

Lebdi F. (1999) *Approche d'Optimisation des Règles de Gestion du Barrage Lakhmès.* Rapport technique. Projet INCO-DC Research N° IC18CT970169, 6 p.

Oweis, T., Rodrigues, P.N. and Pereira, L.S. (2002) Simulation of supplemental irrigation strategies for wheat in Near East to cope with water scarcity. (This issue).

Pereira, L.S., Teodoro, P.R., Rodrigues, P.N. and Teixeira, J.L. (2002) Irrigation scheduling simulation: the model ISAREG. (This issue).

Rodrigues, P.N., Pereira, L.S. and T. G. Machado (2000) KCISA, a program to compute time averaged crop coefficients. Application to field grown vegetable crops. In: M.I. Ferreira and H.G. Jones (Eds.) *Proceedings of the Third International Symposium on Irrigation of Horticultural Crops.* ISHS, Acta Horticulturae Nr. 537, Leuven, vol. I, pp. 535-542.

Rodrigues, P.N, Machado, T., Pereira, L.S., Teixeira, J.L, El Amami, H. and Zairi, A. (2002) Feasibility of deficit irrigation with center-pivot to cope with limited water supply in Alentejo, Portugal. (This issue).

Teixeira, J.L. and Pereira, L.S. (1992) ISAREG - An irrigation scheduling model. *ICID Bulletin.* 41(2): 29-48.

Teixeira, J.L., Fernando, R.M. and Pereira, L.S. (1995) Irrigation scheduling alternatives for limited water supply and drought. *ICID Journal.* 44(2): 73-88.

Zairi, A., Nasr, Z., Ben Mechlia, N., Achour, H. and Oueslati, T. (1996) Réponse à l'eau d'une culture de blé sous conditions hydriques restrictives. *Annales de l'INRGREF*, numéro spécial: 77-89.

Zairi, A., Nasr, Z., Ben Mechlia, N., Pereira, L.S., Achour, H., Derouiche, A., Slatni, A. and Rodrigues, P.N. (1998) Optimisation de l'irrigation du blé en conditions de disponibilités en eau limitées. *Gestion de l'Eau en Conditions de Penurie.* Actes des Journées scientifiques de l'INRGREF (Hammamet, Tunisie, 29-30 Octobre). IRESA, Ministère de l'Agriculture, Tunis.

Zairi, A., Pereira, L.S., El Amami, H., Rodrigues, P.N., Slatni, A. and Teixeira, J.L. (2000) Irrigation scheduling strategies for horticultural field crops under limited water availability. In: M. I. Ferreira and H. G. Jones (Eds.) *Proceedings of the Third International Symposium on Irrigation of Horticultural Crops.* ISHS, Acta Horticulturae Nr. 537, Leuven, vol. I, pp. 503-510.

FEASIBILITY OF DEFICIT IRRIGATION WITH CENTER-PIVOT TO COPE WITH LIMITED WATER SUPPLIES IN ALENTEJO, PORTUGAL

P.N. RODRIGUES[(1)], T. MACHADO[(1)], L.S. PEREIRA[(1)], J.L. TEIXEIRA[(1)], H. EL AMAMI[(2)] and A. ZAIRI[(2)]

[(1)] *Agricultural Engineering Research Center, Institute of Agronomy, Technical University of Lisbon, Portugal*

[(2)] *National Research Institute for Agricultural Engineering, Water and Forestry, Tunisia*

Abstract: The identification of deficit irrigation strategies that provide for a reduced water demand with acceptable impacts on yields is part of preparedness measures to cope with drought and water scarcity. The irrigation scheduling simulation model ISAREG, which has been prior calibrated and validated in several Mediterranean regions, is used to simulate various irrigation schedules under average, high and very high climatic demand conditions to cope with diverse water supply restrictions. The study applies to the Vigia irrigation project in Alentejo, Portugal, which is taken as case study for similar areas in the region. The study concerns the supplemental irrigation of wheat and deficit irrigation of maize and sunflower under center-pivot sprinkler irrigation. The alternative irrigation schedules are evaluated taking into consideration the reduction in demand for irrigation water, the consequent decrease in yield, and the respective impacts on farmers income. Results show that deficit supplemental irrigation strategies for wheat are generally feasible. On the contrary, deficit irrigation strategies for maize are only feasible when the irrigation deficit is limited and water restrictions are low to moderate. Otherwise, the best practice is to limit the area cropped and apply there near optimal irrigation. Results for the sunflower show the feasibility of deficit irrigation because the crop is subsidised, which compensates for the yield losses.

Key words: water scarcity, drought, irrigation scheduling, simulation modelling, economic optimisation, wheat, maize, sunflower.

1. Introduction

Alentejo has a sub-humid climate of Mediterranean type, with most of rainfall occurring in the semester October-March, and very dry and hot summers. Water is typically scarce in Alentejo, where droughts are frequent as analysed in a companion paper (Paulo *et al.*, 2002).

Droughts require pro-active and rational water management to mitigate their impacts. Numerous studies have been devoted to the identification and evaluation of measures to cope with droughts (Batini *et al.*, 1999; Rossi, 2002; Vogt and Somma, 2000) and about the mitigation of drought impacts in irrigated agriculture (Pereira, 1989, 1999, 2002 b). These measures typically resume to those that contribute to reduce the demand for irrigation water, or to adjust the water supply as depending on the water availability. Strategies analysed and proposed in this research concern demand

G. Rossi et al. (eds.), Tools for Drought Mitigation in Mediterranean Regions, 203-222.

management only since they aim at sprinkler irrigation projects where on-demand delivery is applied, and to private farm irrigation systems where the farmer is in control of the water source. However, reduced demand strategies can be considered when simulating the supply and delivery systems (Lamaddalena and Pereira, 1998) to evaluate the functioning of the systems under water scarcity conditions. Further research is being developed along these lines.

Demand management has several facets, including deficit irrigation scheduling, the improvement of farm irrigation systems to make them more performant in terms of reduced water use and efficient application of deficit schedules, as well as the adoption of crop and soil management for water conservation. Studies of sprinkler irrigation performances were developed in another study (Pereira, 2002 a) conducted simultaneously with this research. Other measures to cope with droughts include the reinforcement of the hydraulic infrastructures, the increase of water reuse, the restriction of volumes supplied for irrigation, and graded water prices.

In this research deficit irrigation (English *et al.*, 1990) has been selected, including for supplemental irrigation of cereals. Deficit irrigation means to adopt an irrigation that only partly satisfies the crop irrigation requirements but is able to provide for a yield reduction that allows the economic return of irrigation. Supplemental irrigation is an irrigation practice where irrigation is provided in supplement of rainfall but only during the critical crop stress periods. Otherwise, the crop production could anyway be achieved under rainfed conditions, which is not the case for summer crops under Mediterranean conditions. Good examples of supplemental irrigation of wheat in the Mediterranean region are given by Oweis *et al.* (2000) and Zhang *et al.* (1998).

Maize has a high irrigation requirement, but it is one of the most efficient crops in producing dry matter with the water it uses. Under less than optimal water conditions, maize will use all the water available, and water use efficiency will be closely correlated with yield. But if stress occurs at critical periods such as flowering yields will be greatly reduced and may not be correlated to water use efficiency (Waldren, 1983). The irrigation requirement of wheat is lower than that of maize and sunflower. For that reason wheat is known as water scarcity resistant. However, increasing stomatal resistance that allows the water conservation, the productivity is reduced because of the decreased photosynthesis (Kirkham and Kanemasu, 1983).

In order to prepare for the adoption of appropriate irrigation strategies for grain cereals (wheat and maize) and oil crops (sunflower) when water supply restrictions have to be imposed, the present study was performed to assess how deficit irrigation could be practiced. Thus, the impacts of several deficit irrigation schedules are evaluated in terms of water saving, yield reduction and farmers income reduction. The study applies to the Vigia irrigation project, in central Alentejo region, southern Portugal. Evaluating the alternative irrigation schedules could be helpful to support decision-making by both farmers and irrigation project managers.

2. Case study: Vigia irrigation project

The Vigia irrigation project has 1505 ha irrigated area and is supplied by the Vigia storage dam (15526 hm^3). It is located in Alentejo, southern Portugal, and is part of the Guadiana river basin. The conveyance and distribution system is pressurized and operates on demand. The Alentejo Region represents 45 % of the total cropped area in Portugal (1757000 ha), but irrigated agriculture represents only 7 % (116000 ha) against the national average of 21 % irrigated surface (GPPAA, 2000).

According to the Koeppen classification, this region has a humid mesothermal climate (Csa), with relatively cold winters and hot and dry summers. The annual rainfall ranges from about 300 to 900 mm, erratically distributed inter-annually. The average monthly rainfall, average monthly reference evapotranspiration, ET_0, and average monthly net solar radiation, R_n, for the historical data series of 1941 - 1990 relative to the nearby meteorological station of Évora (36° 31' N, 7° 47' W) are shown in Figure 1. The net solar radiation was computed from observed sunshine hours, and ET_0 was computed with the Penman-Monteith method (Allen *et al.*, 1998).

Data in Figure 1 indicate that rainfall is usually sufficient for the winter crops. On the contrary, the summer crops can not be produced without irrigation because rainfall is too much insufficient during the period when the climatic demand, represented by ET_0, is higher.

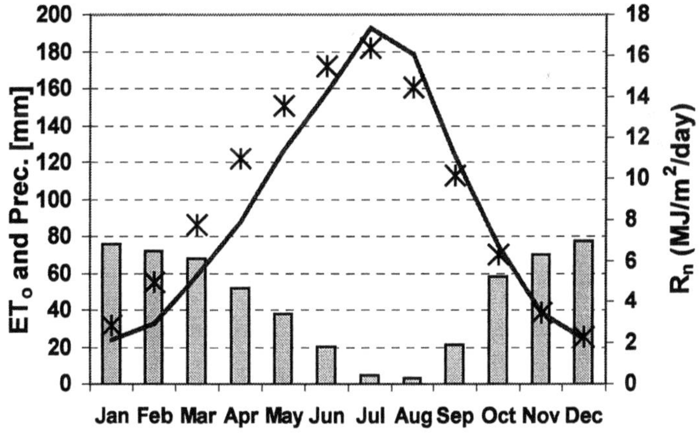

Fig. *1.* *Mean monthly rainfall (▭), mean monthly reference evapotranspiration, ET_0 (—), and mean monthly net radiation, R_n (✱) for Évora, 1941 – 1990*

The average monthly minimum and average relative humidity, and the average monthly wind speed at 2 m height for the same data series are presented in Figure 2. The minimum relative humidity was estimated from the ratio of saturated vapour pressure at minimum and maximum temperature. Results in Figure 2 show that arid conditions prevail during the summer crops growth period.

Main soils are loamy and sandy-loam. In average, the textural classes are 70 % sand, 15 % silt and 15 % clay. The average soil water at field capacity and at wilting point are $\theta_{FC} = 0.34$ mm/mm and $\theta_{WP} = 0.17$ mm/mm. The soil depth varies in the area, but in general is relatively shallow (0.5 m as average).

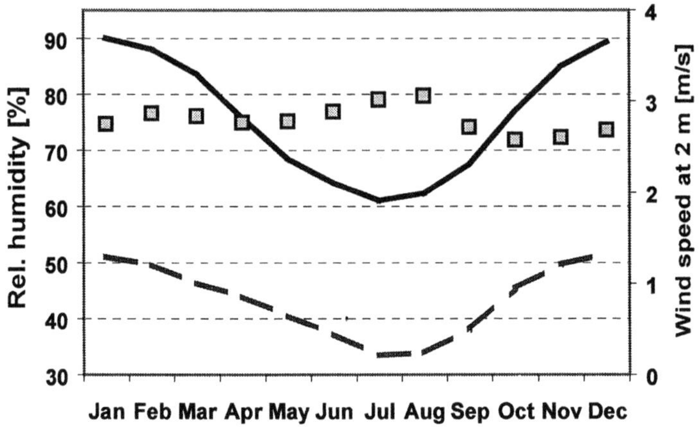

Fig. 2. *Mean monthly minimum (— —) and average (———) relative humidity, and mean monthly wind speed at 2 m height, u_2 (□) for Évora, 1941 - 1990*

The main crops in this irrigation project are grain cereals and industrial crops. Olive trees, forage crops and grapevines are also present but on a limited extend. Sprinkler irrigation, mainly with center pivot lateral systems, is used for the field crops.

3. Simulation modelling

3.1. MODELLING TOOLS

The ISAREG model (Teixeira and Pereira, 1992) has been used to simulate different irrigation management strategies for wheat, maize and sunflower crops. The ISAREG model performs the soil water balance using different options to define and evaluate the irrigation schedules, as described in Teixeira and Pereira (1992) and Liu *et al.* (1998). The reference evapotranspiration is computed with the FAO-Penman-Monteith method (Allen *et al.*, 1998) using the program EVAP56. The crop data parameters are determined with the KCISA program (Rodrigues *et al.*, 2000), which computations follow the updated methodology proposed by FAO (Allen *et al.*, 1998). In particular, the crop coefficients (K_c) and the soil water depletion fraction for no stress (p) have been adjusted to the Vigia climate. Further details are provided in the companion paper by Pereira *et al.* (2002).

Yield losses due to water stress are estimated utilising the yield response factors proposed by Stewart *et al.* (1976). Two approaches are used. In the first, the relationship between the seasonal relative evapotranspiration deficit ($RED = 1 - ET_d / ET_c$) and the relative yield losses ($RYL = 1 - Y_d / Y_c$) is used:

$$(1 - Y_d / Y_c) = K_y (1 - ET_d / ET_c) \quad (1)$$

where ET_d and ET_c are the seasonal crop evapotranspiration (mm) under deficit and full satisfaction, respectively, and Y_d and Y_c are the yields achievable when crop evapotranspiration is ET_d and ET_c, respectively. K_y is the crop yield response factor.

The second approach corresponds to the multiphasic model proposed by Stewart *et al.* (1976), assuming that the impacts on yields vary with the crop development phase where ET is reduced and depend upon the degree of stress occurring in the previous phase. The multiphasic model is described by

$$(1 - Y_d / Y_c) = [\, \Sigma_i K_{y\,i} (ET_{c\,i} - ET_{d\,i}) \,] / ET_c \quad (2)$$

where $ET_{d\,i}$ and $ET_{c\,i}$ are computed for each crop phase i, and $K_{y\,i}$ are the crop yield response factors for the same phases i. The yield response factors used in this study are those proposed by Doorenbos and Kassam (1979), adjusted or validated to local conditions (Alves *et al.*, 1991; Alves and Pereira, 1998).

The gross irrigation depths applied were estimated from the ratio between the net irrigation depths computed with the ISAREG model and the application efficiency estimated for the sprinkler systems. The average 75 % was adopted for the application efficiency based on results of field evaluations (Pereira, 2002 a) and in agreement with performances expected under the prevailing arid environmental conditions (Keller and Bliesner, 1990).

In order to evaluate the impacts of the deficit irrigation strategies on the farmers income, a linear programming model was used as described in a companion paper (Zairi et al., 2002). The activities considered are every one of the irrigation strategies generated for the wheat, maize and sunflower crops, which are characterised by the respective irrigation depths and the yield losses resulting from the irrigation deficits. The objective function is evaluated by the gross margin, a linear function obtained by subtracting the variable costs related with the specified activities to the farm gross product relative to the crops under analysis. The output of the linear problem is shown in terms of percentage of land allocated to each activity after balancing the economic return from water and land (El Amami et al., 2001).

Current prices for land, labour and crop products in Portugal are utilised. Data was obtained through questionnaires and information provided by a governmental agency of the Ministry of Agriculture (unpublished data). Difficulties resulted from a large variation in data relative to crop yields, irrigation costs, and production costs. Average values were adopted despite good farmers achieve higher yields than those considered. The current production and area grants and subsidies for the sunflower crop were also considered.

3.2. WATER SUPPLY STRATEGIES

In the Vigia project, the main irrigation systems are the center-pivot laterals, which are used for all the three crops. The average net irrigation depth I_n = 15 mm is used for the center-pivot systems, corresponding to a rotation completed every 3 days. However, some farmers adopt a higher irrigation frequency.

When water availability is non-limiting, the frequency of irrigation is not restricted and varies along the crop season according to the crop demand since the supply system operates on-demand. Under limited water availability, the supply is made with restrictions in water volumes available during the crop season, and several irrigation scheduling strategies are adopted in practice. The management allowed depletion fraction (MAD), which is a common criterion for irrigation scheduling (Martin *et al.*, 1990) is adopted for defining the alternative schedules. For full irrigation, the MAD fraction equals the soil water depletion fraction for non-stress (p), which is a crop and climate dependent factor, as defined by Allen *et al.* (1998). For water limited schedules, the MAD is increased as described in the companion paper by Pereira *et al.* (2002). Simulations were performed by combining different MAD fractions with different levels of the total available irrigation water for the crop season.

The non-restricted irrigation schedules are only constrained by the system characteristics (SC) and correspond to the achievement of the maximal yield, thus θ_{MAD} equal to θ_p is assumed. The deficit schedules were simulated by decreasing the available water (net season irrigation depth) and decreasing the irrigation threshold below that represented by θ_p. Different MAD are considered for the four crop growth periods defined in Allen *et al.* (1998). For the initial, development, and end season periods θ_{MAD} varies from θ_p to 0.55 θ_p for wheat, 0.50 θ_p for maize, and 0.40 θ_p for sunflower. A lower threshold is considered for sunflower because it is less affected by water deficits. For the mid season period, where water deficits cause higher impacts on yields, θ_{MAD} do not reduces below 0.8 θ_p for maize and 0.70 θ_p for wheat and sunflower.

In parallel to reducing the θ_{MAD}, progressively large restrictions on the water availability are considered. For the wheat crop, the available water decreases from full irrigation requirements to 60 mm when drought is considered. Rainfed wheat is also simulated for average demand conditions, which is the common practice by farmers. For the maize and sunflower crops, the available water decreases from full satisfaction to 390 and 150 mm, respectively.

3.3. CROP PARAMETERS

The average dates for the crop development stages, the crop coefficients (K_c) and the soil water depletion fractions for no stress (p) used in simulation are presented in Table I for the three crops. The crop yield response factors for the considered growing stages and the three crops are given in Table II. The K_{yi} values utilised varied with the intensity of water deficit during the phase under consideration and the precedent phase in the range referred in the table.

Table I. Average crop parameters for wheat, maize and sunflower

Parameters	Crop development stages			
	Initial	Development	Mid season	End season
Wheat:				
Period length (dates)	15 Nov – 24 Jan	25 Jan - 14 Apr	15 Apr -30 May	31 May - 20 Jun
Crop coefficients, K_c	1.12	1.12 - 1.18	1.18	1.18 - 0.25
Depletion fraction, p	0.60	0.60 - 0.57	0.57	0.57 - 0.69
Maize:				
Period length (dates)	20 Apr - 11 May	12 May - 19 Jun	20 Jun - 4 Sep	5 Sep - 30 Sep
Crop coefficients, K_c	0.74	0.74 - 1.27	1.27	1.27 - 0.35
Depletion fraction, p	0.60	0.60 - 0.45	0.45	0.45 - 0.69
Sunflower:				
Period length (dates)	15 Apr - 9 May	10 May - 13 Jun	14 Jun - 14 Jul	15 Jul - 18 Aug
Crop coefficients, K_c	0.68	0.68 - 1.06	1.06	1.06 - 0.35
Depletion fraction, p	0.56	0.56 - 0.40	0.40	0.40 - 0.57

Table II. Crop yield response factors

Crop	Seasonal yield response factor, K_y	Multiphasic yield response factors, K_{yi}		
		Vegetative growth	Flowering and yield formation	Maturation and senescence
Wheat	1.05	0.50 – 0.75	4.30 – 1.25	2.45 – 0.50
Maize	1.25	1.90 – 0.40	4.00 – 2.30	2.00 – 0.50
Sunflower	0.95	1.20 – 0.50	4.00 – 1.00	3.50 – 0.80

4. Net irrigation requirements

The annual net irrigation requirements (NIR) for the wheat, maize and sunflower crops were computed with the ISAREG model by performing a sequential soil water balance for the time series 1941-1990. The average crop parameters presented in Table I were used.

To estimate the available soil water at planting (ASWP) of the wheat crop, expressed as a percentage of the total available soil water (TAW), the simulation started by the end of summer, when the soil water is depleted (assumed ASWP = 5 % TAW), using a bare soil condition until the planting date. Similarly, computations started with a replenished soil water content (ASWP = 95 % TAW) at early January to estimate the ASW at planting for the summer crops.

Results for the ASWP, simulated for the 50 years weather data are shown in Figure 3. Simulations for the wheat crop show that ASW at planting may not be sufficient to provide for appropriate crop establishment in about 25 % of the years. However, because this occurs during the rainy season, farmers generally do not irrigate but often delay seeding until ASW could be enough. For the summer crops, maize and sunflower, irrigation is more often required. Farmers irrigation practices during the initial crop growth period are indicated in Table III.

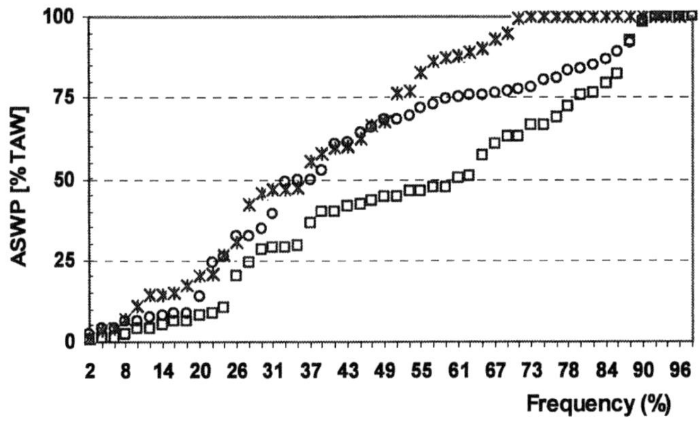

Fig. 3. Frequency distribution of the initial soil water content (ASWP) expressed as a percentage of the total available water (TAW) for wheat (✱), maize (□) and sunflower (o)

Table III. Minimum irrigation events during the initial crop growth period for the summer crops

Crop	Irrigations during the initial period	
	Minimum number	Application depth (mm)
Maize	3	45
Sunflower	2	30

Figure 4 shows the NIR series obtained. Assuming a normal distribution for the computed NIR, the years having crop irrigation requirements corresponding to average, high and very high demand conditions, which correspond to the probability for being exceeded of 50, 20 and 5 %, were identified. The analysis that follows is performed for these three reference years.

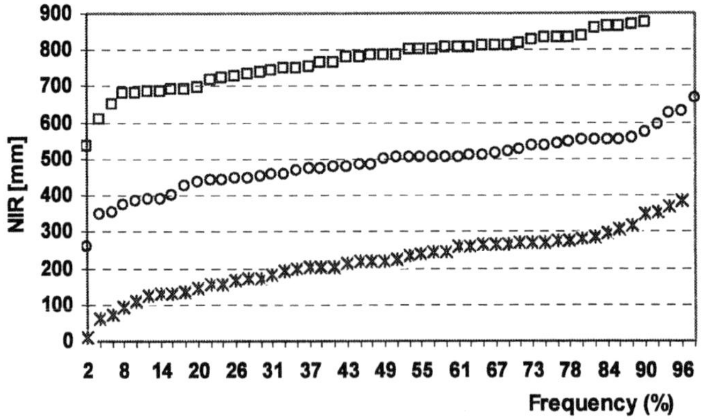

Fig. 4. *Frequency distribution of the net irrigation requirements (NIR) for wheat (✱), maize (□) and sunflower (o)*

The Figures 5 and 6 show the climatic conditions for the mentioned three levels of demand referring to the maize crop. Data in Figure 5 shows that a great variability occurs for precipitation. On the contrary, ET_o is quite stable. It is higher for the drought year, with very high climatic demand. Results in Figure 6 show that the minimum relative humidity follows a trend coherent with that for rainfall. A large variability occurs for wind speed, which is higher for the year representing very high demand. Data in these Figures show that arid conditions prevail during the maize season.

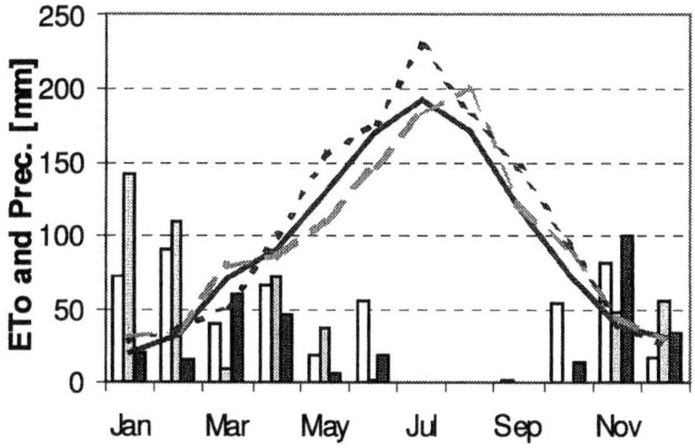

Fig. 5. *Mean monthly precipitation for average (▭), high (▨) and very high (■) climatic demand, and mean monthly ET_o for average (——), high (— —) and very high (- - -) climatic demand*

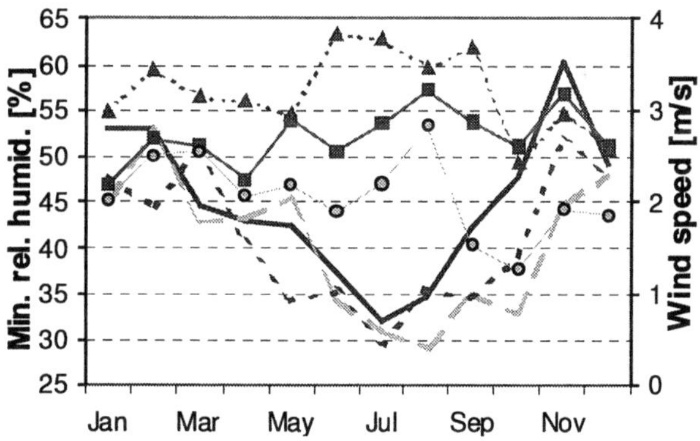

Fig. 6. *Mean monthly minimum relative humidity for average (——), high (— —) and very high (- - -) climatic demand, and mean monthly wind speed for average (—■—), high (··●··) and very high (··▲··) climatic demand*

Table IV summarises for the crops under analysis the available water at planting (ASWP), the season rainfall, the season rainfall excess non used for ET, the season crop evapotranspiration, and the season net irrigation requirements for the average, high and very high climatic demand conditions. To be noted that due to the variability of rainfall during the wheat season, a large fraction of rainfall during the Winter is not used by the crop in the year having high climatic demand.

Table IV. Total available soil water at planting (ASWP) and seasonal rainfall, crop evapotranspiration, and net irrigation requirements (NIR) for the average, high and very high climatic demand conditions for different crops

Crop	Demand condition	ASWP (mm)	Season rainfall (mm)	Non used rainfall (mm)	Season ET (mm)	Season NIR (mm)
Wheat	Average	85.0	334.6	56.5	553	217
	High	53.2	479.4	307.9	472	276
	Very high	76.6	153.1	8.3	558	366
Maize	Average	32.6	96.6	0.0	904	787
	High	6.4	61.4	0.0	891	830
	Very high	36.2	46.0	0.0	1017	967
Sunflower	Average	66.7	76.3	0.0	545	446
	High	69.1	58.4	0.0	580	492
	Very high	43.7	55.9	0.0	620	557

5. Feasibility of deficit irrigation

5.1. WHEAT

For average climatic demand conditions (Table V), the adoption of deficit supplemental irrigation for wheat is easily achievable. Reductions of about 50% in the seasonal irrigation depths, from the optimal 225 mm to 120 mm, produce RYL close to 30%.

Table V. Selected results of the response of the wheat crop to supplemental irrigation strategies in Vigia using center-pivot systems (I_n = 15 mm)

Climatic demand	Strategy	Season irrigation (mm)	Number of irrigation events	Season ET (mm)	Relative ET (%)	RYL (%)
Average	SC	225	15	553.9	100.0	0.0
	E	210	14	549.9	99.3	1.6
	F	180	12	521.8	94.2	12.8
	G	120	8	463.1	83.6	29.3
	H	60	4	388.2	71.0	36.4
	I	30	2	361.7	65.3	37.7
	NI	0	0	343.3	62.0	38.8
High	SC	270	18	475.9	99.5	0.0
	D	240	16	451.8	94.4	12.0
	E	210	14	423.7	88.6	25.1
	F	180	12	394.6	82.5	34.9
	G	120	8	335.2	71.0	42.2
	H	60	4	275.6	57.6	47.6
	I	30	2	245.6	51.3	49.5
Very high	SC	345	23	547.1	99.7	0.0
	B	300	20	514.3	93.7	12.3
	C	270	18	486.5	88.7	21.3
	D	240	16	457.5	83.4	31.2
	E	210	14	427.9	78.0	38.5
	F	180	12	398	72.5	42.7
	G	120	8	338.3	61.7	47.2
	H	60	4	278.3	57.0	49.8

This happens because the season available rainfall satisfies a large fraction of the crop water requirements, so limited yield losses occur when adopting deficit irrigation. In fact, farmers generally do not irrigate under such climatic demand conditions because the rainfed crop is economically feasible, even with a RYL near 40 % (Table VI). The few farmers that irrigate wheat under average climatic conditions use center-pivot systems, which show to adapt well to provide light irrigations during critical crop stress periods.

Under high climatic demand, reducing the seasonal irrigation depth from 270 to 120 mm would produce a RYL close to 42 %, thus higher than that for the rainfed crop under average demand (Table V). Such RYL (39 %) would be obtained when applying 150 mm. Under very high climatic demand, the same RYL corresponds approximately to a season application depth of 210 mm.

Simulations for the application of 120 mm for the three demand conditions are shown in Figure 7. They indicate that irrigation starts earlier, by March, before heading, for the drought case. The water deficit during the end season is smaller for average demand due to larger spring rainfall, which is negligible in case of drought.

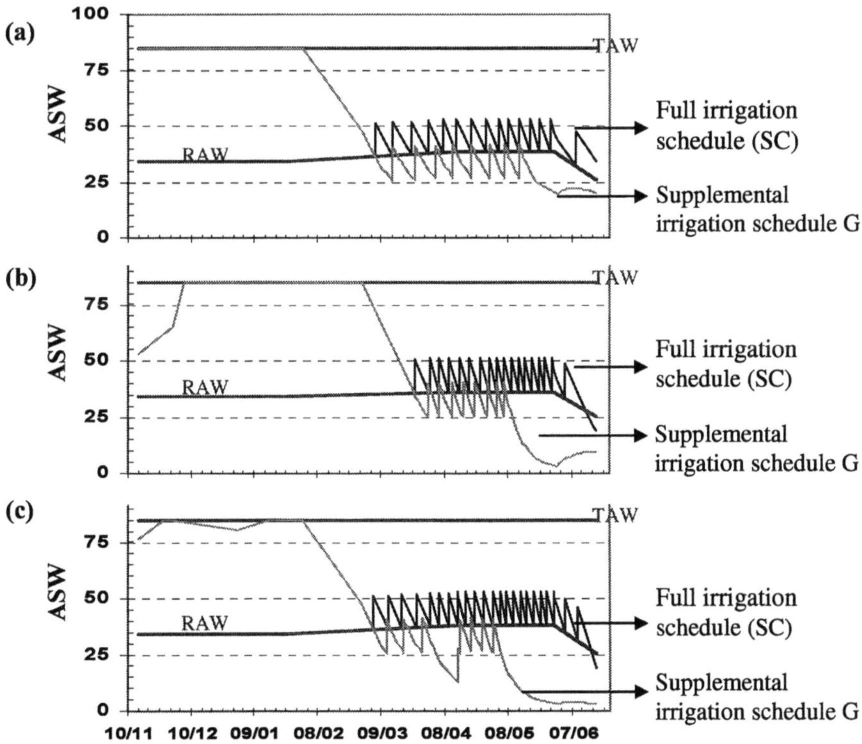

Fig. 7. *Simulated available soil water (ASW) when comparing the full irrigation schedule (SC) for wheat with those for application of 120 mm for (a) average, (b) high, and (c) very high climatic demand*

The gross margins per unit volume of water (GM/m^3) show to increase for all cases from full irrigation to the heavier deficit, while the gross margins per unit land area (GM/ha) decrease but remaining positive, as shown in Figure 8. The GM/m^3 grows steadily from 0.38 USD/m^3 to 1.76 USD/m^3 under average climatic demand, but follows a different trend for high and very high demand. First it slowly decreases from 0.36 to 0.32 USD/m^3 in case of high demand (strategy F, when 180 mm are applied), and later sharply increases up to 1.57 USD/m^3. Similarly, for very high demand it varies from 0.30 USD/m^3 down to 0.27 USD/m^3 (strategy E, with 210 mm target depth) and then increase to 0.82 USD/m^3 (Table VI).

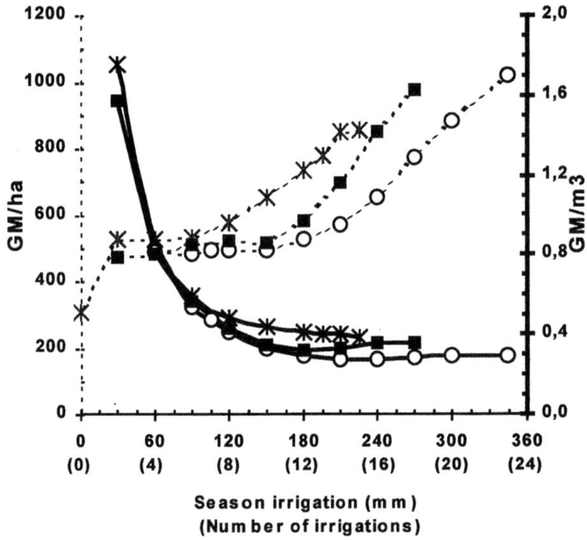

Fig. *8.* *Variation of the gross margins per hectare cropped (---) and per m^3 of water applied (—) relative to the wheat crop when irrigation depths decrease from full irrigation to heavy deficits for average (—✱—), high (-■-) and very high (—○—) climatic demand conditions*

This behaviour indicates that the increase in water productivity for the mild deficit schedules is not able to valorise the water applied when compared with the land, to which correspond relatively high gross margins per hectare cropped (GM/ha). For higher deficits, the water productivity becomes high enough to assure the feasibility of the irrigated crop under deficit irrigation.

Results in Table VI show that all deficit irrigation strategies are feasible as 100 % of the surface is allocated to the crop. However, the optimal strategies often correspond to a combination of two irrigation strategies that, together, have the same demand as the strategy not selected. For instance, the case for application of 120 mm (Figure 7) would have as optimal solution a combination of the E and I strategies for average demand, of SC and I for high demand, and SC and H for very high demand. This means that the practical solution to apply 120 mm, the strategy G, is sub-optimal. Results are similar to those obtained by Zairi *et al.* (2002).

Table VI. Economic responses of the wheat crop to deficit irrigation: gross margins per unit volume of water applied (GM/m^3) and percentage of surface allocated to each irrigation strategy

Strategy	Target depth	Average climatic demand		High climatic demand		Very high climatic demand	
		GM/ m^3	Surface allocated to each strategy	GM/ m^3	Surface allocated to each strategy	GM/ m^3	Surface allocated to each strategy
	(mm)	(USD)	(%)	(USD)	(%)	(USD)	(%)
SC	345	–	–	–	–	0.30	100/SC
SC	270	–	–	0.36	100/SC	–	–
SC	225	0.38	100/SC	–	–	–	–
B	300	–	–	–	–	0.29	84/SC + 16/H
C	270	–	–	–	–	0.29	74/SC + 26/H
D	240	–	–	0.35	88/SC + 12/I	0.27	63/SC + 37/H
E	210	0.40	100 E	0.33	75/SC + 25/I	0.27	53/SC + 47/H
F	180	0.41	83 E+17 I	0.32	63/SC + 37/I	0.29	42/SC +58/H
G	120	0.48	50/E + 50/I	0.43	37/SC + 63/I	0.41	21/SC + 79/H
H	60	0.88	17/E + 83/I	0.80	12/SC + 88/I	0.82	100/H
I	30	1.76	100/I	1.57	100/I	–	–
NI	0	–	100/NI	–	–	–	–

5.2. MAIZE

Simulations for maize produced results very different from those for wheat. Results in Table VII show that RYL grow much faster then the reduction in applied irrigation. Reducing the season irrigation depth for the average, high and very high climatic demand conditions, respectively from 795, 825, and 930 mm to 660 mm produces water savings equal to 17, 20 and 29 %, respectively, but causing RYL equal to 24, 36, and 47%. Therefore, when relatively high water savings are considered, the impacts on yields are very important.

Table VII. Selected results of the response of the maize crop to deficit irrigation strategies in Vigia using center-pivot systems (I_n = 15 mm)

Climatic demand	Strategy	Season irrigation (mm)	Number of irrigation events	Season ET (mm)	Relative ET (%)	RYL (%)
Average	SC	795	53	907.8	99.6	0.0
	C	780	52	897.6	98.5	3.6
	D	720	48	846.0	92.8	15.1
	E	660	44	787.6	86.4	24.4
	F	600	40	728.1	79.9	34.3
	G	540	36	668.4	73.3	41.9
High	SC	825	55	878.8	98.9	0.0
	C	780	52	843.0	94.9	18.0
	D	720	48	786.3	88.5	27.1
	E	660	44	727.0	81.8	36.1
	F	600	40	667.3	75.1	46.2
Very high	SC	930	62	998.0	99.6	0.0
	B	825	55	903.0	95.0	14.2
	C	780	52	856.2	86.1	21.6
	D	720	48	796.6	81	34.0
	E	660	44	736.7	74.0	47.3

These results show that for severe to very severe drought conditions, to maintain a RYL near to that estimated for average climatic demand (24 %) would require that the net seasonal application depths be 720 mm and 780 mm for the high and very high climatic demand conditions, respectively. These season irrigation depths are close to that required for no stress under average climatic demand (795 mm). This indicates that the adoption of deficit irrigation for maize, like for other summer crops such as tomato (Zairi *et al.*, 2002), is difficult and does not produce the high demand reduction that is desired to cope with droughts. These difficulties are very well known by the farmers, who generally reduce the irrigation area and apply there a near to optimal irrigation scheduling.

Results for the GM/ha of the maize crop (Figure 9) show to decrease steadily, from values near 1100 USD/ha when the crop is fully irrigated, to near 400 USD/ha when relatively high irrigation deficits are considered. For the GM/m^3, on the contrary to the wheat crop, they steadily decrease from the full irrigation to the higher deficit schedules. The GM/m^3 are much smaller than those for wheat, thus indicating that for the case study under consideration the crop gross margins are too small to valour the high volumes of water required. These low values for the gross margins are not expected. They are probably due to the fact that center-pivot irrigation is relatively new in the area and irrigation is yet poorly managed. Results from field evaluations of irrigation systems (Pereira, 2002 a) have shown quite poor irrigation performance in this area.

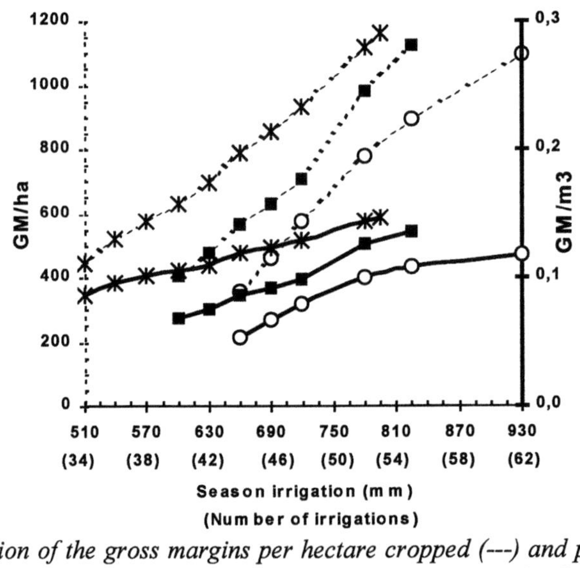

Fig. 9. *Variation of the gross margins per hectare cropped (---) and per m^3 of water applied (—) relative to the maize crop when irrigation depths decrease from full irrigation to heavy deficits for average (—✳—), high (-■-) and very high (—○—) climatic demand conditions*

As expected from the results relative to the variation of GM/m^3, which decrease despite the water productivity increases, the optimization model proposes to progressively reduce the cropped area adopting there the full irrigation SC strategy (Table VIII). When the water supply decreases, the percentage of area reduction depends on climatic conditions, being higher for the high and very high demand.

Results generally agree with those obtained by Zairi *et al.* (2002) for tomato. Both crops, tomato in Tunisia and maize in Portugal are cultivated during the dry season and therefore rely on irrigation only, *i.e.* rainfall and soil water reserve play a very minor role in satisfying crop requirements. Therefore, any reduction in irrigation water produces an heavy impact on the yield, which may negatively affect the farmer income when gross margins are not high enough to accommodate that yield decrease. Results do not agree with those presented by Zand-Parsa and Sepaskhah (2001) for Iran, where mild deficit irrigation shows to be feasible. Reasons for this difference in behaviour related to costs of production, including irrigation costs, and yield prices.

Table VIII. Economic responses of the maize crop to deficit irrigation: gross margins per unit volume of water applied (GM/m3) and percentage of surface allocated to each irrigation strategy

Strategy	Target depth	Average climatic demand		High climatic demand		Very high climatic demand	
		GM/ m^3	Surface allocated to each strategy	GM/ m^3	Surface allocated to each strategy	GM/ m^3	Surface allocated to each strategy
	(mm)	(USD)	(%)	(USD)	(%)	(USD)	(%)
SC	930	–	–	–	–	0.12	100/SC
SC	825	–	–	0.14	100/SC	–	–
SC	795	0.15	100/SC	–	–	–	–
B	825	–	–	–	–	0.11	89/SC
C	780	0.14	98/SC	0.13	95/SC	0.10	84/SC
D	720	0.13	91/SC	0.10	87/SC	0.08	77/SC
E	660	0.12	83/SC	0.09	80/SC	0.05	71/SC
F	600	0.11	75/SC	0.07	73/SC	–	–
G	540	0.10	68/SC	–	–	–	–

Summarizing, the economic evaluation indicates that for the conditions analysed deficit irrigation is not feasible for the maize crop. This conclusion is in agreement with farmers practice of reducing the maize cropped area or replacing maize by a drought resistant crop such as sunflower. However, results can not be interpreted as indicating that deficit irrigation of maize is absolutely unfeasible. In fact the model searches an optimal solution under a large number of constraints in farm resource utilization, including those for price of inputs and grain yield. Other near optimal solutions may be feasible but searching these solutions was not considered. Economic results for other maize producing areas may be different and lead to also different conclusions about feasibility of deficit irrigation.

5.2. SUNFLOWER

Results for irrigation scheduling simulations with sunflower are presented in Table IX. Although the sunflower crop demands less water than other summer crops such as maize, it requires considerable amounts of water to fully satisfy its crop water requirements.

Simulations show that it is possible to significantly reduce the seasonal irrigation from 435, 510 and 540 mm to 180 mm only, *i.e.*, producing RYL equal to 39, 43 and 44 %, respectively for average, high and very high climatic demand conditions. In other words, reducing the irrigation water by more than 60 %, the yield losses are near 40 %. The simulated ASW for these three cases is presented in Figure 10. It shows that two irrigations at planting are required for the three cases. Irrigation start later for average demand (Figure 10 a), which makes that the stress at the last crop phase is less than for high and very high demand. Nevertheless, the crop is heavily stressed during the last growth stage.

Table IX. Selected results of the response of the sunflower crop to deficit irrigation strategies in Vigia using center-pivot systems (I_n = 15 mm)

Climatic demand	Strategy	Season irrigation (mm)	Number of irrigation events	Season ET (mm)	Relative ET (%)	RYL (%)
Average	SC	435	29	539.1	99.9	0.0
	E	420	28	529.0	98.0	6.6
	G	330	22	452.7	83.9	24.0
	I	240	16	375.0	68.6	33.4
	J	180	12	311.5	57.7	39.3
	K	150	10	281.7	52.2	40.5
High	SC	510	34	594.8	99.8	0.0
	E	420	28	526.7	88.4	20.8
	G	330	22	439.9	73.8	31.4
	I	240	16	351.5	59.0	38.8
	J	180	12	291.8	49.0	43.1
	K	150	10	262.0	44.0	44.9
Very high	SC	540	36	616.0	99.8	0.0
	B	510	34	594.7	97.2	8.3
	E	420	28	516.5	84.4	26.2
	G	330	22	428.2	70.0	34.7
	I	240	16	339.3	55.5	40.9
	J	180	12	279.4	45.7	43.7

For the sunflower crop, the economic evaluation of alternative irrigation schedules produced the results shown in Table X and in Figure 11 considering actual subsidies.

Table X. Economic responses of the sunflower crop to deficit irrigation: gross margins per unit volume of water applied (GM/m3) and percentage of surface allocated to each irrigation strategy

Strategy	Target	Average climatic demand		High climatic demand		Very high climatic demand	
	depth (mm)	GM/ m^3 (USD)	Surface allocated to each strategy (%)	GM/ m^3 (USD)	Surface allocated to each strategy (%)	GM/ m^3 (USD)	Surface allocated to each strategy (%)
SC	540	–	–	–	–	0.12	100/SC
SC	510	–	–	0.13	100/SC	–	–
SC	435	0.16	100/SC	–	–	–	–
B	510	–	–	–	–	0.12	92/SC + 8/J
C	480	–	–	0.12	92/SC + 8/K	0.11	83/SC + 17/J
E	420	0.15	95/SC + 5/K	0.13	75/SC + 25/K	0.12	67/SC + 33/J
G	330	0.17	63/SC + 37/K	0.15	50/SC + 50/K	0.14	42/SC + 58/J
I	240	0.23	32/SC + 68/K	0.21	25/SC + 75/K	0.20	17/SC + 83/J
J	180	0.29	11/SC + 89/K	0.28	8/SC + 92/K	0.28	100/J
K	150	0.36	100/K	0.33	100/K	–	–

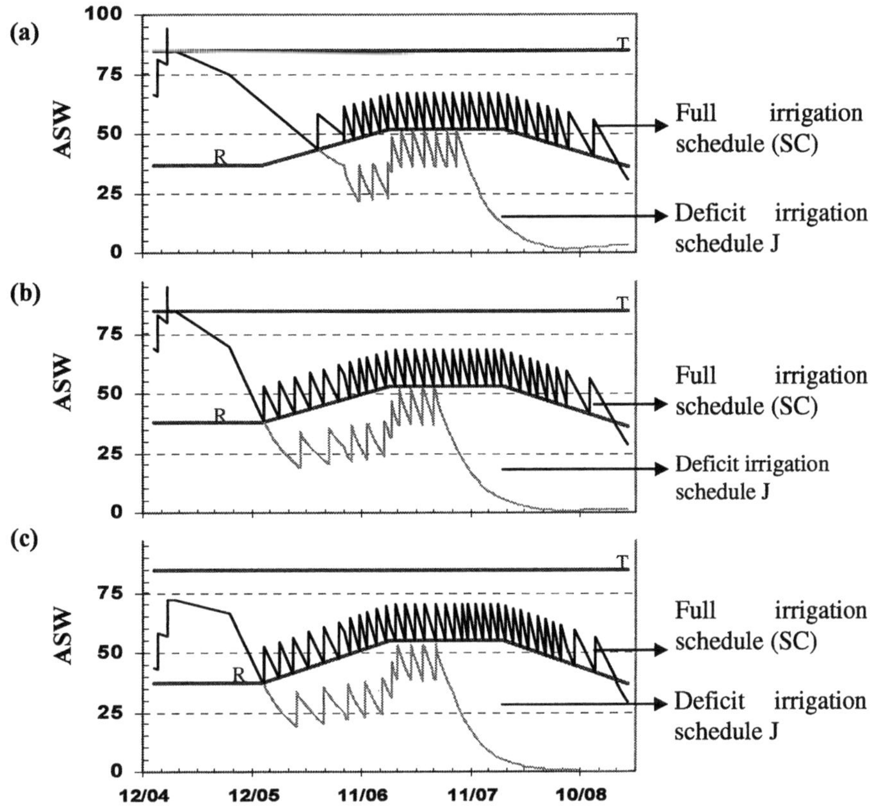

Fig. 10. Simulated available soil water (ASW) for sunflower adopting irrigation schedules for no stress (SC) and for a target depth of 180 mm under (a) average, (b) high, and (c) very high climatic demand

The GM/m^3 for sunflower crop grows up as water deficits become more severe (Figure 11). This trend is similar to that for wheat but values are smaller than those for this cereal crop. They range from 0.16 to 0.29 USD/m^3 for average demand and from 0.12 to 0.28 USD/m^3 for very high climatic demand. However, this positive trend is due to the crop subsidies that add to the value of the yield.

The GM/ha decrease from full irrigation to high irrigation deficits (Figure 11). However, they are much lower than those for maize when full water requirements are satisfied. This fact justifies the farmers behaviour, which do not crop sunflower but maize out of drought years.

As for the wheat crop, the mathematical model selects for each applied depth a combination of the full irrigation schedule SC with the most severe deficit schedule, the respective percentages depending upon the water available (Table X). This indicates that the application of the target depths relative to each strategy is feasible but the deficit schedules simulated are sub-optimal.

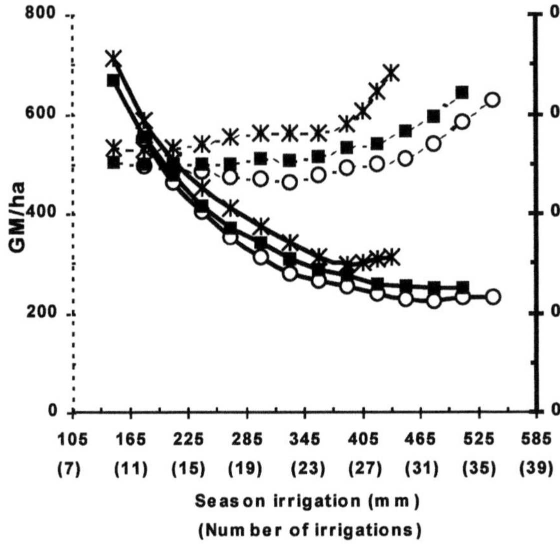

Fig. *11. Variation of the gross margins per hectare cropped (---) and per* m^3 *of water applied (—) relative to the sunflower crop when irrigation depths decrease from full irrigation to heavy deficits for average (—✱—), high (-■-) and very high (—○—) climatic demand conditions*

6. Conclusions

This study shows the usefulness of using simulation models to generate alternative irrigation schedules in relation to the available seasonal irrigation water and to evaluate the feasibility of their adoption using an economical model.

Results show that responses of crops to deficit irrigation varied with the climatic demand (average, high and very high) and from crop to crop. For the wheat crop, traditionally rainfed, it is economically feasible to adopt deficit irrigation, including when severe restrictions to water availability are considered under severe drought demand conditions. The sunflower crop despite being a summer crop, behaves similarly to wheat. Deficit irrigation is feasible in every case and allows for very large reductions in irrigation demand.

For the maize crop, large reductions in water available for irrigation lead to important yield losses, so making it not feasible to achieve water saving by adopting deficit irrigation. Therefore, when facing water restrictions as under drought the best economic solution for the maize crop is to fully irrigate a reduced cropped area.

The results of this study constitute a first step in a more extensive analysis to better understanding the crops economic patterns and to evaluate the main factors likely to affect farmers decisions. An in-depth sensitivity analysis of the impacts of water and crop prices is required, as well as relative to crop yields, irrigation systems and irrigation performances.

Acknowledgements

This research is funded by the research contract ERBIC-18CT-970169 with the European Union, the Contract Interreg IIC 99.09.5942.1 with the Portuguese Ministry of Agriculture, and the Portuguese-Tunisian joint research program ICCTI/SERST (423).

References

Allen, R.G., Pereira, L.S., Raes, D. and Smith, M. (1998) *Crop Evapotranspiration. Guidelines for Computing Crop Water Requirements.* FAO Irrig. Drain. Pap. 56, FAO, Rome, 300p.

Alves, I., Pereira, L.S. (1998) Relações água-produção. Caso da rega em condições de carência hídrica. *Comunicações IICT, Série Ciências Agrárias* nº20: 37-41.

Alves, I.L., Fontes, J.C. and Pereira, L.S. (1991) Water-yield relations for corn. *Planning, Operation and Management of Irrigation Systems for Water and Energy Conservation* (Proc. ICID Techn. Session). Chinese Nat. Com. ICID, Beijing, vol. I-A, pp. 154-161.

Batini, G., Rossi, G., Benedini, M. and Monacelli, G. (Eds.) (1999) *Territorial Planning and Coping with Effects of Droughts* (Proc. Int. Workshop, Taormina, Italy) Presidenza del Consiglio dei Ministri, Dip. Servizi Tecnici Nazionali, Roma. (CD-ROM).

Doorenbos, J. and Kassam, A.H. (1979) *Yield Response to Water.* FAO Irrig. Drain. Pap. 33, FAO, Rome.

El Amami, H., Zairi, A., Pereira, L. S., Machado, T., Slatni, A. and Rodrigues, P.N. (2001) Deficit irrigation of cereals and horticultural crops: economic analysis. *Agricultural Engineering International,* Vol. III, Manuscript LW 00 007b (http://agen.tamu.edu/cigr/volume3.html).

English, M.J., Musick, J.T. and Murty, V.V.N. (1990) Deficit irrigation. In: G.J. Hoffman, T.A. Howell and K.H. Solomon (Eds.) *Management of Farm Irrigation Systems.* ASAE, St. Joseph, MI, pp: 631 - 663.

GPPAA (2000) *Agricultura do Alentejo: Principais indicadores.* Ministério da Agricultura, do Desenvolvimento Rural e das Pescas, Lisboa.

Keller, J. and Bliesner, R. (1990). *Sprinkle and Trickle Irrigation.* Van Nostrand Reinhold, New York.

Kirkham, M.B. and Kanemasu, E.T. (1983) Wheat. In: I.D. Teare and M.M. Peet (Eds.) *Crop-Water Relations.* John Wiley & Son, Inc., pp. 187 - 211.

Lamaddalena, N. and Pereira, L.S. (1998) Performance analysis of on-demand pressurised irrigation systems. In: L.S. Pereira and J.W. Gowing (Eds.) *Water and the Environment: Innovation Issues in Irrigation and Drainage* (1st Inter-Regional Conf. Environment-Water, Lisbon), E&FN Spon, London, pp. 247-255.

Liu, Y., Teixeira, J. L., Zhang, H.J. and Pereira, L.S. (1998) Model validation and crop coefficients for irrigation scheduling in the North China Plain. *Agric. Water Manag.* 36: 233-246.

Martin, D.L., Stegman, E.C, Fereres, E. (1990) Irrigation scheduling principles. In: G.J. Hoffman, T.A. Howell and K.H. Solomon (Eds.) *Management of Farm Irrigation Systems.* ASAE, St. Joseph, MI, pp. 155 - 203.

Oweis, T., Zhang, H. and Pala, M. (2000). Water use efficiency of rainfed and irrigated bread wheat in a Mediterranean environment. *Agron. J.* 92(2): 231-238.

Paulo, A.A., Pereira, L.S., and Matias, P.G. (2002) Analysis of local and regional droughts in southern Portugal using the theory of runs and the standardised precipitation index. (This issue).

Pereira, L.S. (1989) Mitigation of droughts. 2 - Irrigation. *ICID Bulletin* 38: 16 -34.

Pereira, L.S. (1999). Management of irrigation systems under drought. In: G. Battini, G. Rossi, M. Benedini and M. Monacelli (Eds.) *Territorial Planning and Coping with Effects of Droughts* (Proc. Int.

workshop,Taormina, Italy) Presidenza del Consiglio dei Ministri, Dip. Servizi Tecnici Nazionali, Roma. (CD-ROM).

Pereira, L.S. (Coordinator) (2002 a) *Diagnóstico dos Sistemas de Rega em Pressão.* Final Report of a PEDIZA Project, CEER and DER, ISA, Lisbon.

Pereira, L.S. (2002 b) Irrigation demand management to cope with drought and water scarcity. (This issue).

Pereira, L.S., Teodoro, P.R., Rodrigues, P.N. and Teixeira, J.L. (2002) Irrigation scheduling simulation: the model ISAREG. (This issue).

Rodrigues, P.N., Pereira, L.S. and Machado, T. (2000) KCISA, a program to compute time averaged crop coefficients. Application to field grown vegetable crops. In: M.I. Ferreira and H.G. Jones (Eds.) *Proceedings of the 3rd International Symposium on Irrigation of Horticultural Crops.* ISHS, Acta Horticulturae Nr. 537, Leuven, vol. I, pp. 535-542.

Rossi, G. (2002) An integrated approach to drought mitigation in Mediterranean regions. (This issue).

Stewart, J.L., Hagen, R.M. and Pruit, W.O. (1976) *Water production functions and predicted irrigation programs for principal crops as required for water resources planning and increased water use efficiency.* University of California, Dep. Land, Air and Water Resources, Davis, and USDI/Bureau of Reclamation, Denver, CO.

Teixeira, J.L. and Pereira L.S. (1992) ISAREG - An irrigation scheduling model. *ICID Bulletin.* 41(2): 29-48.

Vogt, J.V. and Somma, F. (Eds.) (2000) *Drought and Drought Mitigation in Europe.* Kluwer Acad. Publ., Derchrecht.

Waldren, R.P. (1983) Corn. In: I.D. Teare and M.M. Peet (Eds.) *Crop-Water Relations.* John Wiley & Son, Inc., pp. 187 - 211.

Zairi, A., El Amami, H., Slatni, A., Pereira, L.S., Rodrigues, P.N. and Machado, T. (2002) Coping with drought: deficit irrigation strategies for cereals and field horticultural crops in Central Tunisia. (This issue).

Zand-Pasra, Sh. and Sepaskhah, A.R. (2001) Optimal applied water and nitrogen for corn. *Agric. Water Manag.* 52(1): 73-85.

Zhang, H., Oweis, T.Y., Garabet, S. and Pala, M. (1998). Water-use efficiency and transpiration efficiency of wheat under rain-fed conditions and supplemental irrigation in a Mediterranean-type environment. *Plant and Soil* 201: 295-305.

IRRIGATION SCHEDULING SIMULATION FOR CITRUS IN SICILY TO COPE WITH WATER SCARCITY

I. ALBA [(1)], P.N. RODRIGUES [(2)] and L.S. PEREIRA[(2)]
[(1)] *Department of Civil and Environmental Engineering, University of Catania, Italy*
[(2)] *Agricultural Engineering Research Center, Institute of Agronomy, Technical University of Lisbon, Portugal*

Abstract: The successful cultivation of citrus under the semi-arid conditions of the Mediterranean area depends upon the adequate supply of good quality water for summer irrigation. Under limited water availability, the adoption of appropriate irrigation scheduling practices could reduce yield losses and avoid death of trees and loss of capital investments. With this purpose, the irrigation scheduling simulation model ISAREG has been validated with data collected in citrus experimental plots located in the Catania plain irrigation district. Then, different deficit irrigation strategies have been simulated aiming at selecting those that minimise the water demand with smaller impacts on yields. Simulations were performed for different levels of climatic demand: average, high and very high. The impacts due to the existing farm reservoirs have been considered. Results show the usefulness of simulation models to identify and evaluate strategies for deficit irrigation, and made evident the importance of farm reservoirs to allow farmers to adopt the more appropriate irrigation schedules under water scarcity.

Key words: simulation modelling, drought, water scarcity, citrus, deficit irrigation, soil water balance.

1. Introduction

The impacts of long and severe droughts on agriculture, mainly the irrigated agriculture, have been very heavy in the last decades in several Mediterranean regions. Under drought and water scarcity conditions, the principal objective of irrigation management is to optimise a limited water supply. Criteria for timing and depth of applications can be obtained by using several approaches including soil water measurements, crop stress indicators and soil water balance models (Pereira, 1989; Tarjuelo and de Juan, 1999).

Under limited water resources, due to the variability of rainfall, both in frequency and quantity, irrigation scheduling becomes challenging for irrigation managers. The adoption of appropriate irrigation scheduling practices should lead to increased yields and higher profits for farmers, appreciable water savings, controlled environmental impacts of irrigation, and improved conditions for the sustainability of irrigated agriculture (as reviewed in several papers in Smith *et al.,* 1996).

The selection of irrigation strategies that minimise the water demand and produce less impacts on yields need to be performed taking into account different levels of crop demand conditions: average, high and very high, the latter corresponding to the

G. Rossi et al. (eds.), Tools for Drought Mitigation in Mediterranean Regions, 223-242.

occurrence of severe drought. However, while the selection of such strategies is relatively common for annual crops (El Amami *et al.*, 2001), for fruit trees it is more difficult because a drought can reduce yields for several years, even after it ended. Actually, the yield loss due to drought is not the worst damage when compared with death of trees and consequent loss of capital investments. Thus when severe droughts occur, it is essential to select a strategy in which yields could be sustained for future and tree survival is not compromised.

The success of citrus production in the Mediterranean region depends upon an adequate supply of good quality water for irrigation during the rainless summers. Several authors (Levy and Syverstsen, 1981; Sardo and Germaná, 1990, 1994) have demonstrated that citrus trees growing in diverse environmental conditions are capable of modifying their physiological responses to water. However, other authors have shown that stressed trees did not show any osmotic adjustment and, therefore, yields decrease when evapotranspiration is not fully satisfied by irrigation (Castel and Buj, 1990; Gonzalez-Altozano and Castel, 2000). Nevertheless, mild regulated deficit irrigation did not significantly influence the number of flowers, the post bloom and June drop, or the final number of fruits per tree. Yield decreases are often due to smaller fruit weight (Castel and Buj, 1990).

It is generally accepted that proper irrigation is of major importance during flowering and fruit setting. Although seasonal conditions may cause citrus to bloom at various times, the main blooming period of citrus trees in Mediterranean climate is the Spring. A water deficit during this period leads to abscission of flowers and young fruits, so to subsequent yield losses. Proper water management at this time may indeed reduce flowers to fall. A second critical period is the time of rapid fruit growth, which is variable according to the variety but generally occurs from end June to end July, when a water deficit would induce smaller fruit size. Water stress has also effects on yield after fruit formation and at the end of season (Germanà *et al.*, 1985a; Germaná and Sardo, 1986; Gonzalez-Altozano and Castel, 2000). Throughout the season, the irrigation regimes influence fruit quality as well as vegetative growth and root development. Controlled water deficits may favour juice quality (Cruse *et al.*, 1982). However, the ability of citrus trees to resist to water stress is variable according to the species and cultivars.

On-farm irrigation performances play a fundamental role in making efficient the use of water in irrigated agriculture. To mitigate the impacts produced by water scarcity due to droughts, the adoption of appropriate irrigation scheduling practices could lead to increased yields and/or water saving. However, quite often, the farmer is constrained by the district water delivery schedules, so he is not free to apply the most appropriate irrigation scheduling.

The technological evolution of irrigation systems and the advances in know-how about the soil-water-plant-atmosphere relationships during the last thirty years have changed concepts on irrigation management. Results of studies on citrus irrigation in experimental plots irrigated by under-tree sprinkler irrigation (Alba and Torrisi, 1976) and by microsprinkler systems (Alba, 1979; Germanà, 1992) have shown that irrigation uniformity is not a major factor because citrus have large root systems. Since soil water uptake by the roots is generally concentrated in the top 60-90 cm of the soil, a regular wetting of deeper soil layers may promote considerable deep percolation. Not wetting the entire root zone may be advantageous for crop production (Shmueli *et al.*, 1973). However, to adopt irrigation systems such as microsprinklers that produce a large percentage of root area wetted is essential to both growth and yield of citrus trees (Zekri and Parsons, 1988; 1989). In addition, the frequency of irrigation treatments may influence both water savings and plant physiological responses. A proper irrigation scheduling help to manage citrus irrigation with less water (Goell, 1988; Goell and Cohen, 1989).

The objective of the present study is to establish practical rules for a suitable irrigation management of citrus under limited water supply. The paper includes the validation of the irrigation scheduling model ISAREG, and the analysis of different deficit irrigation strategies for citrus in terms of water saving and consequent yield reduction.

2. The Catania-Plain irrigation district

The Catania Plain Irrigation District is located in the Eastern part of Sicily, near the city of Catania (Figure 1).

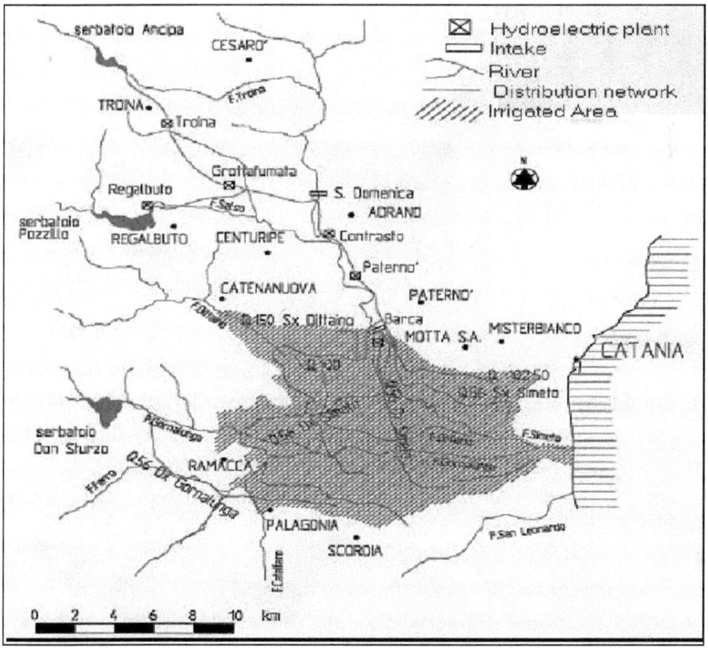

Fig. 1. The Catania plain district

It is one of the most important irrigation districts in Southern Italy. The total irrigable area is about 47000 ha, but the irrigated area is 18000 ha only. Citrus orchards cover more than 90 % of the irrigated area, the other cultivated crops being fodder crops, artichokes, watermelons and vegetables.

The climate is typically Mediterranean with mild winters, and rainless and hot summers. The rainy period is usually from October to March. For the period investigated in this study (1959-1996), the average annual precipitation (Figure 2) is 473 mm, with a maximum of 863 mm (1969) and a minimum of 195 mm (1981). In the same period, the monthly minimum temperature range from 1.4 °C in January to 25.6 °C in July, while the monthly maximum temperature varies from 9.1° C to 35.8 °C for the same months.

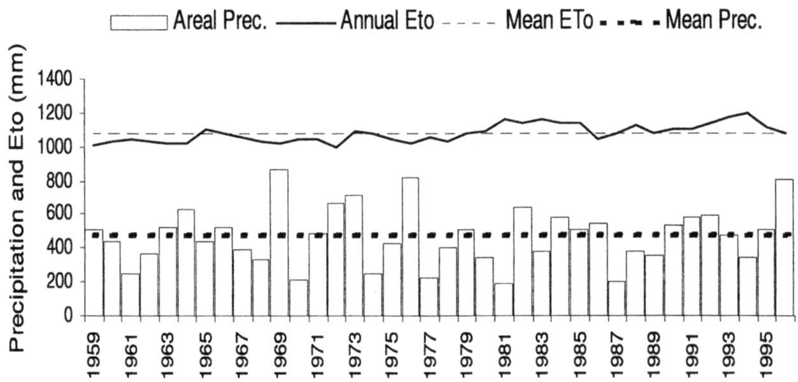

Fig. 2. Annual areal precipitation and reference evapotranspiration in Catania plain district (1959-1996)

The soils are of loamy type, mainly clay-loam, loam and sandy-loams, and have a medium to good level of fertility. The total available water (TAW) generally ranges from 150 mm/m to 210 mm/m.

A land reclamation institution, the "Consorzio di Catania", under the control of the authorities of the Sicily Region, is responsible for the operation, management and maintenance of the irrigation district. The distribution network is mainly supplied by surface water from the Salso-Simeto system (Figure 1), including two reservoirs (Pozzillo and Ancipa), three river intakes and five hydroelectric plants (Barbagallo and Rossi, 1993). The Pozzillo reservoir has a storage capacity of about 123 million m^3, mainly for irrigation purposes. Open channels supply about 65 % of the irrigated area while a pressurised pipe network serves the remaining. The operation of the irrigation district is based on a rotational delivery schedule. Water is delivered to the farms with fixed irrigation intervals and volumes. Commonly, the flow rate is 12 l/s.

As far as the irrigation method is concerned, gravity methods are almost abandoned. Field vegetable crops, fodder crops, and cereals are generally irrigated by sprinkling. Microirrigation is used in citrus groves, more often two 180° sprayers installed on both sides of the trunk, using a T in polyethylene pipe (16 mm) and placed at 40-50 cm height (Alba, 1979). However, despite citrus orchards irrigation is performed by modern systems, the irrigation delivery remain unchanged.

The Catania Plain district is particularly drought-prone (Cancelliere and Rossi, 2002). It experiences frequent periods of water scarcity due to significant reduction of precipitation and runoff in the Simeto river basin. The regulatory mechanism used when the water supply is limited is to give priority to the irrigation of fruit trees. During the most severe drought (1987-90), the emergency measures taken by the public administration consisted of loans, grants and tax reduction to farmers. The response of farmers, frightened by the perspective of total yield losses and death of the grove itself, was rather effective: on the one hand, they increased the groundwater use, both drilling new wells and better exploring the existing ones; on the other hand, they constructed farm ponds to store runoff and canal water available during the winter season to overcome the problems due to the insufficient frequency of deliveries, thus providing for intermediate regulation storage when water is scarce (Barbagallo *et al.*, 1992).

3. Validation of the ISAREG model for citrus in the Catania Plain

A preliminary analysis and application of several irrigation scheduling models - CROPWAT (Smith, 1992), ISAREG (Teixeira and Pereira, 1992; Liu *et al.*, 1998) and SOWABAMO (Mannocchi and Mecarelli, 1995) – was performed during 1998. The results pointed out that ISAREG is the more suitable model since it is more able to fit the observations of the real system (Privitera, 1998). Afterwards, the validation of the model ISAREG was carried out in a representative citrus farm of the Catania irrigation district (Sole Greco, 1999).

The experimental farm is located at Fontanazza, Catania province, 13 km Southeast from the city (latitude 37° 30', longitude 15°. 06', altitude 65 m a.s.l.). The citrus crop is a 10-year old plantation, orange Cv. Tarocco Nucellare, with an average height of 2.5 m. The trees are planted in rows spaced by 7 m, with 6 m spacing between trees in each row (Figure 3).

The citrus plantation is equipped with a micro-irrigation system constituted of two micro-sprinklers by tree, which flow rate is 106 l/h at 1 bar. The average wetted area by tree is about 20 m^2, thus the wetted area fraction is 0.5. There is no active under tree ground cover during the irrigation season because the soil is tilled for weeds control and water conservation at the end of the winter, by March. On the contrary, weeds develop from October to March, so a full ground cover protects the soil during the rainy season.

Fig. 3. Citrus experimental microirrigated plots.

Two schedules have been evaluated in the four selected plots where the farmer decided application depths:

a) the Irrigation District schedule, as it is currently practiced in the area, with 21-day interval between irrigations (plots A1 and B1)
b) an improved schedule, as practiced by the farmer, using canal water and groundwater (plots A2 and B2).

Climatic data were available from an automated meteorological station (AF88, Figure 4) located in the same farm, belonging to the Sicilian Regional Department of Agriculture. Daily data on temperature, relative humidity, radiation and wind speed were used for computing the reference evapotranspiration (ET_o) by the FAO-Penman-Monteith method (Allen *et al.*, 1998).

Fig. 4. Automated weather station AF88

The soil water content was measured by the gravimetric method. Two soil samples per plot at 0-30 cm and 30-60 cm soil layers were used. Observations were performed before and after irrigations, and with variable frequency between irrigation events. As far as the crop coefficients (K_c) for citrus are concerned, four different sets of values were tested in simulations carried out with the ISAREG model. The best results were obtained using the K_c computed from an experimental investigation carried out from 1975 to 1985 on irrigation of citrus, variety Tarocco, in the Catania Plain (Germanà *et al.*, 1985b; Sardo and Germanà, 1990).

The ISAREG model was validated by comparing the simulated and the observed soil water contents considering 60 cm soil depth (Figure 5). Results from comparing the simulated and observed soil water content (Figure 6) show a regression coefficient b =1.015, very close to 1, but a relatively low determination coefficient, $R^2 = 0.49$. The large dispersion around the regression line is due to the less appropriate field methodology because soil sampling was performed until 60 cm depth only but the active root zone is deeper. However, results show that the ISAREG model is able to predict soil water in irrigated citrus orchards.

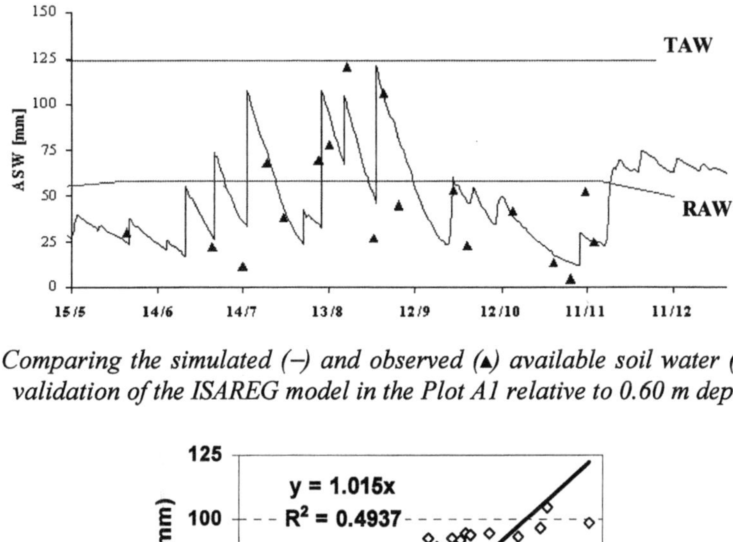

Fig. 5. Comparing the simulated (–) and observed (▲) available soil water (ASW) for validation of the ISAREG model in the Plot A1 relative to 0.60 m depth

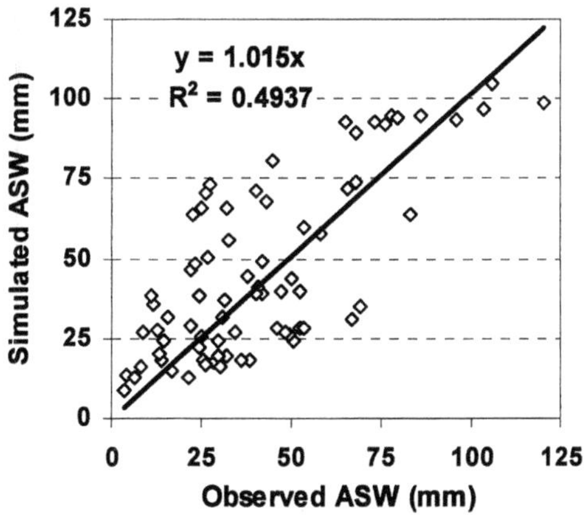

Fig. 6. Observed vs. simulated available soil water (ASW) for all experimental plots

4. Irrigation scheduling strategies

4.1. INPUT DATA

Four daily rainfall data series (1959-1996) observed at Catenanuova, Motta S. Anastasia, Ramacca and Lentini were used. The areal precipitation was calculated using the Thiessen method. The weights used are the following: Catenanuova (10.73 %), Motta S. Anastasia (41.63 %), Ramacca (31.27 %) and Lentini (16.30 %).

The reference evapotranspiration (ET_o) was computed using the daily temperature data of Catania station for the same time period. The Hargreaves method has been adopted. Observations performed in wide climatic conditions have shown that the Hargreaves method do not largely deviate from the FAO-PM method (Allen *et al.*, 1998). The FAO-Penman-Monteith method was used with monthly data only because several periods of daily data other than temperature were missing.

Crop data are shown in Table I. The crop coefficients (K_c) utilised have been derived from several experimental studies carried out in the Catania Plain with three orange cultivars, Cv. Tarocco, Cv. Moro and Cv. Navelina, over the irrigation season May-October (Sardo and Germanà, 1994; 1990) as referred before. For January-April and November-December, the K_c values obtained by Castel *et al.* (1987) are used.

Table I. Crop data used for water balance simulations

STAGES	A	B	C	D	E	F
First day of development stages	01/01	01/03	15/04	15/05	15/11	31/12
Soil water depletion, p	0.55	0.50	0.45	0.45	0.55	0.55
Root depth (m)	0.90	0.90	0.90	0.90	0.90	0.90

A – Starting date; B - Beginning of vegetative growth; C - Beginning of flowering; D - Beginning of yield formation; E - Beginning of ripening; F - End date

The yield response factor (K_y) was utilised to estimate the yield reductions due to water stress. Doorenbos and Kassam (1979) suggest variable values, ranging from 0.72, for mild evapotranspiration deficits, up to 1.20 for high water deficits. The K_y values used are listed in Table II for different ranges of relative evapotranspiration deficit (RED), that is defined by $(1 - ET_d / ET_c)$, where ET_d and ET_c are the seasonal crop evapotranspiration (mm) under deficit and full satisfaction, respectively. The yield reductions were analysed by the relative yield loss (RYL), which is defined by $(1 - Y_d / Y_c)$, where Y_d and Y_c are the yields achievable when crop evapotranspiration is ET_d and ET_c respectively.

Table II. Factors K_y for citrus as a function of RED

RED (%)	K_y
RED ≤ 15	0.72
15 < RED ≤ 25	0.83
25 < RED ≤ 35	0.90
35 < RED ≤ 45	0.98
45 < RED ≤ 50	1.20

Soil data have been obtained from a soil map and soil survey of the Catania Plain, the latter carried out by the SOAT, Regional Agricultural Department. Adopting the Thiessen method, the weighted average TAW used in the simulations with the ISAREG model is 202 mm/m.

4.2. IRRIGATION REQUIREMENTS

Using the above mentioned input data, the annual net irrigation requirements (NIR) for citrus have been computed with the ISAREG model. Results are presented in Figure 7. The average annual NIR is 383 mm, and the minimal and maximum values are respectively 196 and 614 mm. These results make evident a very large inter-annual variation in relation with rainfall variability.

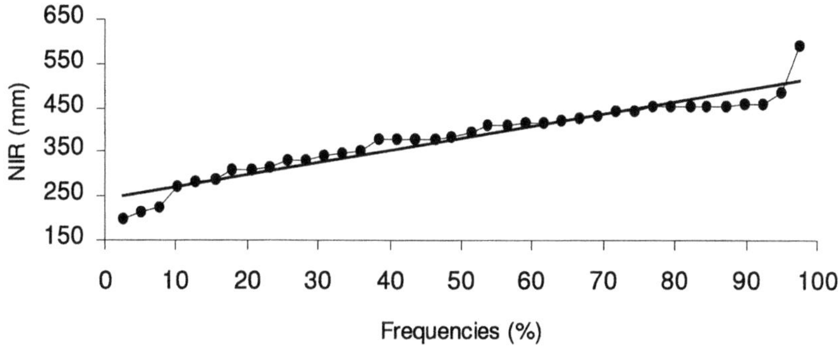

Fig. 7. Frequencies for NIR of citrus by using daily areal climatic data (DT)

The computed 38 years (1959-1996) data set for NIR was analysed using an empirical frequency analysis. Consequently, the years corresponding to the average (f ≈ 50 %), high (f ≈ 80 %) and very high (f ≈ 95 %) climatic demand were selected to perform the irrigation scheduling simulation. Conditions for f ≈ 80 % correspond to moderate droughts while those for f ≈ 95 % concern severe droughts. The main climatic data and irrigation requirements referring to those years are reported in Table III.

Table III. Main data characterising the climatic demand conditions

Climatic demand and probability for not being exceeded	Total available soil water at planting (mm)	Season Rainfall (mm)	Season ET_c (mm)	Season NIR (mm)
Average (50 %)	164	548	713	395
High (80 %)	164	444	757	452
Very high (95 %)	164	204	711	484

4.3. IRRIGATION SCHEDULING STRATEGIES TO COPE WITH DROUGHT

The irrigation scheduling strategies simulated are based on the actual rules of water management applied in the irrigation district. The irrigation season usually starts between May and April and continues until the end of the summer season according to the climate conditions and the available water resources.

As pointed out earlier, many reservoirs (small ponds) have been constructed in the citrus farms for intermediate storage regulation. Therefore, simulations are performed for two scenarios:

- *Rigid schedules* for farms without reservoirs, where the farmers are constrained to schedule the irrigations according to the delivery schedules decided by the district managers. Alternative schedules analysed are:
  a) Irrigation intervals of 21-day with restrictions on the initiation of the irrigation season, 29/04 or 20/05. Only the net irrigation depth $I_n = 50$ mm is considered.
  b) As above but irrigation schedules reduced to 12-day frequency, adopting both irrigation depths $I_n = 25$ mm and $I_n = 50$ mm.
  c) As above but irrigation schedules reduced to 7-day frequency, using only the irrigation depth $I_n = 25$ mm.
- *Flexible schedules* for farms with reservoirs, where farmers can decide their own irrigation scheduling independently of the delivery schedules adopted. Therefore, the following deficit irrigation strategies using the application depths $I_n = 25$ mm and $I_n = 50$ mm are considered:
  a) No limitations in water volumes or irrigation timings, adopting variable irrigation frequency dictated by different irrigation thresholds varying from $\theta_{MAD} = \theta_p$ (no stress) to $\theta_{MAD} = 0.4\ \theta_p$ (heavy stress). An explanation on irrigation thresholds is given in a companion paper by Pereira *et al.* (2002).
  b) No irrigation after September, adopting variable frequency using the irrigation thresholds $\theta_{MAD} = \theta_p$ (no stress) to $\theta_{MAD} = 0.9\ \theta_p$ (light stress).
  c) As for b) but without irrigation after August.
  d) As for b) but without irrigation after July.

5. Analysis of deficit irrigation schedules

5.1. RIGID SCHEDULES

Rigid schedules with fixed time intervals between irrigations of 7, 12 and 21-day are applied in farms not having ponds or reservoirs. In Table IV are compared the relative ET and relative yield deficits (RYL) for different available water supplies under several scenarios for climatic demand: average, moderate drought (high demand) and severe drought (very high demand). The comparison refers to net application depth $I_n = 25$ mm with 7 and 12-day time intervals when irrigation starts by 20/05.

Simulation results show that, when water is non limited, the best results for the relative ET and RYL are obtained with 7-day irrigation intervals (Table IV). Adopting 12-day intervals, water use do not attain those maximal values unless the irrigation thresholds is maintained at θ_p or above (results not shown). Adopting the 12-day frequency when the available water is moderately restricted, *e.g.* 350 mm, causes higher water stress and RYL than for 7-day irrigation intervals. This is due to the fact that soil water is already below the optimal θ_p threshold when irrigation starts since the 12-day intervals do not allow to keep the soil water above that threshold. On the contrary, using the 7-day irrigation frequency, the soil water content remains above θ_p during the critical water stress periods. Therefore, the RYL are smaller and the crop ET are near the maximal ET.

Table IV. Selected results of citrus response to deficit irrigation adopting rigid schedules (7 and 12-day irrigation frequency) and net irrigation depths In = 25 mm when irrigation starts by 20/05

Season irrigation (mm)	7-day irrigation intervals Season ET (mm)	Relative ET (%)	RYL (%)	12-day irrigation intervals Season ET (mm)	Relative ET (%)	RYL (%)
Average demand						
400	713.1	100.0	0.0	–	–	–
350	702.9	98.6	1.0	658.5	92.3	5.5
300	664.7	93.2	4.9	656.6	92.1	5.7
250	620.2	87.0	9.4	625.2	87.7	8.9
200	582.5	81.7	15.1	581.8	81.6	15.2
Moderate drought						
450	755.9	99.8	0.1	–	–	–
400	740.5	97.8	1.6	–	–	–
350	706.3	93.3	4.9	636.3	84.0	13.2
300	664.5	87.7	8.8	635.4	83.9	13.3
250	618.6	81.7	15.1	609.7	80.5	16.1
200	571.0	75.4	20.3	568.8	75.1	20.5
Severe drought						
450	694.9	97.7	1.7	–	–	–
400	662.9	93.2	4.9	–	–	–
350	620.6	87.2	9.2	596.6	83.9	13.3
300	574.7	80.8	15.9	565.7	79.5	16.9
250	527.0	74.1	23.3	524.3	73.7	23.7
200	478.2	67.2	29.5	477.6	67.1	29.6

When water is limited to 250 mm or 200 mm only, then the irrigation period is reduced in proportion to the available water and differences between both schedules become negligible. A quite heavy water deficit occurs for both cases as shown in Figure 8 for severe drought conditions.

It can be therefore concluded that when small net irrigation depths I_n = 25 mm are adopted there is no advantage in enlarging the time intervals between irrigations. On the contrary, using the fixed 7-day schedule allows better crop responses when water is not limited or restrictions are smaller. Responses to both 7 and 12-day frequency are similar only when very high limitations in water deficits are imposed.

If the date to start irrigations is anticipated to 29/04, there is a slight improvement in relative ET and RYL for the 12-day frequency when limitations in water supplies are small (Table V).

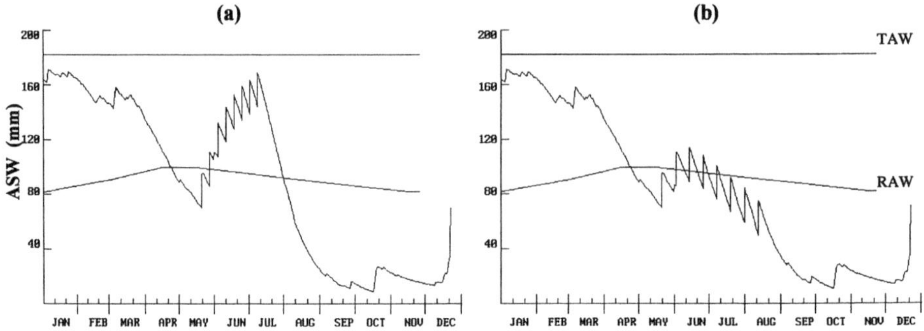

Fig. 8. Simulated available soil water (mm) for the severe drought scenario when irrigation depths are I_n = 25 mm, the irrigation starting date is 20/05 and the irrigation frequency are: (a) 7-day and (b) 12-day

Table V. Selected results of citrus response to deficit irrigation adopting rigid schedules (7 and 12-day irrigation frequency) and net irrigation depths In = 25 mm when full irrigation is anticipated to 29/04

Season irrigation (mm)	7-day irrigation intervals Season ET (mm)	Relative ET (%)	RYL (%)	12-day irrigation intervals Season ET (mm)	Relative ET (%)	RYL (%)
Average demand						
475	713.2	100.0	0.0	–	–	–
400	685.4	96.1	2.8	–	–	–
350	642.3	90.1	7.2	681.9	95.6	3.2
300	602.8	84.5	12.8	677.7	95.0	3.6
250	562.2	78.8	17.5	628.5	88.1	8.6
200	523.6	73.4	23.9	582.5	81.7	15.1
Moderate drought						
500	757.4	100.0	0.0	–	–	–
425	724.8	95.7	3.1	–	–	–
350	664.4	87.7	8.8	661.0	87.3	9.2
300	618.4	81.6	15.1	649.9	85.8	10.2
250	570.7	75.3	20.3	615.1	81.2	15.5
200	525.6	69.4	27.5	570.5	75.3	20.4
Severe drought						
525	704.1	99.0	0.7	–	–	–
450	654.4	92.0	5.8	–	–	–
375	587.3	82.5	14.4	621.4	87.3	9.1
300	519.5	73.0	24.3	570.7	80.2	16.3
250	478.3	67.1	29.6	526.2	74.0	23.4
200	439.5	61.5	37.7	478.3	67.2	29.5

However, when the available water is heavily limited, *e.g.* 250 or 200 mm, the relative ET and RYL for the 7-day frequency become much higher than those computed for the starting date by 20/05. For the severe drought scenario, RYL increases from 29.5 % to 38 %. On the contrary, differences are negligible for the 12-day frequency. This is explained by the fact that anticipating the date when irrigation starts and using the same total volume of water it results that the last irrigation is also anticipated. The long water deficit period that occurs in summer initiates earlier for the 7-day frequency (Figure 9 a) because ET is higher during the irrigation period when compared with the 12-day frequency (Figure 9 b).

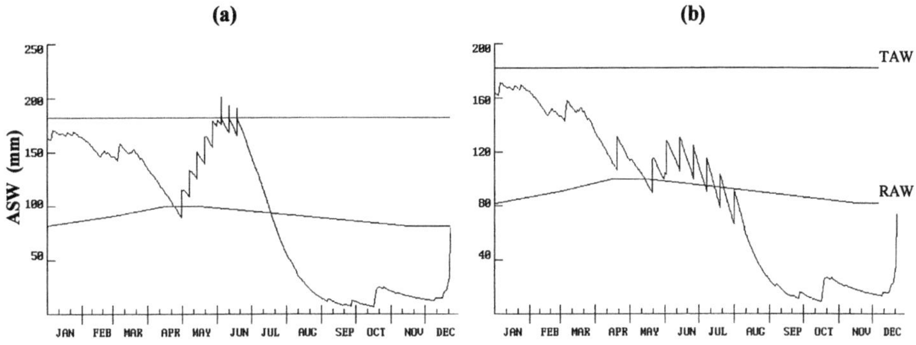

Fig. 9. Simulated available soil water (mm) for citrus for the severe drought scenario when irrigation depths are $I_n = 25$ mm, the irrigation starting date is 20/05 and the irrigation frequency are: (a) 7-day and (b) 12-day

Results from the simulations above indicate that there is no advantage in anticipating the date for the first irrigation when water supplies are drastically reduced because it would produce a longer and heavier water deficit during summer. Under these conditions, crop responses to irrigation deficits are better for the 12-day frequency when available water is heavily reduced.

Comparing the 12 and 21-day intervals between irrigations using a net depth $I_n = 50$ mm, when the irrigation starts by 20/05, it can be observed (Table VI) that when there are no limitations in water availability only the 12-day schedule is flexible enough to fully satisfy the crop demand. These results are similar to those analysed above to compare 7 and 12-day frequencies when $I_n = 25$ mm.

On the contrary, when high restrictions on available water are enforced, the relative ET is higher and the RYL are smaller for the 21-day frequency. The simulated ASW are shown in Figure 10 for moderate and severe drought conditions adopting both schedules when seasonal available water is limited to 200 mm. It can be observed for both drought scenarios that the large summer deficit starts earlier for the 12-day frequency, thus producing a larger water stress period with consequent heavy reduction in ET and yield. This results are similar when comparing dates 29/04 and 20/05 to start irrigation (results not shown), *i.e.* impacts of water stress are increased, mainly for the 12-day frequency, when the first irrigation is anticipated and the available water remains unchanged. This confirms the appropriateness of the present practice of starting the irrigation season by late May.

Table VI. Selected results of citrus response to deficit irrigation adopting rigid schedules and net irrigation depths In = *50 mm when irrigation starts by 20/05*

Season irrigation	12-day irrigation intervals			21-day irrigation intervals		
	Season ET	Relative ET	RYL	Season ET	Relative ET	RYL
(mm)	(mm)	(%)	(%)	(mm)	(%)	(%)
Average demand						
500	713.1	100.0	0.0	–	–	–
400	687.4	96.4	2.6	693.5	97.2	2.0
300	613.4	86.0	10.1	671.0	94.1	4.3
200	543.6	76.2	19.6	582.5	81.7	15.1
Moderate drought						
500	755.7	99.8	0.2	–	–	–
400	717.5	94.7	3.8	681.6	90.0	7.2
300	632.7	83.5	13.6	656.4	86.7	9.6
200	548.6	72.4	24.8	570.7	75.3	20.3
Severe drought						
500	681.3	95.8	3.1	–	–	–
400	609.6	85.7	10.3	644.0	90.5	6.8
300	526.5	74.0	23.4	572.9	80.5	16.1
200	455.0	63.9	35.3	478.2	67.2	29.5

Figure 10 also shows some essential differences between the moderate and severe drought analysed herein: (1) the soil water deficit is higher when irrigation starts and, (2) the recovery of soil water occurs much later under severe drought. This explains differences in Tables IV, V and VI.

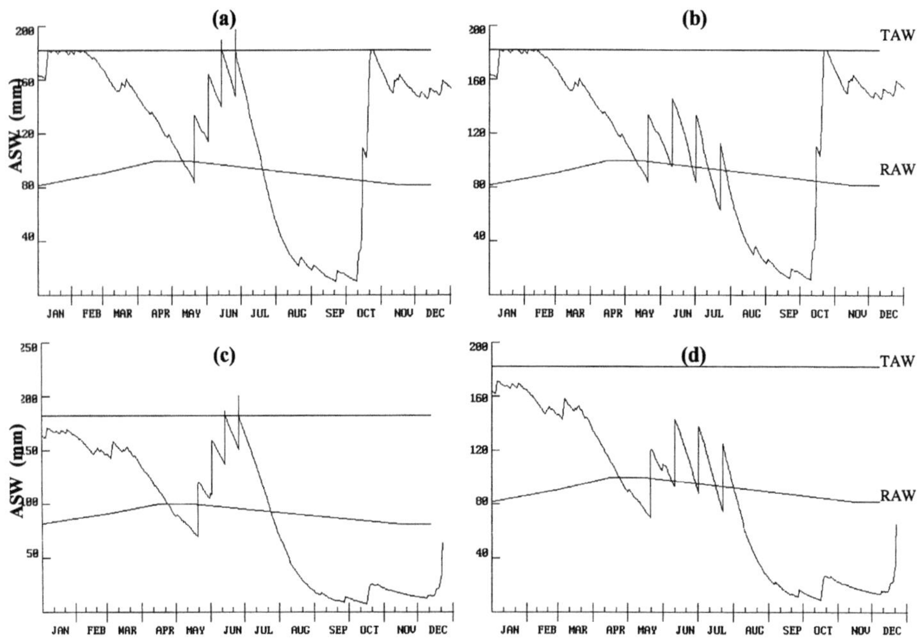

Fig. 10. Simulated available soil water (mm) when irrigation depths are I_n = 50 mm, the irrigation starting date is 20/05 under moderate drought with (a) 12-day and (b) 21-day irrigation frequency and under severe drought with (c) 12-day and (d) 21-day frequency

5.2. FLEXIBLE SCHEDULES

Farms having intermediate storage regulation reservoirs may adopt flexible schedules. Simulations were performed adopting different irrigations thresholds θ_{MAD} or diverse dates for ending the irrigation season.

When θ_{MAD} is progressively reduced, ET and yield consequently decrease. Simulation results relative to both drought scenarios (Table VII) show that crop responses are better when $I_n = 50$ mm than those for $I_n = 25$ mm when adopting lower θ_{MAD} thresholds. This is due to the fact that smaller irrigation depths do not allow refilling the soil water to the readily available water (RAW) corresponding to θ_p (Figure 11). However more water is used when application depths are higher. Thus, to achieve the same restriction in seasonal water different θ_{MAD} thresholds have to be selected, e.g. to apply 300 mm under moderate drought the thresholds 0.6 θ_p has to be selected for $I_n = 25$ mm and 0.5 θ_p when $I_n = 50$ mm. Similarly, to apply 250 mm under severe drought the thresholds 0.5 θ_p and 0.4 θ_p are required for the application depths 25 and 50 mm, respectively (Table VII).

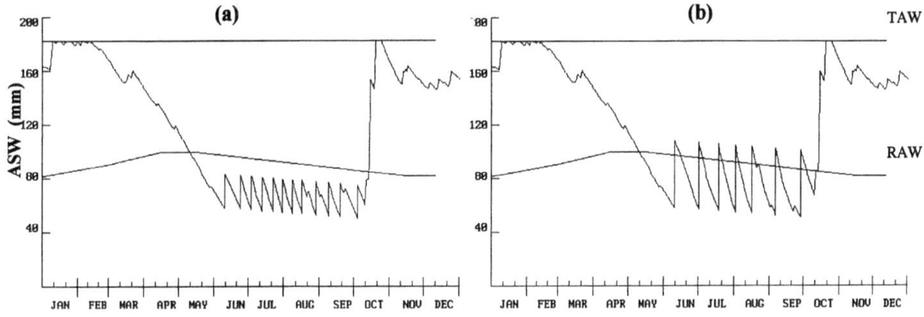

Fig. 11. Simulated available soil water (mm) for $\theta_{MAD} = 0.6\ \theta_p$ when using (a) $I_n = 25$ mm and (b) $I_n = 50$ mm under moderate drought

Simulation results do not show evident differences in RYL when comparing both application depths to achieve a given seasonal applied depth. Therefore preferring one or another is not evidenced. However, adopting $I_n = 25$ mm provides for more flexibility in irrigation management.

When scheduling irrigations taking as criteria the dates for ending the irrigation (Table VIII), seasonal irrigation volumes can be reduced similarly to those adopting progressively low thresholds θ_{MAD}. Results are similar using 25 or 50 mm irrigation depths. Impacts of water deficits increase when the last irrigation is progressively anticipated (Figure 12) since the summer deficit period then increases. However, a large RYL is predicted only when the very high demand corresponding to severe drought is considered.

Table VII. Selected results of the response of the citrus crop to deficit irrigation strategies adopting flexible schedules using net irrigation depth I_n = 25 and 50 mm and with θ_{MAD} varying from p to 0.4 θ_p

θ_{MAD}	Net irrigation depths I_n = 25 mm				Net irrigation depths I_n = 50 mm			
	Season irrigation (mm)	Season ET (mm)	Relative ET (%)	RYL (%)	Season irrigation (mm)	Season ET (mm)	Relative ET (%)	RYL (%)
Average demand								
θ_p	375	713.2	100.0	0.0	400	713.2	100.0	0.0
0.8 θ_p	325	679.4	95.3	3.4	350	637.2	97.8	1.6
0.7 θ_p	275	634.6	89.0	7.9	300	663.8	93.1	5.0
0.6 θ_p	225	594.3	83.3	13.8	–	–	–	–
0.5 θ_p	175	548.2	76.9	19.1	250	597.4	83.8	13.4
0.4 θ_p	150	515.2	72.2	25.0	200	547.5	76.8	19.2
Moderate drought								
θ_p	475	757.4	100.0	0.0	500	757.4	100.0	0.0
0.8 θ_p	400	718.7	94.9	3.7	450	737.3	97.3	1.9
0.7 θ_p	350	673.7	88.9	8.0	400	708.1	93.5	4.7
0.6 θ_p	300	628.0	82.9	14.1	350	671.9	88.7	8.1
0.5 θ_p	250	579.8	76.6	19.3	300	628.6	83.0	14.0
0.4 θ_p	200	537.1	70.9	26.2	250	580.0	76.6	19.3
Severe drought								
θ_p	500	711.5	100.0	0.0	500	711.5	100.0	0.0
0.8 θ_p	450	668.2	93.9	4.4	500	689.1	96.9	2.3
0.7 θ_p	400	618.5	86.9	9.4	450	660.8	92.9	5.1
0.6 θ_p	325	565.2	79.4	17.0	400	617.2	86.7	9.5
0.5 θ_p	275	516.6	72.6	24.7	350	568.0	79.8	16.6
0.4 θ_p	225	467.5	65.7	30.9	250	511.2	71.8	25.3

Table VIII. Selected results of the response of the citrus crop to deficit irrigation strategies adopting flexible schedules using net irrigation depth I_n = 25 and 50 mm and θ_{MAD} = 0.9 θ_p with different endings of irrigation season

End of irrigation season	Net irrigation depths I_n = 25 mm				Net irrigation depths I_n = 50 mm			
	Season irrigation (mm)	Season ET (mm)	Relative ET (%)	RYL (%)	Season irrigation (mm)	Season ET (mm)	Relative ET (%)	RYL (%)
Average demand								
Free	375	705.0	98.9	0.8	400	708.9	99.4	0.4
30/09	350	703.6	98.7	1.0	350	706.2	99.0	0.7
31/08	300	673.1	94.4	4.0	300	673.7	94.5	4.0
31/07	175	558.3	78.3	17.9	200	582.4	81.7	15.1
Moderate drought								
Free	450	747.7	98.7	0.9	450	752.3	99.3	0.5
30/09	425	746.3	98.5	1.1	450	752.3	99.3	0.5
31/08	350	704.8	93.1	5.0	350	705.7	93.2	4.9
31/07	225	594.8	78.5	17.7	200	618.6	81.7	15.1
Severe drought								
Free	475	700.3	98.4	1.1	500	706.2	99.3	0.5
30/09	400	662.5	93.1	5.0	400	663.5	93.3	4.9
31/08	325	598.0	84.0	13.2	350	621.0	87.3	9.2
31/07	200	478.3	67.2	29.5	200	478.3	67.2	29.5

Comparing the relative yield losses for rigid schedules (Table IV and VI) with those for flexible schedules (Tables VII and VIII), it may be concluded that the RYL are easily controlled when flexible schedules are adopted in case that seasonal volumes are not limited or restrictions are not very heavy. When the available water is reduced to 200 mm only, then RYL close to 30 % are predicted for both rigid and flexible schedules. However, because flexible schedules are feasible only in farms having intermediate storage, in case of heavy restrictions in delivery these farms may use other sources of water and not depend only on the irrigation system deliveries. It is enough to add additional 50 mm in storage to significantly decrease RYL (Table VII). This is the main advantage for adopting intermediate storage.

Comparing both flexible schedules simulated by using a θ_{MAD} threshold or limiting the last irrigation date, there is a trend that adopting $\theta_{MAD} = 0.9\,\theta_p$ and ceasing irrigation as early as the water available is restricted may be the best approach. However, to optimise flexible schedules would require information on the weather and the soil water variables in real time and use a simulation model to timely predict the irrigation dates.

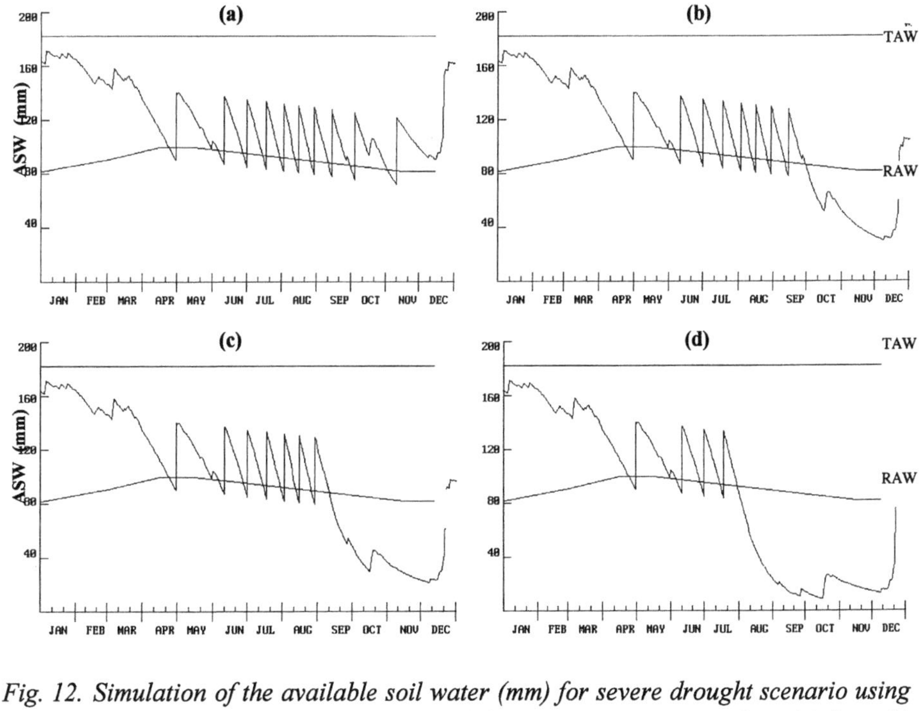

Fig. 12. Simulation of the available soil water (mm) for severe drought scenario using I_n = 50 mm with $\theta_{MAD} = 0.9\,\theta_p$ and the irrigation season ending (a) free, (b) by 30/09, (c) 31/08 and (d) 31/07

6. Conclusions

The present study shows that the application of deficit irrigation strategies for citrus should be feasible when a minimum season net depth of 200 mm is available. Further, experimental data would be welcome to verify the assumptions resulting from the simulations despite the ISAREG model has been validated previously to be used in this study. However, the validation was not performed for severe water restrictions.

Using the ISAREG model for simulating a large number of alternative irrigation schedules allowed comparing rigid and flexible schedules, several criteria for irrigation and impacts of irrigation depths.

Adopting flexible schedules shows to have the potential to reduce the water deficits relative to rigid schedules. However, differences are relatively small when large restrictions on the seasonal water availability are imposed because the summer deficits are then very large for all the scenarios evaluated.

Results show that when 25 mm application depths are used and rigid schedules are adopted, the 7-day irrigation frequency may lead to reduced impacts on yields relative to the 12-day frequency. However, when very heavy restrictions on the available water are enforced, the resulting RYL are similar for both irrigation frequencies. When large application depths are used, as $I_n = 50$ mm, the 21-day irrigation frequency looks to be best adapted to limit impacts on yields. Therefore, it is advisable to adopt a delivery scheduling with 7-day intervals with application depths $I_n = 25$ mm when limitations in water supply are not severe. On the contrary, under severe drought, large intervals between deliveries such as 12 or 21-day may be adopted. The application depths should be then 25 mm and 50 mm respectively.

Simulations did not show advantage in anticipating the date for starting the irrigation. In case of drought and when the available water is drastically limited, anticipating the start of the irrigation season would lead to more severe impacts on yields.

In case flexible irrigation schedules can be adopted in farms having intermediate storage reservoirs, the best results are achieved when the irrigation threshold is kept relatively high ($\theta_{MAD} = 0.9\ \theta_p$). Then, the date for the last irrigation has to be anticipated according to the water availability.

When comparing flexible and rigid schedules, simulations do not show an evident advantage for the first when very heavy restrictions on water availability are imposed. However, because farms adopting flexible schedules may store and use water from other additional sources, they have available large seasonal depths than farms fully depending on the irrigation system deliveries. These additional volumes increase the flexibility of irrigation and lead to lower impacts on yields.

Further investigations are required to test with field observations the assumptions above and to validate the yield response factor K_y for the citrus varieties being used in the area. Moreover, economic information relative to the yield deficits produced would be required to better select the most adequate practices to cope with water scarcity.

Acknowledgements

This research was funded by the research contract ERBIC-18CT-970169 with the European Union. Particular thanks are due to the personnel of the SOAT of Catania for their help in collecting climatic data; the personnel of the SOAT of Acireale for the physical and soil moisture content analysis; Dr. Francesco Marziani for providing information on the Catania plain district; and to Fabio Privitera and Giancarlo Sole Greco for their participation in experiments and data collection.

References

Alba, I.(1979) Experiences de microirrigation dans les agrumes en Sicile. *L'irrigazione.* 1(2): 21-27.

Alba, I. and Torrisi, S. (1976) Influenza di diversi regimi irrigui sull'accrescimento dei frutti di arancio *cv.* Tarocco. *L'Irrigazione.* 1(2): 25-35.

Allen, R.G., Pereira, L.S., Raes, D. and Smith, M. (1998) *Crop Evapotranspiration. Guidelines for Computing Crop Water Requirements.* FAO Irrig. Drain. Pap. 56, FAO, Rome, 300p.

Barbagallo, S. and Rossi, G. (1993) Effetti della siccità in agricoltura e misure di mitigazione. Il ruolo dell'ingegneria per l'agricoltura del 2000. *V° Convegno nazionale A.I.G.R.*., Jun. 1993, pp. 133-142.

Barbagallo, S., Rossi, G. and Tamburino, V. (1992) Emergency measures in water supply systems for alleviating the effects of the 1987-90 drought in Sicily. *ICID XVI European Regional Conference*, Budapest, pp. 7-23

Cancelliere, A. and Rossi, G. (2002) Droughts in Sicily and comparison of identified droughts in mediterranean regions.(This issue).

Castel, J.R. and Buj, A. (1990) Response of Salustiana oranges to high frequency deficit irrigation. *Irrig. Sci.* 11: 121-127.

Castel, J.R., Bautista, L., Ramos, C. and Cruz, G. (1987) Evapotranspiration and irrigation efficiency of mature orange orchards in Valencia (Spain). *Irrigation and Drainage Systems.* 3: 205-217.

Cruse, R.R., Wiegand, C.L. and Swanson, W.A. (1982) The effects of rainfall and irrigation management on citrus juice quality in Texas. *J. Amer. Soc. Hort. Sci.* 107(5): 767-770.

Doorenbos, J. and Kassam, A.H. (1979) *Yield Response to Water*, FAO Irrig, and Drain, Paper Nr. 33, FAO, Rome.

El Amami, H., Zairi, A., Pereira, L.S., Machado, T., Slatni, A. and Rodrigues, P.N. (2001) Deficit irrigation of cereals and horticultural crops: economic analysis. *Agricultural Engineering International* Vol. III, Manuscript LW 00 007b. (http://agen.tamu.edu/cigr/volume3.html)

Germanà, C. (1992) Increasing water use efficiency through irrigation management. *Proc. Int. Soc. Citriculture.* 2: 638-642.

Germanà, C. and Sardo V. (1986) Misura del grado di sofferenza idrica e della sua influenza su alcuni parametri fisiologici degli agrumi. In: *Agrumicoltura* (Proc. Cong. 29/4-3/5 1986), Cagliari.

Germanà, C., Sardo, V. and Cutore, L. (1985a) Scheduling irrigation on orange trees, In: *Third International Drip-Trickle Irrigation Congress* (Nov. 17-21, 1985), Fresno, California, USA.

Germanà, C., Sardo, V. and Cutore, L. (1985b) Esperienze sull'irrigazione dell'arancio, *L'irrigazione.* 4: 17-22,

Goell, A. (1988) Citrus irrigation in Israel. Making it with less water. *Proceedings of 6th International Citrus Congress* (Mar. 6-11, 1988), Tel Aviv.

Goell, A. and Cohen, A. (1989) Analysis of fruit growth in grapefruit as affected by irrigation intervals. *Sci. Horticulturae.* 39: 223-233.

Gonzalez-Altozano, P., and Castel, J.R. (2000) Effects of regulated deficit irrigation on "Clementina de Nules" citrus trees growth, yield and fruit quality. In: M.I. Ferreira and H.G. Jones (Eds.) *Proceedings of the 3rd International Symposium on Irrigation of Horticultural Crops*. ISHS, Acta Horticulturae Nr. 537, Leuven, vol. II, pp. 749-758.

Levy, Y. and Syverstsen, J.P. (1981) Water relations of Citrus in climates with different evaporative demands, *Proc. Int. Soc. Citriculture*. 2: 501-503.

Liu, Y., Teixeira, J.L., Zhang, H.J. and Pereira, L.S. (1998) Model validation and crop coefficients for irrigation scheduling in the North China Plain. *Agric. Water Manag.* 36: 233-246.

Mannocchi, F. and Mecarelli, P. (1995) A soil-water-crop model for large and small scale applications. In: L.S. Pereira, B.J. van den Broek, P. Kabat and R.G. Allen (Eds.) *Crop-Water-Simulation Models in Practice*. Wageningen Pers, Wageningen, pp. 105-120.

Pereira, L.S. (1989) Mitigation of droughts. 2. Irrigation. *ICID Bulletin*. 38: 16 -34.

Pereira, L.S., Teodoro, P.R., Rodrigues, P.N. and Teixeira, J.L. (2002) Irrigation scheduling simulation: the model ISAREG. (This issue).

Privitera, E. (1998) *Modelli di Programmazione dell'Intervento Irriguo in Condizioni di Siccità: Applicazione al Comprensorio della Piana di Catania*, tesi di Laurea, Università di Catania.

Sardo, V. and Germanà, C. (1990) Irrigation of orange trees. Crop coefficient and yield response. In: *XXIII mt. Hort. Congress* (27 Ago.- 1 Sep.), Firenze, pp. 290.

Sardo, V. and Germanà, C. (1994) Ulteriori esperienze sull'irrigazione dell'arancio. *Irrigazione e Drenaggio* 41(4): 8-16.

Shmueli, E., Bielorai, H., Heiler, J. and Manteli, A. (1973) Citrus water requirement experiments conducted in Israel during the 1960's. In: A. Hadas *et al.* (Eds.) *Physical Aspects of Soil Water and Salts in Ecosystems.* Ecological Studies 4. Springer Verlag, Berlin, pp, 339-350.

Smith, M. (1992) *CROPWAT, a Computer Program for Irrigation Planning and Management.* FAO Irrig. Drain. Pap. 46, FAO, Rome.

Smith, M., Pereira, L.S., Berengena, J., Itier, B., Goussard, J., Ragab, R., Tollefson, L. and Van Hoffwegen, P. (Eds.) (1996) *Irrigation Scheduling: From Theory to Practice*. Water Report 8, FAO, Rome, 384p.

Sole Greco, G. (1999) *Modelli di Simulazione del Bilancio Idrologico del Terreno Irrigato. Verifica Sperimentale per Colture Agrumicole nella Piana di Catania.* Tesi di Laurea, Università di Catania.

Tarjuelo, J.M. and de Juan, J.A. (1999) Crop water management. In: H.N. van Lier, L.S. Pereira and F.R. Steiner (Eds.) *CIGR Handbook of Agricultural Engineering, vol. I: Land and Water Engineering*, ASAE, St. Joseph, MI, pp. 380-429.

Teixeira, J.L. and Pereira, L.S. (1992) ISAREG, an irrigation scheduling model. *ICID Bulletin*, 41(2): 29-48.

Zekri, M. and Parsons, L.R. (1988) Grapefruit leaf and fruit growth in response to drip, microsprinkler, and overhead sprinkler irrigation. *J. Amer. Soc. Hort. Sci.* 113(6): 819-823.

Zekri, M. and Parsons, L.R. (1989) Water relations of grapefruit trees in response to drip, microsprinkler, and overhead sprinkler irrigation. *J. Amer. Soc. Hort. Sci.* 114(1): 25-29.

IRRIGATION MANAGEMENT DYNAMICS IN THE JORDAN VALLEY UNDER DROUGHT CONDITIONS

M. SHATANAWI[(1)], Y. AL-ZU'BI[(2)] and O. AL-JAYOUSSI[(3)]
[(1)]*Faculty of Agriculture, University of Jordan, Amman, Jordan*
[(2)]*General Corporation for the Environmental Protection, Amman, Jordan*
[(3)]*Applied Science University, Amman, Jordan*

Abstract: During the last twenty years, Jordan has witnessed many drought events, which put constraints on irrigated agriculture. These conditions urge the government to take necessity measures to alleviate the impact of drought. This paper aims at evaluating the effect of different irrigation dynamics on yield and water consumption through the application of deficit irrigation. By using meteorological records and soil and crop data for the Jordan Valley, ISAREG model as well as EVAPOT and KCISA programs were applied to estimate irrigation water requirements and reduction in yield. Water productivity under different levels of water deficits was determined. It was concluded that wheat and citrus are the crops that best perform under drought using economic criteria. ISAREG model proved to be a valuable and helpful tool to be used for irrigation management and estimation of yield reduction.

Key words: Irrigation management, drought, ISAREG model, water productivity, Jordan, multi-criteria analysis.

1. Introduction

Jordan is classified among few countries of the world with limited water resources where demands are far exceeding supplies. The total area of Jordan is 89.4 km2 where 92% of it is located in the arid to semi arid region. Most of the lands in this region are arable land if water resources can be made available. Water in Jordan is considered not only the main factor of production but also a very crucial factor of survival and social development.

Comparing the volume of natural renewable resources of 940 million cubic meters (MCM) to the current population estimate of 5.0 million people (DOS, 2000) would result in a per capita share of water of 188 m^3 per year. On the other hand, considering allocation of water for domestic and industrial purposes, the agricultural per capita share of water will be about 100 m^3 per year. This amount represents about 12% of the individual need for water to provide enough food and fibre.

At the present, the current annual water consumption is estimated at 935 MCM; 610 MCM of them are allocated for agriculture, 270 MCM are consumed for domestic purposes, and an estimated 65 MCM are allocated to industrial and other purposes. The amount of water consumed has been derived from different sources as: 520 MCM from ground water, 370 MCM of surface water and 65 MCM as treated wastewater. It is clear that water consumption is heavily dependent on groundwater where about 80

G. Rossi et al. (eds.), Tools for Drought Mitigation in Mediterranean Regions, 243-258.

MCM of it comes from the non-renewable resources and about 440 MCM comes from the already exploited renewable groundwater aquifers. The rate of abstraction from these aquifers is 160 MCM above their safe yield, which is estimated at 280 MCM.

Surface water resources in Jordan comes from different sources such as the Yarmouk River flow, the flow of the side wadies, release of the dams, and the share of Jordan from River Jordan. Out of the 600 MCM as potential surface water only about 370 MCM have been utilized due to different reasons mainly the lack of enough storage facilities and the difficulty in catching desert floods.

This amount varies from one year to another depending on the amount of precipitation. The last three years, for example, have witnessed severe drought conditions where surface water amounts have significantly dropped affecting irrigated agriculture. Therefore, water allocation for farmers was reduced from 4 mm/day on the average during normal wet years to 2.0 mm/day and 1.7 mm/year for the drought years of 1999 and 2000, respectively.

This paper aims at determining crop irrigation requirements and yield under different drought management dynamics. Irrigation management scenarios under different drought measures are evaluated in terms of yield reduction and water productivity. The model ISAREG coupled with EVAPOT and KCISA programs are used to determine such variables. Water productivity as a new measure of water use efficiency is determined as the ratio of gross value of the crop to different levels of water consumption.

2. Irrigation development in Jordan and in the Jordan Valley

2.1. GENERAL

In Jordan, water not land is the determinate factor for increasing the size and the productivity of the irrigated areas. Historically, up to 1940, agriculture production in Jordan depended on rainfed agriculture while scattered irrigated areas could be found around springs and along the beds of the side wadis. Traditional surface irrigation was the general practice and the farmers themselves used to manage their water in terms of distribution and scheduling. As early as 1940, Jordan started the irrigation development through the establishment of irrigation schemes along the valley beds of the natural streams and around major springs through the provision of constructing hydraulic structures such as weirs, concrete canals, diversion boxes and control structures. The major development in irrigation has started in the 1950s aiming at integrated rural development of the Jordan Valley utilizing Jordan share of the water resources of the River Jordan system. At the present, irrigation is practiced in two main areas: the Jordan Rift Valley (JRV) and the Upland.

In the late sixties and early seventies, the government began developing pilot projects in the desert and the upland of Jordan using groundwater. The expansion in irrigation began in eighties and early nineties by the private sectors through utilizing the groundwater. These activities concentrated on the major basins of Azraq, Amman-Zarka, upper Yarmouk, and the Dead Sea basins. The uncontrolled pumping from these aquifers has exceeded their safe yield to about 150%. Irrigation has also been extended to utilize the non-renewable resources of Dissi and Mudwarah area where about 70 MCM are being pumped annually from the aquifers and are used for agriculture. Recent studies have estimated that it would be possible to pump about 120 MCM annually from the non-renewable aquifers of Dissi for about 100 years.

The total irrigated area in the upland and southern Desert has reached 42000 ha according to the Ministry of Agriculture estimates. Therefore the total irrigated area in Jordan is about 72000 ha.

2.2. IRRIGATION IN THE JORDAN RIFT VALLEY

The Jordan Rift Valley area extents from the lake Tabarieh in the North to the Red Sea in the South passing through the Dead sea. The area north of the Dead Sea is called the Jordan Valley while the area south to it is called southern Ghors and Wadi Araba.

A major study for the integrated development of the Jordan Valley was conducted in 1953 (Baker, 1953) for the development of the land and the water resources in the Jordan River System. This study showed that it would possible to put under irrigation an area of about 360000 dunums (36000 hectares) in the Jordan Valley. As of 1957 the government of Jordan began implementing the study at different stages, but due the abstraction of the flow of the River Jordan and its tributaries, the above figure was never achieved. The area, which has been developed and could receive irrigation water is 23000 ha, plus 6000, which has been equipped by irrigation distribution system but no water allocation. The irrigation development during this period includes building a main canal with a head capacity of 20 m^3/s in addition to a diversion weir on the Yarmouk river and a 3-km tunnel. The canal length has reached 110 km. Other major development activities were the construction of five dams with a capacity of 160 MCM. Future plans include the construction of Wahda Dam on the Yarmouk River with a capacity of 225 MCM and utilizing of about 20 MCM of the treated effluent in a closed area north of the Jordan Valley. Works on improving the conveyance and diversion efficiency of the side wadis are in progress.

The main concern in the Jordan valley is the decline of the available water resources due to drought, upstream uses and the overuse of groundwater in the tributary basins. The available resource has dropped from an average of 370 MCM per year during the last 2-3 decades to about 250 MCM per year during 1997-2000 periods.

Irrigation projects in the southern Ghors utilize the base flow of the Dead Sea side wadis by regulating their flow through construction of the weirs and irrigation canals. At the first stage, an area about 4800 ha was put under irrigation. After construction of two major dam, the irrigated area in the second stage will reach 6300 ha. There is a potential to irrigate about 2000 ha in Wadi Araba using the groundwater resources in that area.

The Jordan Valley irrigation project is a public scheme controlled and managed by the Jordan Valley Authority (JVA). It was established in 1977 and is responsible of land and water resources development in the Jordan Rift Valley. After land reform, the land under irrigation is divided into farm units of 3-4 hectares each. Each farm unit is provided with water through a farm intake. Water is conveyed from different sources through the main canal (King Abdullah Canal, KAC). For more than 90% of the area, water is delivered to the farm in pressurized pipes. Pressure head is provided from pumping stations adjacent to the main canal or the natural head of some dams at higher elevation

The most common on-farm irrigation system is micro irrigation, which covers about 64% of the Jordan Valley area. Very limited number of farms use sprinkler irrigation while the rest (about 35%) are still using the conventional surface irrigation that is practiced in citrus and banana farms (Shatanawi, 1994)

The Division of Operation and Maintenance of JVA is in charge of irrigation water management. Farmers in the Jordan Valley submit their irrigation request for their units early in the morning to ten stage offices distributed along the Valley. The requests are processed according to supplies fed from KAC or other sources. Demands are compared with supplies and irrigation delivery schedules are issued to ditch riders who open the turnouts of the open channel system, or farm turnouts, or intakes of piped irrigation systems.

Releases from dams are determined taking into consideration the target level until the end of the season. In the daily water balance, uncontrolled resources (flows from Yarmouk & others side Wadis) are first used starting from south to north, then controlled resources according to pre-defined schedules also from south to north. Excess water in KAC is stored in Wadi Arab Dam and pumped to Amman for municipal and industrial uses.

In the case of severe water shortages, contingency policies are adopted to reduce water allocations with more care given not to harm permanent licensed crops (banana, citrus, ect.).

To operate such a system while the resource is not only scarce but also fluctuating and sometimes unpredictable, with an ever-increasing demand, becomes year after a more critical issue, and thus, a great challenge for the JVA staff. The issues of this challenge include: (1) to build a water allocation strategy so as to make the best possible use of the water resources available, and (2) to provide each user the "right amount of water at the right time", so as to make the best use of the water allocated to each project and to substantially upgrade the efficiency of the whole scheme.

A requirement necessary to meet the above objectives for both water resources and water distribution is the global management at the scale of the entire scheme. Such a management may only be obtained when all the information related to the water situation is gathered efficiently and be readily available to the management staff. This requirement led the JVA to the implementation of the Management Information System (MIS).

The scarcity of water in Jordan is not only due for its being located in the arid to semi arid region of the world, but other factors increases the severity of water shortage. First, the population growth rate is high so that the per capita share of water from the renewable water resources is always decreasing. Most of the country's water resources are shared with its neighbouring countries. The abstraction of surface water and springs and the over pumping of groundwater at the upper parts of the watersheds has affected the base and flood flow of major watercourses downstream. A major factor is the frequent occurrence of drought events during the last three decades. For example, the historical average annual flow of the Yarmouk River was about 438 MCM up to 1980. But due to upstream uses and other factors mentioned above, this flow has dropped to about 300 MCM/year. This has affected the diverted water from Yarmouk to the Jordan Valley from traditional 125 MCM during 1980-1985 to about 100 MCM/year in the years 1997 and 1998. For the drought years of 1999 and 2000 the diverted flow from the Yarmouk to the Jordan Valley has dropped to 65 MCM/year and 55 MCM/years respectively. It will drop further during 2001 due to the continuous drought. Table I shows the available water resources to the Jordan Valley during the last four years.

Table I. Water resources in the Jordan valley for the period 1997-2000.

<table>
<tr><th rowspan="2">Source</th><th colspan="4">MCM/Year</th></tr>
<tr><th>1997</th><th>1998</th><th>1999</th><th>2000</th></tr>
<tr><td>Yarmouk to canal</td><td>99.7</td><td>100.4</td><td>62.9</td><td>54.6</td></tr>
<tr><td>Tabaria</td><td>47.9</td><td>55.9</td><td>41.9</td><td>54.5</td></tr>
<tr><td>Mukheibeh</td><td>15.1</td><td>15.02</td><td>16.3</td><td>18.0</td></tr>
<tr><td>Arab Dam</td><td>0</td><td>-</td><td>0.7</td><td>0.3</td></tr>
<tr><td>Wadi Arab</td><td>-</td><td>0.2</td><td>0.1</td><td>0</td></tr>
<tr><td>Ziglab Dam</td><td>0</td><td>0</td><td>0</td><td>0</td></tr>
<tr><td>Wadi Ziglab</td><td>3.9</td><td>6.2</td><td>6.2</td><td>5.2</td></tr>
<tr><td>Wadi Taibeh</td><td rowspan="4">6.5</td><td>0</td><td>0</td><td>0</td></tr>
<tr><td>Wadi Abu Zaid</td><td>0.7</td><td>0.4</td><td>4.5</td></tr>
<tr><td>Wadi Jurum</td><td>4.9</td><td>2.7</td><td>0.6</td></tr>
<tr><td>Wadi Yabis</td><td>2.9</td><td>0.1</td><td>5.0</td></tr>
<tr><td>Wadi Kufranja</td><td>-</td><td>8.9</td><td>2.0</td><td>5.0</td></tr>
<tr><td>Wadi Rajib</td><td>-</td><td>6.5</td><td>1.7</td><td>3.5</td></tr>
<tr><td>Zarqa River</td><td>71.5</td><td>73</td><td>67.1</td><td>77.3</td></tr>
<tr><td>Wadi Mallah</td><td>-</td><td>0</td><td>0</td><td>1.7</td></tr>
<tr><td>Wadi Shuib</td><td>5.7</td><td>4.2</td><td>4.3</td><td>4.7</td></tr>
<tr><td>Wadi Kafrein</td><td>13.0</td><td>11.2</td><td>6.5</td><td>8.0</td></tr>
<tr><td>Wadi Hisban</td><td>1.7</td><td>2.2</td><td>1.7</td><td>1.5</td></tr>
<tr><td>Total</td><td>265.0</td><td>292</td><td>215</td><td>240</td></tr>
<tr><td>To Amman</td><td>38.6</td><td>30.4</td><td>41.4</td><td>41.2</td></tr>
</table>

The frequent occurrence of drought especially during recent years has put many constraints on irrigated agriculture in the Jordan Valley. The JVA has taken certain management practices and measures to ease the problem. Long-term measures to assure sustainable irrigated agriculture include: (1) Increase the volume and number of storage facilities by constructing new dams and enlarging the capacities of existing ones; (2) Plan for a dynamic cropping pattern by limiting the area of perennial crops and fruit trees to less than 30% to allow flexible cropping pattern during drought years; (3) Encourage the introduction of water saving technology like the use of micro irrigation, mulches, and protected agriculture; (4) Augment water resources through the use of treated wastewater and ground water during drought years.

In addition to the above, the following short-tem measures are taken during any drought year: (1) Limit the area planted with annual crops and vegetables during the period of high consumption, especially during summer months. In the years of severe drought, planning is imposed on summer plantation; (2) Reduce the acreage of irrigation area through reducing water allocation but allowing the irrigation of fruit trees; (3) Enforce conservation measures thought the control of water wastes, spills and illegal uses; (4) Practice deficit irrigation to the extend that to keep fruit trees growing or surviving while risking the yield; (5) Some farmers exploit groundwater wells to combat shortages of water supply in a drought year.

The general practice during normal years is to allocate irrigation water to farmers to satisfy crop water requirements at rates of 4-5 mm/day. During drought years the rate is reduced to 2-3 mm/day, and even less in case of severe drought. This is the case during the years of 1997, 1998, 1999 and 2000 where water was given at rates of 4, 3, 2 and 1.7 mm/day; respectively. This study aims to evaluate crop responses to these measures.

3. Methodology

3.1. PROCEDURE

In this paper, the ISAREG model (Teixeira and Pereira, 1992) was used to evaluate the effect of different level of water deficits on irrigation scheduling, evapotranspiration, and yield reduction. Water productivity was determined as the ratio between gross value of the crop and water use.

The meteorological data from Deir-Alla agricultural research station in the central Jordan Valley during the period 1980-1999 (JMD, 1999) were used as input to EVAPOT program. EVAPOT uses Penman-Montieth equation to calculate the reference evapotranspiration. The EVAPOT program follows the updated methodology of FAO (Allen et al., 1998) to compute the reference evapotranspiration with the Penman Monteith method. Data of maximum and minimum air temperature, relative humidity, wind speed, and sunshine duration or solar radiation are used as input to EVAPOT. The crop coefficients were determined using the KCISA program (Rodrigues, 1999), which is based on the procedure described by FAO 56 (Allen et al.,

1998) for the selected crops, namely; citrus, tomatoes, potatoes and wheat. The soil moisture depletion fraction for no stress (p) and the effective rooting depth (Zr) for each crop development stages are also results of computations with KCISA.

The results of EVAPOT and KCISA as well as irrigation modes, effective precipitation and soil data are used as input to ISAREG model. ISAREG model calculated actual crop water requirements under no stress and for several water deficit levels. The model provides the user with irrigation scheduling options, including the percent reduction in yield under imposed schedules and water deficits. A DSS is used as the general forum within which irrigation management variables and models can be combined to incorporate the necessary interacting components which play a role in the decision making process. Figure 1 describes the input variable and data to ISAREG model and DSS as well the outcome of both ISAREG and DSS to calculate relative yield and increasing water productivity, and thus select the best management options.

3.2. DSS AND ANALYTICAL HIERARCHY PROCESS (AHP)

The analytic hierarchy process (AHP) is incorporated in the DSS. The AHP utilizes multi-criteria analysis (MCA), a meta decision-support system that leverages the comprehensive expertise of to enhance strategic decisions through a process that provides structured clarity, communication, and synthesis. Basically speaking, the AHP is a method of breaking down a complex, unstructured situation into its component parts, assigning numerical values to subjective judgments to determine which variables have the highest priority and should be acted upon to influence the outcome of the situation. The necessity of assigning a numerical value to each variable of the problem helps decision makers to maintain cohesive thought patterns and to reach a conclusion (Al-Jayyousi and Shatanawi, 1995). AHP represents a powerful tool in the hand of the system analysis and a valuable help in solving problems, which include conflicting criteria (Al-Zu'bi, 2000).

The values for the gross margin from each of the four dominant crops are calculated using equations published by McLennan (2000). Water productivity is the gross value of the crop divided by its water consumption. The first required variable for the gross margin calculations is variable cost. The values for the variable costs for each crop have already been calculated and reported by the Agricultural Credit Cooperation, ACC (1998). Variable costs are those that will occur only if production actually takes place and tend to vary with the level of production. They are 3653, 4312, 4834 and 449 JD/ha for citrus, tomatoes, potatoes and wheat, respectively.

The gross margin of profit is calculated as GM = TR – VC, where GM is gross margin, TR is total return and VC is variable cost (all units JD/ha). The water economic productivity (WEP) in JD/m^3 is the gross margin per hectare divided by the water consumption (WC) in m^3/ha. In order to achieve the selection of the best economic performing crop under drought condition using DSS, analytical hierarchy process was achieved. The model for decision making used is the Analytical Hierarchy Process (AHP) in the DSS "expert system" software Expert Choice (Saaty, 2000). The expert system has a major advantage: ease of use by non-specialists. It has been applied successfully in many aspects of water resource management (Bender and Simonovic, 1993).

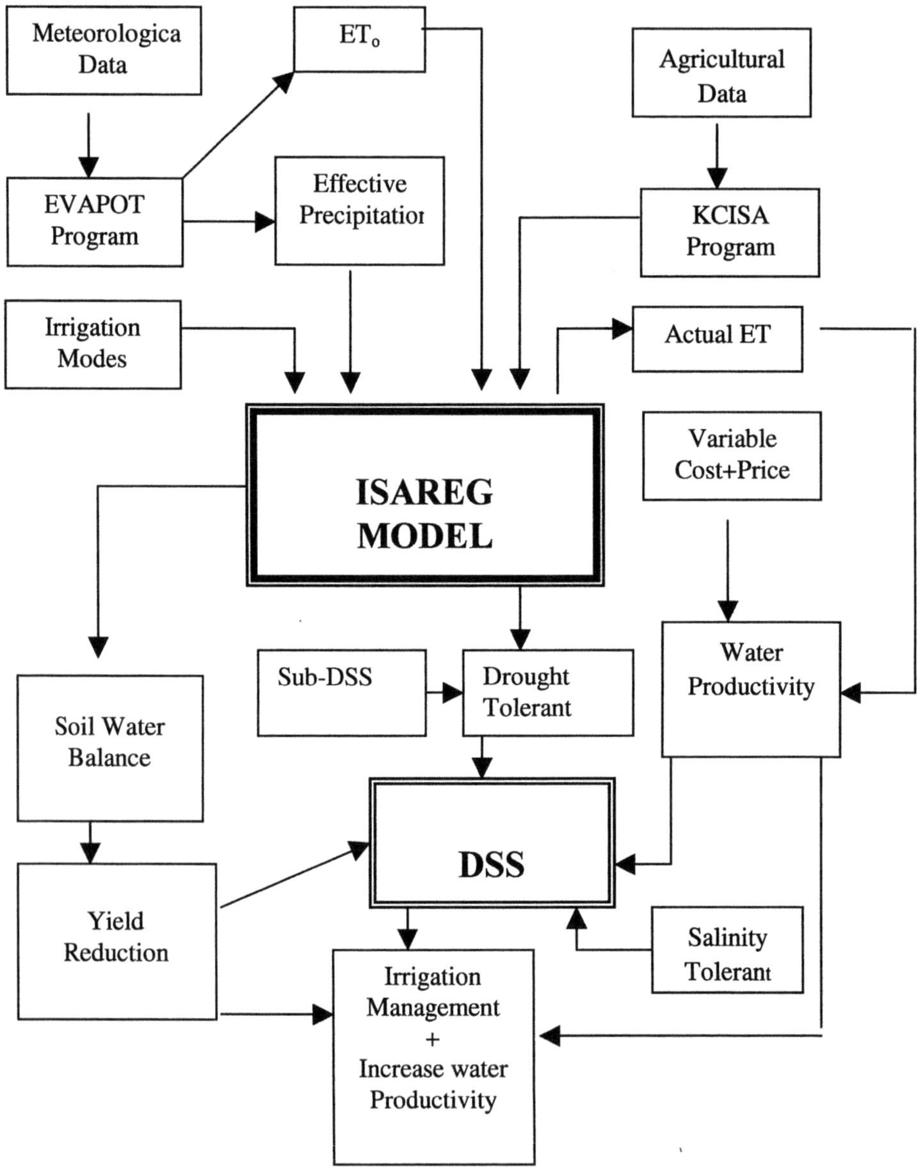

Fig. 1. The flow chart describing the input and output data and the procedure to obtain water productivity and irrigation scheduling strategy in the Jordan Valley under drought conditions

The problem was broken into different levels to construct a hierarchy tree. The hierarchy tree shown is composed of three levels. The first level defines the 'goal' to achieve the best crop under drought conditions.

The second level outlines four scenarios; based upon them we can synthesize judgments to achieve that goal. These are to irrigate with 4mm/day, 3mm/day, 2mm/day, and 1.7mm/day. The third level lists the alternatives, which consist from the main crops grown in Jordan Valley: Citrus, Tomato, Potato, and Wheat.

Based on the goal (level 1), the four scenarios priorities are defined. Different approaches are followed to achieve the priorities for the alternatives (level 3) with respect to the scenarios indicated in level 2. Drought tolerancy was evaluated based on the reduction in yield for each crop under different scenarios using ISAREG model.

3.3. GENERAL DSS: ANALYTICAL HIERARCHY PROCESS UNDER DROUGHT CONDITIONS

The output of the above mentioned decision process to select the best crop under drought condition is utilized to construct the main DSS System for increasing water economics productivity (WEP) in Jordan Valley To construct this DSS, similarly the problem was broken into three levels: the first level defines the goal to be achieved, which is to increase WEP in Jordan Valley. The second level concerns the objectives, which are: (1) to save water, (2) to achieve high economic return, (3) to select the most tolerant crops to salinity as described in (Doorenbos and Kassam, 1979) and (4) to achieve the most tolerant crops under drought condition. The third level lists the alternatives, which consist of the main crops grown in Jordan Valley: Citrus, Tomato, Potato, and Wheat. Based on the goal (level 1), the objective priorities were defined. Different approaches were followed to achieve the priorities for the alternatives (level 3) with respect to the objectives indicated in level 2. Drought tolerancy was evaluated based on the reduction in yield for each crop under different scenarios using the ISAREG model as previously mentioned. Also crop irrigation requirements were calculated by ISAREG model. Based on water consumption and an other variables indicated in the methodology described in calculation of economic return, the fourth objective 'to achieve high economic return' was defined.

4. Results

4.1. ACTUAL EVAPOTRANSPIRATION

The average monthly values for the reference evapotranspiration (ETo) for Jordan Valley (Deir Alla station) during the period 1980-1999 are presented graphically in Figure 2. The results demonstrated that the maximum mean monthly ETo is 205.7 mm during July, while the minimum mean monthly ETo was 72.4 mm during January.

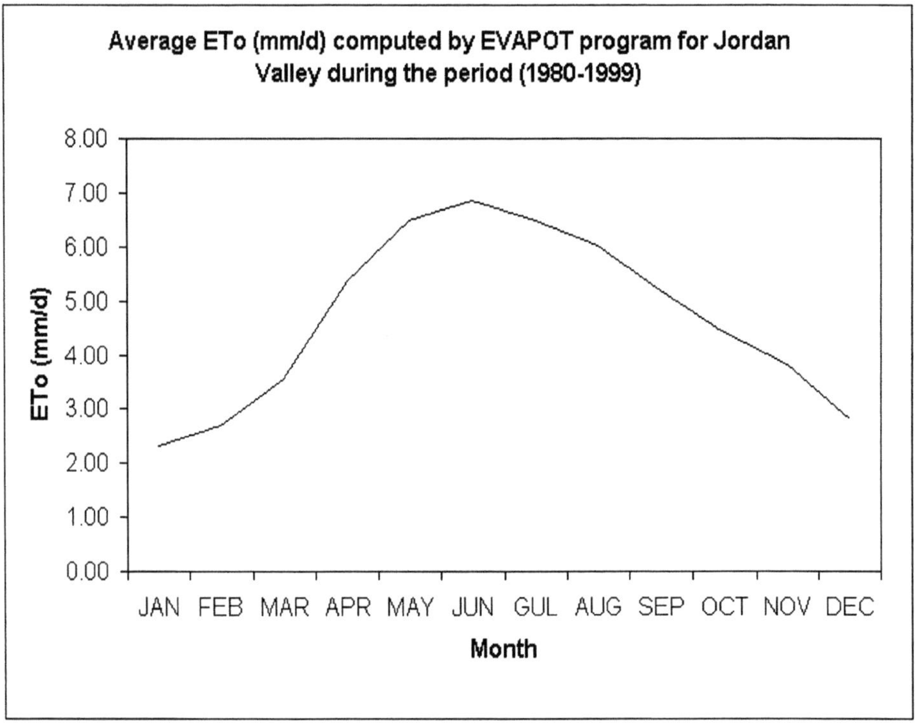

Fig. 2. Average ETo (mm/d) computed by EVAPOT program for Jordan Valley during the period (1980- 1999).

Actual ET under no stress conditions for the four crops under study are shown in Table II. The yearly average actual ET is 1064 mm, 591, 303 mm and 528 mm for citrus, tomatoes, potatoes and wheat respectively. Table III shows actual ET under the four levels of deficit irrigation for the four selected crops.

Table II. Crop growth stages and actual crop ET for the major and important crops grown in the Jordan Valley.

Crops	Initial Stage (Days)	Dev. Stage (Days)	Mid Stage (Days)	Late Stage (Days)	Total Growth period (Days)	Plant Date	Water Request mm./Season
Citrus	47	73	123	91	334	Jan.	1087
Tomato	31	40	41	25	137	Mar.	591
Potato	20	25	40	25	110	Jan.	303
Wheat	30	66	40	25	161	Dec.	528

Table III. Actual crop evapotranspiration (mm) under four levels of Deficit Irrigation.

Crop	Maximum ET	ET at 4 mm/day	ET at 3 mm/day	ET at 2 mm/day	ET at 1.7 mm/day
Citrus	1087	1022 (3.2)	819 (22.2)	568 (37.3)	496 (42.7)
Tomato	591	521 (12.3)	383 (37.0)	260 (58.8)	221 (66.0)
Potato	303	303 (0.0)	245 (3.2)	211 (33.7)	178 (45.8)
Wheat (**)	537	537 (0.0)	448 (16.5)	347 (35.4)	274 (41.7)

* Numbers between brackets are percent reduction in ET.

** The average daily ET for wheat is 3.52 mm/day under standard conditions.

4.2. POTENTIAL YIELD REDUCTION AND ECONOMIC RETURN

It is difficult to report acceptable figures for the potential yield for the selected crops. The statistics of the Ministry of Agriculture, Department of General Statistics and the Annual Bulletin of the Central Bank are not accurate enough to calculate the maximum yield because they consider the total areas and the volume of exports and local consumption. For the purpose of this study, the authors made every possible effort to report the potential yield after reviewing the results of research and interviewing many farmers. The best estimate values of potential yield as well as the retail prices are presented in Table IV.

Table IV. The potential yield and retail price of selected crops.

Crops	Potential Yield Ton/ ha	Retail price JD/Ton	Total Return JD/ ha
Citrus	60	150	9000
Tomatoes	60	100	6000
Potatoes	40	150	6000
Wheat	5	170	850

The yield reduction as a fraction of the potential yield was calculated by ISAREG model based on the linear crop-water production function.

$$\left(1-\frac{Ya}{Ym}\right)=Ky\left(1-\frac{ETa}{ETc}\right)$$

Where K_y is yield response factor, ETa is actual ET under stress condition, ETc is ET for standard conditions (no water stress), Ya is yeld when ET=Eta and Ym is potential yield, when ET=ETc.

Table V shows the percentage reduction in yield for the four selected crops. It is clear that tomatoes are the most sensitive crop to JVA management followed by potatoes. The most tolerant crop is wheat where only 41% reduction in yield was noticed versus 51% reduction in ETo (values are indicative only).

Table V. Reduction in yield in percentage for the four selected crops growing in Jordan Valley under deficit irrigation, computed with ISAREG model.

Scenarios Crops	4mm/day	3mm/day	2mm/day	1.7mm/day
Citrus	3.2	22.2	37.3	42.7
Tomatoes	12.3	37.0	58.8	67.8
Potatoes	0.0	3.2	33.7	45.0
Wheat	0.0	16.5	35.4	40.7

4.3. WATER ECONOMIC PRODUCTIVITY

Table VI show the water economic productivity under maximum potential yield and under different levels of water stress.

Table VI. WEP and respective components when crop ET is 4 mm/day and maximum yield for the selected crops.

Crop	TR (JD/ha)	VC (JD/ha)	GM (JD/ha)	WC (m3/ha)	WEP (JD/m3)
Crop ET = 4 mm/day					
Citrus	9000	3653	5347	10640	0.5
Tomato	6000	4312	1680	5910	0.28
Potato	6000	4834	1160	3030	0.38
Wheat	850	4486	4014	5280	0.08
Crop ET = 3 mm/day					
Citrus	7020	3578	3442	8190	0.42
Tomatoes	3780	4250	-470	3860	-0.12
Potatoes	5810	4817	993	2450	0.41
Wheat	710	421	289	4480	0.065
Crop ET= 2 mm/day					
Citrus	5643	3483	2160	5680	0.38
Tomatoes	2508	4213	-1705	2600	-0.65
Potatoes	3762	4806	-1044	2110	-0.49
Wheat	549	388	161	3470	0.046
Crop ET= 1.7 mm/day					
Citrus	4797	3476	321	4960	0.065
Tomatoes	204 0	420	-2160	2210	-0.97
Potatoes	3264	4797	-1533	1780	-0.86
Wheat	491	370	121	2740	0.045

TR: Total Return, VC: Variable Cost, GM: Gross Margin, WC: Water Consumption; WP: Water Productivity.

It is clear that the WEP of citrus is the highest under standard and stressed conditions. The WEP of tomato and potato crops are highly affected by water scarcity. Wheat and citrus are the most tolerant to deficit irrigation in terms of water stress while tomato and potato are very sensitive. Therefore it is not advisable to plant tomato or potato (thus vegetables in general) under drought conditions.

4.4. SELECTION OF THE BEST CROP UNDER DROUGHT

Results for the DSS AHP model to select the best crop under drought conditions, are given in Table 7. This Table summarizes the results of the selection of the best crop under drought condition for four scenarios that will determining the selection of the "best" crop, based on the following scenarios: (1) Eta = 4mm/day; (2) Eta = 3mm/day; (3) Eta = 2mm/day, and (4) Eta = 1.7mm/day.

The highest-ranking crop in the "4mm" scenario was potato followed by wheat. In the "3mm" category wheat was ranked highest, followed by citrus. Potato and wheat crops ranked highest in the "2mm" scenario, followed by citrus. The overall ranking "wheat" comes in the highest place, followed by citrus. Based on the goal (level 1), to select the best crop under drought, wheat scores high for overall ranking (0.46) followed by citrus (0.25) and potato (0.24), while tomato scores the lowest (0.05).

Data in Table VII illustrates the relative confluence or divergence of desired scenarios for each crop. For example, tomatoes receive low rankings for all four scenarios, while citrus tend to range above average for all four scenarios. Potato shows broader distribution of high and low scores, while wheat receives high rankings for the three scenarios of 3mm, 2mm and 1.7 mm.

Table VII. Priorities of alternative crops growing in Jordan Valley based on the Analytical Hierarchy Process analysis under different scenarios and with respect to overall goal to select the best crop under drought.

Scenario / Alternative	Scenario1 (4mm)	Scenario2 (3mm)	Scenario3 (2mm)	Scenario4 (1.7 mm)	Over all
Citrus	0.15	0.29	0.22	0.27	0.25
Tomato	0.05	0.06	0.06	0.04	0.05
Potato	0.5	0.16	0.36	0.18	0.24
Wheat	0.3	0.5	0.36	0.51	0.46

4.5. CROP SELECTION FOR HIGHER WEP

The general DSS AHP for sustainable irrigation management in Jordan Valley was used for selection of the best crop under drought.

The results from the objectives and alternatives prioritization are tabulated in Table VIII. The weights donated to the objectives with respect to the goal 'to increase water productivity' were based on the consultation with the staff of Faculty of Agriculture in the University of Jordan. The highest priority objective is objective 2 (40%) followed by objective 1 (36%) and objective 4 (17%), while objective 3 received the lowest score (7%).

The AHP ranked each alternative with respect to the four objectives (Table VIII). The highest-ranking crop with respects to objective 1 "to save water" was potato followed by wheat. With respect to objective 2 "to achieve high economic return" citrus was ranked highest, followed by tomato. Wheat crop ranked highest with respect to objective 3 "to select the most tolerant crops to salinity " followed by tomato. Overall ranking for each of the alternatives across the 4 objectives show higher overall values for alternative 3, potato, followed by alternative 1, citrus, while alternative 4, wheat, comes in the third place, and alternative 2, tomato ranked the lowest. The variation in weighting among the four objectives across each objective is obvious. But it is interesting to notice the relative lack of variation in weighting under overall ranking for the three crops; potato, citrus and wheat. With comparison, there is a consistency between the overall results of both the sub-DSS and the general DSS, with tomato receiving the lowest score.

Table VIII. Priorities of alternative crops growing in Jordan Valley based on the Analytical Hierarchy Process analysis with respect to objectives and to overall Goal ' to increase water economic productivity'.

Objectives Alternatives	To save Water (0.36)	Economic return (0.40)	Salinity tolerant (0.07)	Drought Tolerant (0.17)	overall
Citrus	0.06	0.52	0.1	0.18	0.27
Tomato	0.12	0.27	0.25	0.06	0.18
Potato	0.57	0.15	0.1	0.2	0.3
Wheat	0.26	0.06	0.56	0.56	0.25

5. Discussion and conclusions

In light of the above results, it is apparent that there is the need to define economic criteria to optimize water use for irrigation. The contribution of one cubic meter in crop yield was estimated to provide a guideline for sustainable agriculture in the Jordan Valley. The above results reveal that the economic return for citrus was 0.5 JD/m^3 while, the lowest economic return, for wheat, was 0.08 JD/m^3. These values are reasonable and reflect market mechanisms in the Jordan Valley. In fact, farmers responded to the high value of citrus in the market by shifting from low-value crops (like wheat) to higher valued crops (e.g. citrus).

It should be noted that the comparative advantage of certain crops due to climatic conditions and water availability encouraged farmers and agricultural policy makers to be market responsive. Specifically, the drought conditions forced farmers to adopt various mitigation measures such as to apply deficit irrigation, adopt the irrigation of high-valued crops, and to limit irrigation to only a portion of land.

Due to the fact that water is the major constraint in agricultural production, there is an uncertainty and variability of yield. Four irrigation schemes are practiced, i.e., 4 mm/day, 3 mm/day, 2 mm/day, and 1.7 mm/day. These represent drought severity scenarios, for low, medium, high, and extreme severity respectively. The ISAREG model shows that the reduction in yield, for citrus as an example, under the above four scenarios was 2%, 16.5%, 35.4%, and 35.4% respectively.

In order to achieve the main goal of "sustainable irrigation management in Jordan Valley" it is recommended that the current policy in use must not continue which will deteriorate the water quality and deplete existing quantities. The objectives considered in the main DSS allow for water use without compromising supply.

The highest priority crop with respect to the goal of sustainable irrigation management is the potato crop. This is due to high relative yield, market prices and less water consumption attributed to the short growing period, which is only 110 days. The problem with Potato is the sensitivity to salinity. In comparison, wheat scored highly under both objective 3 (Salinity tolerant) and objective 4 (drought tolerant), which implies that wheat is the most tolerant crop to in Jordan Valley.

The different alternative ranking results demonstrated that a DSS allows flexibility and can answer questions specific to management needs.

The following can be given:

1. The highest priority crop with respect to the goal of increasing water economic productivity in the Jordan Valley is potato followed by citrus under standard conditions.
2. The highest priority crop with respect to the goal of selecting the best crop under drought condition in Jordan Valley is wheat followed by citrus.
3. The ISAREG model is a valuable and helpful tool that can be used for best irrigation management and estimation of yield reduction to increase water productivity.
4. The analytical Hierarchy Process AHP represents a powerful tool in system analysis and a valuable help in solving problems, which include conflicting criteria. ISAREG along with AHP proved to provide an enabling DSS framework for the analysis of drought management.

References

ACC. (1998) *A guide for Agricultural Costs and Returns.* Agricultural Credit Corporation, Amman, Jordan.

Allen, R.G., Pereira, L.S., Raes, D. and Smith, M. (1998) *Crop Evapotranspiration. Guidelines for Computing Crop Water Requirements.* FAO Irrig. Drain. Pap. 56, FAO, Rome, 300p.

Al-Jayyousi, O.R. and Shatanawi, M.R. (1995) An Analysis of Future Water Policies in Jordan Using Decision Support Systems. *J. Water Resour. Develop.* 11: 315-330.

Al-Zu'bi, Y.A. (2000) *Decision Support System for Sustainable Management of Water Resources in the Azraq Basin.* Ph. D. thesis, University of Jordan.

Baker, M.J. (1953) *Princes idea project.* Water Authority, Amman, Jordan, 105p.

Bender, M.J. and Simonovic, S.P. (1993) Interactive Computer Graphic for Expert- System Verification. *J. Water Res. Plann. Manag.* 119: 518- 530

DOS (2000) *Yearly Report.* Department of Statistics, Amman, Jordan.

Doorenbos, J. and Kassam, A.H. (1979) *Yield Response to Water.* FAO Irrig. and Drain. Paper N°33, FAO, Rome.

JMD (1999) *Time Series of Monthly Means for Some Weather Elements in Jordan.* Jordan Meteorological Department, Amman, Jordan.

McLennan, W. (2000) *Water Account for Australia.* Australian Bureau of Statistics.

Rodrigues, P.A.N. (1999) *KCISA program Users Guide.* Departamento de Engenharia Rural, Instituto Superior de Agronomia, Universidade Teccnica de Lisboa.

Saaty, T. (2000) *Expert Choice Tutorials.* Expert Choice, Inc. UAS.

Shatanawi, M.R. (1994) *Irrigation Management and Water Quality in the Central Jordan Valley.* Water and Environment Research and Study Center, University of Jordan, Amman, Jordan.

Teixeira, J.L. and Pereira, L.S. (1992) ISAREG-An irrigation-scheduling model. *ICID Bulletin 41(2):* 29-48.

SIMULATION OF SUPPLEMENTAL IRRIGATION STRATEGIES FOR WHEAT IN NEAR EAST TO COPE WITH WATER SCARCITY

T. OWEIS [(1)], P.N. RODRIGUES [(2)] and L.S. PEREIRA [(2)]
[(1)] *International Center for Agricultural Research in Dry Areas, Aleppo, Syria*
[(2)] *Agricultural Engineering Research Center, Institute of Agronomy, Technical University of Lisbon, Portugal*

Abstract: The development of supplemental irrigation strategies that reduce water demand with acceptable impacts on yields is part of preparedness measures to cope with drought and water scarcity. The irrigation scheduling simulation model ISAREG, after appropriate validation, is used to simulate various irrigation schedules for wheat under average, high and very high climatic demand conditions to cope with limited water supplies. Surface and set sprinkler irrigation are considered. The study was performed with field data collected at the Tel Hadya research station of ICARDA, Aleppo, Syria. The alternative irrigation schedules are evaluated through the achieved reduction in demand for irrigation water and the consequent yield decrease. Results show that water saving strategies for supplemental irrigation of wheat are generally feasible.

Key words: drought, water saving, deficit irrigation, irrigation scheduling, simulation modelling, North Syria.

1. Introduction

The present general practice in irrigated agriculture is to maximize crop yield per unit land. However, for several crops maximizing yield is at the account of the water productivity, *i.e.* the harvestable yield per unit volume of water used. For small grain cereals, water productivity drops at high yield levels under full irrigation (Zhang and Oweis, 1999). Maximum water productivity is attained at lower than potential yields. In areas where water, not land, is the most limiting resource to production, maximizing water productivity is more profitable to the farmer than maximizing crop yield. This is because water saved by deficit irrigation can be used to irrigate additional land, since land is not the liming factor, at much higher overall water productivity. In northern Syria it was found by ICARDA that applying 50% of full supplemental irrigation requirements would reduce yield by only 15% under average climatic conditions. Applying the saved 50% water to lands that would otherwise be rainfed could increase total farm production up to 38% (Oweis, 2001). Such relation is important to determine the proper strategies for irrigation in areas where water, not land, is the most limiting factor. This is of particular importance when planning supplemental irrigation to cope with droughts.

Water management under drought requires measures and policies, which are common with aridity, such as those to avoid water wastes, reduce demand and make water use more efficient (Pereira, 2002). Demand management for irrigation under water scarcity includes practices and management decisions of agronomic, economic,

G. Rossi et al. (eds.), Tools for Drought Mitigation in Mediterranean Regions, 259-272.

and technical nature. Objectives include: reduced irrigation requirements, adopted practices leading to water savings in irrigation, reduced demand for water at the farm, and increased yields and income per unit of water used, *i.e.*, an increase in water productivity as it is intended for supplemental irrigation (Oweis, 1997; Oweis *et al.*, 1999).

Supplemental irrigation is an important practice in rainfed areas for crops normally producing without irrigation. Under low and/or varying rainfall, rainfed yields are usually low and highly fluctuating from one season to the other. Supplemental irrigation provides limited amount of water during times when precipitation fails to satisfy basic crop needs. Research conducted at ICARDA and elsewhere shows substantial increases in field crop yields by applying small amounts of irrigation water in times when precipitation is not adequate (Oweis, 1997; Zhang and Oweis, 1999). Water productivity of supplemental irrigation water is very high and increases for deficit supplemental irrigation (Oweis *et al*, 1999). The frequent occurrence of drought and the dependency of supplemental irrigation management on the variable rainfall make the efficient scheduling of this practice difficult.

To assist the development of supplemental irrigation strategies for coping with droughts and water scarcity, both field experiments and model simulation are used. The first provides a solid basis for using a model, and the latter extend results to wider climatic demand and supply conditions as described in this study.

2. Modelling approach

2.1. CLIMATE AND SOIL

The wheat supplemental irrigation under limited water availability in North Syria was studied in the Block B of ICARDA's Tel Hadya research farm, located 30 km North of Aleppo. The study used historical meteorological daily data from September 1984 to August 2000 recorded in the weather station installed at Tel Hadya (latitude 36° 01'N, longitude 36° 56'E, elevation 284 m above sea level).

The average monthly rainfall, average monthly reference evapotranspiration (ET_0), and average monthly minimum and maximum temperatures for that period are shown in Figure 1. ET_0 was computed with the Penman-Monteith method (Allen *et al.*, 1998). Results in Figure 1 show that rainfall is insufficient during the critical wheat stages in April and May. The average monthly minimum and mean relative humidity, and the average monthly wind speed at 2 m height for the same period are presented in Figure 2. The minimum relative humidity is estimated from the ratio of saturated vapor pressure at minimum and maximum temperature. Results in Figure 2 show that arid conditions start prevailing after March, thus increasing the crop water demand during the critical growth stage of wheat.

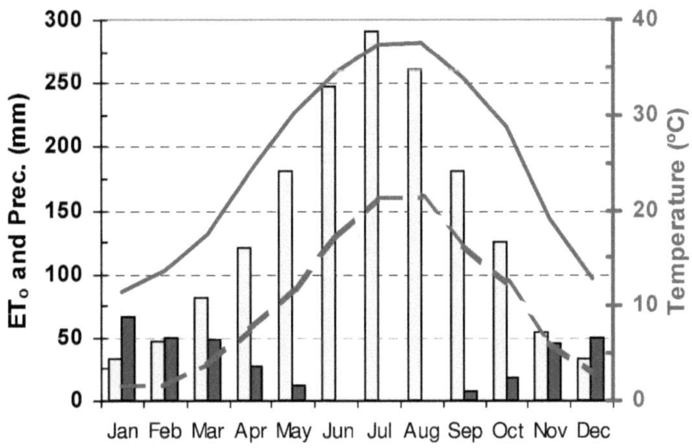

Fig. 1. *Average monthly rainfall (▬), average monthly reference evapotranspiration, ET_o (▭), and average monthly minimum (— —) and maximum (——) temperature at Tel Hadya, 1984 – 2000*

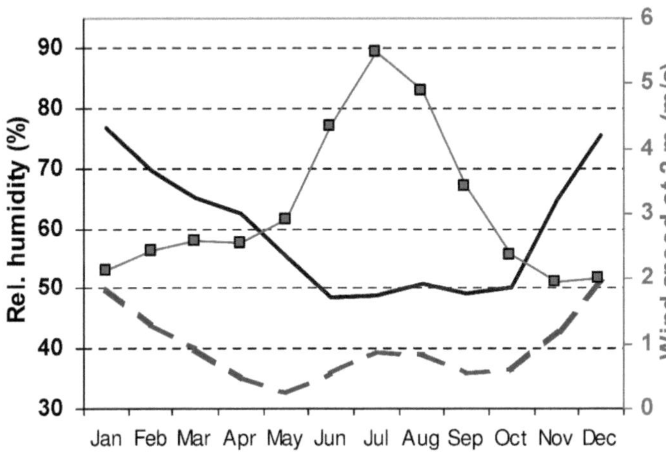

Fig. 2. *Average monthly minimum (— —) and mean (——) relative humidity, and average monthly wind speed at 2 m height, u_2 (—■—) at Tel Hadya, 1984 – 2000*

The main soil type at Tel Hadya is a Calcixerollic Xerochrept. On average, the textural classes are 15 % sand, 25 % silt and 60 % clay. The volumetric soil water content at field capacity (θ_{FC}) and at the permanent wilting point (θ_{WP}) are 40 % and 24 %, respectively. The soil depth is 1.0 to 1.8 m.

2.2. MODELLING TOOLS

The ISAREG model (Teixeira and Pereira, 1992) has been used to simulate different irrigation water management strategies. The model performs the soil water balance using different options to define and evaluate the irrigation schedules, as described by Pereira *et al.* (2002). The reference evapotranspiration is computed with the FAO-Penman-Monteith method (Allen *et al.*, 1998) using the program EVAP56. The crop data parameters are determined with the KCISA program (Rodrigues *et al.*, 2000), using the methodology proposed by FAO (Allen *et al.*, 1998). Thus, the crop coefficients (K_c) and the soil water depletion fraction for no stress (p) are adjusted to the Tel Hadya climate. More details on the modelling tools used are provided in a companion paper by Pereira *et al.* (2002).

The average dates for the crop development stages, the crop coefficients (K_c) and the soil water depletion fractions for no stress (p) are presented in Table I for wheat in Tel Hadya. The maximum effective rooting depth was estimated to be 1.5 m.

Table I. Average crop parameters for wheat

Parameters	Crop development stages			
	Initial	Development	Mid season	End season
Wheat:				
Period length (dates)	20 Nov – 29 Dec	30 Dec - 9 Apr	10 Apr -14 May	15 May – 8 Jun
Crop coefficients, K_c	0.87	0.87 - 1.19	1.19	1.19 - 0.25
Depletion fraction, p	0.60	0.60 - 0.53	0.53	0.53 - 0.67

Yield losses due to water stress are estimated by using the yield response factors proposed by Stewart *et al.* (1976) and Doorenbos and Kassam (1979). The crop yield response factors for the considered growth stages are given in Table II. The K_{yi} values utilized are not constant but vary with the intensity of water deficit in each phase and in the precedent phases in the range referred in that Table.

Table II. Crop yield response factors

Crop	Seasonal Yield response factor, K_y	Multiphasic yield response factors, K_{yi}		
		Vegetative growth	Flowering and yield formation	Maturation and senescence
Wheat	1.05	0.50 – 0.75	4.30 – 1.25	2.45 – 0.50

2.3. WATER SUPPLY STRATEGIES

In the Tel Hadya research farm, several irrigation methods are used. In this study surface irrigation and set sprinkler systems are considered. The average net irrigation depth I_n = 80 mm is adopted for surface irrigation and I_n = 40 mm for sprinkling.

When water availability is non-limiting, the frequency of irrigation is not restricted and varies along the crop season according to the crop demand. Under limited water availability, the supply is made with restrictions in the available water volumes. Several irrigation scheduling strategies are applied in practice. For simulation purposes, the management allowed depletion fraction, MAD (Martin *et al.*, 1990) is adopted. For full irrigation, the MAD fraction equals the soil water depletion fraction for non-stress (p), which is shown in Table I.

The supplemental water saving schedules were simulated by imposing restrictions on the irrigation period, decreasing the available water (net season irrigation depth) and increasing water stress represented by progressively lower thresholds $\theta_{MAD} < \theta_p$ as referred by Pereira *et al.* (2002). The corresponding irrigation strategies analyzed in this study are summarized in Table III. The alternative irrigation schedules are evaluated as for other companion papers (Rodrigues *et al.*, 2002; Zairi, *et al.*, 2002).

Table III. Constraints for wheat irrigation scheduling simulation

Surface irrigation (I_n = 80 mm)			Set-sprinkler systems (I_n = 40 mm)		
θ_{MAD} (1) (% θ_p)	Available water (mm) Nov. - Mar.	Apr. - May	θ_{MAD} (1) (% θ_p)	Available water (mm) Nov. - Mar.	Apr. - May
100	No restrictions	No restrictions	100	No restrictions	No restrictions
90	80	320	90	80	260
80	80	240	90	80	220
70	80	160	85	80	200
60	80	80	75	80	160
55	0	80	70	80	120
Non irrigated	0	0	65	80	80
			60	40	80
			55	40	40
			55	0	40
			Non irrigated	0	0

(1) For the mid season period, $\theta MAD \geq 75\% \ \theta p$

3. Simulation of supplemental irrigation

3.1. MODEL VALIDATION IN NORTH SYRIA

The ISAREG model was validated for wheat using field experimental data observed at the Tel Hadya research station. Environmental conditions there are representative of North Syria, including soils and wheat varieties. Data of field trials with surface irrigation using variable irrigation depths for 3 irrigation seasons (1991/92, 1992/93 and 1993/94) were used. For the first two seasons, data on treatments with no irrigation are also analyzed.

The ISAREG model validation is performed comparing the soil water content (% in volume) observed in field experiments throughout the crop season with those simulated by the model. Figure 3 presents the results of the model simulation for wheat at Tel Hadya in 1992/93 for the irrigated and rainfed treatments. The figure also includes the rainfall events for the same crop season. It shows that the model appropriately describes the soil water content along the entire crop season.

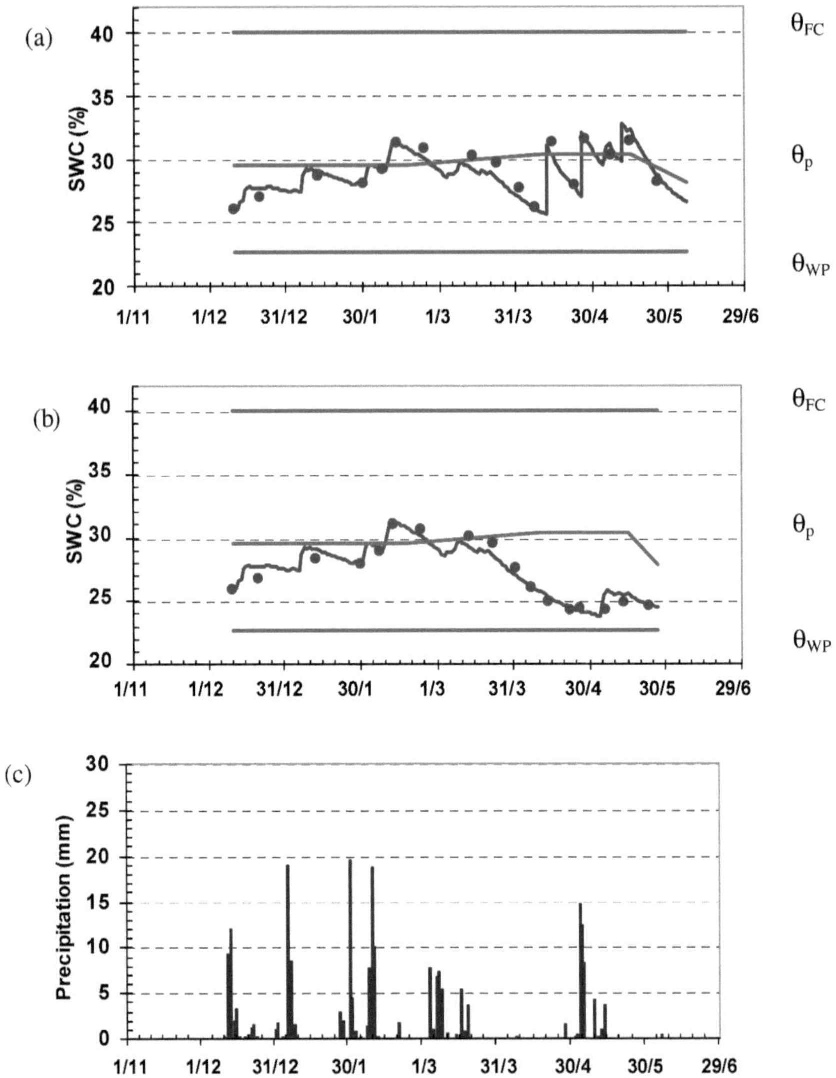

Fig. 3. Validation of the ISAREG model with Tel Hadya wheat data for 1992-93: simulated (——) and observed (●) soil water content (expressed in % volume) for: (a) irrigated and (b) rainfed wheat, and (c) daily rainfall data

The Figure 4 concerns the regression through the origin relative to the simulated vs. observed soil water content for the same experiments. The regression coefficients are very close to 1 and the determination coefficients are high, indicating that the model appropriately predicts the soil water content in both cases.

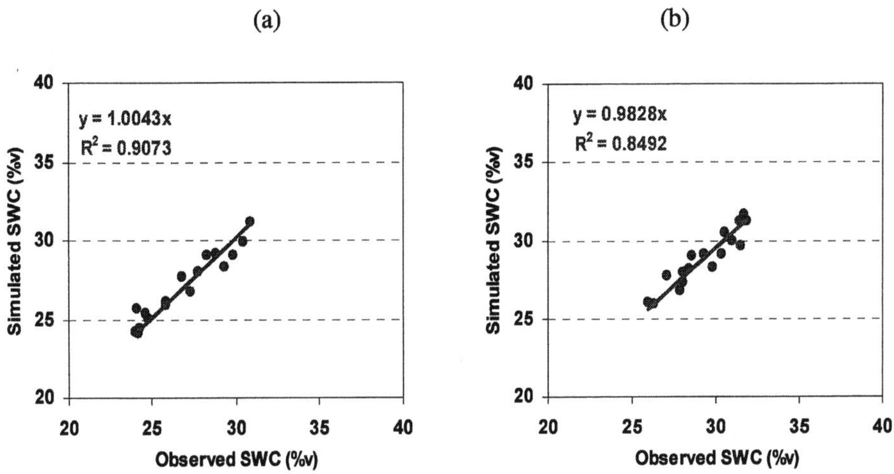

Fig. 4. Simulated vs. observed soil water content (expressed in % volume) for the irrigated (a) and rainfed wheat treatments (Tel Hadya, 1992/93)

Figure 5 compares the simulated and observed soil water content (% in volume) for all field trials analyzed at Tel Hadya. It shows a regression coefficient b =1.00 but a relatively low determination coefficient, $R^2 = 0.60$, due to a quite large dispersion around the regression line, particularly for the high soil water content values. This may be due to the fact that the location of the weather station is distant from the field trials sites, which can cause errors in the rainfall estimates when local storms pass over Tel Hadya. In fact, for 1991/92 a negative difference between simulated and observed soil water content was verified following one rainfall event that causes a nearly constant model underestimation from there on. On the contrary, for 1993/1994 the model steadily overestimates the soil water after another rainfall event. These facts may indicate that rainfall was underestimated for that particular event in 1991/92, and was overestimated for that day in 1993/94. For verification of this hypothesis, a correction in the rainfall depths for both events were simulated that produced results similar to those for 1992/93 (results not shown). However, because it was not possible to appropriately correct the rainfall data, results are kept as originally observed, so with the discrepancies identified in Figure 5.

Nevertheless, the fact that the regression coefficient for all data is b = 1.0 indicates that the prediction capability of the model is satisfactory. Thus, the validation results are appropriate and show that the ISAREG model is able to predict the soil water content for irrigated and rainfed wheat in Tel Hadya. Therefore, it can be used to simulate alternative water saving supplemental irrigation schedules for similar conditions in the region.

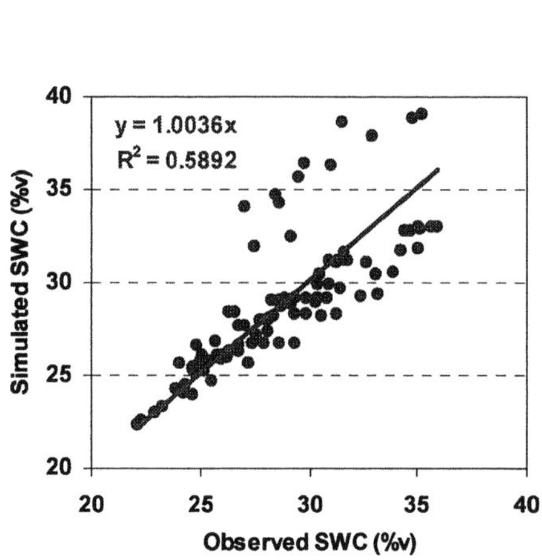

Fig. 5. Simulated vs. observed soil water content (% volume) for all wheat trials analyzed (Tel Hadya, 1991/92, 1992/93 and 1993/94)

3.2. NET IRRIGATION REQUIREMENTS

The annual net irrigation requirements (NIR) and the available soil water at planting (ASWP) for the time series 1984-2000 were computed with the ISAREG model by performing a sequential soil water balance. Results are shown in Figure 6. The ASWP is estimated performing a bare soil water balance starting in the end of summer with an assumed soil water content of 5 % of total available water (TAW) and finishing in the wheat planting date. Results in Figure 6 show that the ASWP is often insufficient to ensure appropriate crop establishment. The irrigation requirements for average, high and very high demand conditions, which correspond to the probabilities of 50, 20 and 5 % for being exceeded calculated from the empirical distribution of NIR (Figure 6), were identified. The analysis that follows is performed for these three reference years.

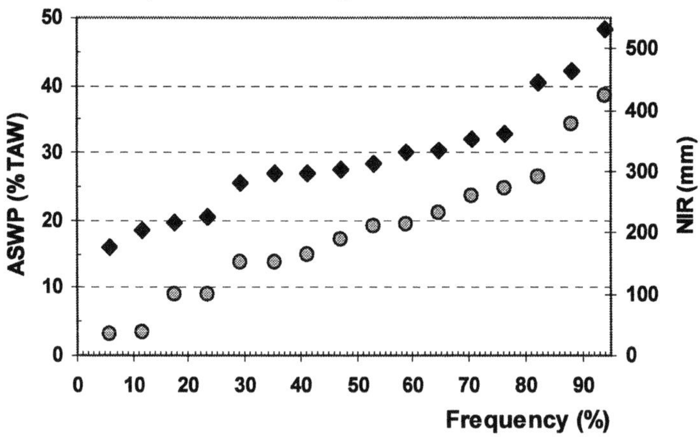

Fig. 6. Frequency distribution of the available soil water at planting, ASWP (●) and the net irrigation requirements, NIR (◆) for wheat (Tel Hadya, 1984-2000)

Table IV summarizes the net irrigation requirements, the available soil water at planting, the season rainfall and the season crop evapotranspiration for high and very high climatic demand conditions.

Table IV. Net irrigation requirements (NIR), available soil water at planting (ASWP), seasonal rainfall, and seasonal crop evapotranspiration (ET_c) for average, high and very high climatic demand conditions

Demand condition	Season NIR (mm)	ASWP (mm)	Season rainfall (mm)	Season ET_c (mm)
Average	312	44	275	525
High	445	14	221	554
Very high	532	50	152	682

3.3. YIELD RESPONSES TO SUPPLEMENTAL IRRIGATION

The wheat water requirements in North Syria are about 320, 400 and 560 mm for the average, high and very high climatic demand conditions, respectively. Model results are presented in Table V for surface irrigation and in Table VI for set sprinkler systems. Results show that it is recommended to apply one irrigation of 80 mm at planting when surface irrigation is practiced, or two irrigations of 40 mm in case of sprinkling, because rainfall is generally insufficient to ensure appropriate crop establishment. In alternative, planting may be delayed (Oweis *et al.*, 2000).

Table V. Selected results of wheat crop responses to supplemental irrigation strategies in Tel Hadya using surface irrigation (I_n = 80 mm)

Climatic demand condition	Season Irrigation (mm)	Number of irrigation events					Season ET (mm)	Relative ET (%)	RYL (%)
		Season	20/11 to 30/12	30/12 to 10/4	10/4 to 15/5	15/5 to 8/6			
Average	320	4	1	-	2	1	523.9	99.9	0.0
	240	3	1	-	2,	-	507.0	96.6	9.4
	160	2	1	-	1	-	453.4	86.4	23.5
	80	1	-	-	1	-	383.6	73.1	31.3
	0	-	-	-	-	-	310.7	59.1	34.6
High	400	5	1	1	2	1	553.4	99.8	0.0
	320	4	1	1	2	-	516.7	93.2	14.9
	240	3	1	1	1	-	455.9	82.2	30.0
	160	2	1	-	1	-	385.5	69.5	40.0
	80	1	-	1	-	-	308.0	55.6	46.0
Very high	560	7	1	3	2	1	681.6	99.9	0.0
	400	5	1	2	2	-	580.5	85.1	26.2
	320	4	1	2	1	-	508.7	74.6	37.0
	240	3	1	1	1	-	430.9	63.2	43.5
	160	2	1	1	-	-	354.5	52.0	47.9

Table VI. Selected results of wheat crop responses to supplemental irrigation strategies in Tel Hadya using set-sprinkler systems (I_n = 40 mm)

Climatic demand condition	Season irrigation (mm)	Number of irrigation events: Season	20/11 to 30/12	30/12 to 10/4	10/4 to 15/5	15/5 to 8/6	Season ET (mm)	Relative ET (%)	RYL (%)
Average	280	7	2	-	3	2	523.7	99.8	0.1
	240	6	1	-	4	1	501.7	95.6	12.8
	200	5	1	-	3	1	478.3	91.2	20.3
	160	4	1	-	3	-	451.3	86.0	22.5
	120	3	1	-	2	-	419.8	80.0	26.2
	80	2	1	-	1	-	385.2	73.4	28.1
	40	1	1	-	-	-	349.0	66.5	31.6
	0	-	-	-	-	-	310.7	59.2	34.6
High	400	10	2	2	4	2	553.1	99.8	0.1
	360	9	2	1	5	1	539.6	97.3	6.7
	320	8	2	1	5	-	517.8	93.4	13.9
	280	7	2	1	4	-	486.8	87.8	21.0
	240	6	2	1	3	-	454.7	82.0	25.5
	200	5	2	1	2	-	420.6	75.9	29.9
	160	4	2	-	2	-	384.8	69.4	36.0
	120	3	1	1	1	-	347.0	62.6	40.3
	80	2	1	1	-	-	309.2	55.8	42.4
	40	1	1	0	-	-	270.8	48.8	45.2
Very high	560	14	1	6	5	2	681.6	99.9	0.1
	360	9	1	5	3	-	545.2	79.9	28.4
	320	8	1	5	2	-	508.9	74.6	34.1
	280	7	1	5	1	-	470.8	69.0	36.9
	240	6	1	4	1	-	430.5	63.1	38.6
	200	5	1	3	1	-	392.8	57.6	42.2
	160	4	1	3	-	-	354.4	52.0	45.8
	120	3	1	2	-	-	314.8	46.2	48.6

Under average climatic demand conditions, the adoption of supplemental irrigation for wheat is easily achievable. In fact, the water-yield characteristics of the wheat crop (Kirkham and Kanemasu, 1983) favor such strategies. Reducing the seasonal irrigation from the full irrigation strategy to 160 mm *i.e.* reducing irrigation demand to 50%, produces a relative yield loss (RYL) of about 23 % (23.5 % in surface irrigation and 22.5 % in case of sprinkler irrigation) Irrigations would be required both at planting and by April-May, when final yield is formed. Results of simulation of the soil water content for those irrigation schedules for both irrigation methods are shown in Figure 7. In this figure are also shown the total available water (TAW) characterizing the Tel Hadya soils, and the readly available water (RAW) corresponding to the depletion fraction p indicated in Table I. In case of sprinkler irrigation, a mild water stress would be produced after emergency and irrigation would be required earlier than for surface irrigation, by April.

Larger reductions (75 %) in the seasonal irrigation depths, from full irrigation to only 80 mm, would produce RYL near 31% and 28% respectively for surface irrigation and sprinkling (Tables V and VI). Differences in RYL among both irrigation methods are due to the fact that irrigation at planting would then be kept for sprinkling but would not be considered for surface irrigation, thus inducing water deficits at earlier crop stages. Nevertheless, there is limited impact on yield due to such large reduction in the season irrigation depth because the seasonal rainfall is enough to ensure yield, including under rainfed conditions. To highly increase water productivity is achievable under average climatic demand conditions. Therefore, when supplemental irrigation is applied under average climatic conditions it is recommended to be practiced at planting to guarantee an appropriate crop establishment and by April, during the mid crop growth period, because it is then the most sensitive crop stage to water deficits since it controls grain formation and filling.

(a) (b)

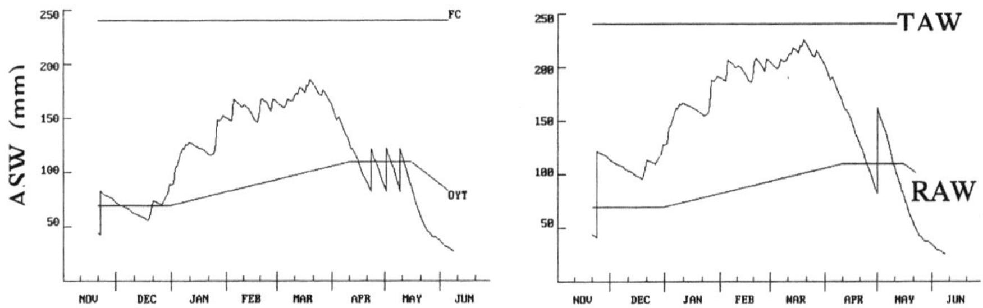

Fig. 7. Simulated available soil water (mm) when 160 mm supplemental irrigation is applied during the wheat crop of season under average climatic demand conditions using: (a) surface irrigation; and (b) set-sprinkler systems. TAW and RAW indicate the total and readly available soil water

For high climatic demand conditions, which correspond to mild or moderate droughts, an irrigation at planting has to be considered to provide for adequate soil water at crop establishment. Winter rains are then insufficient to assure appropriate soil water conditions at flowering and grain forming stages, thus spring irrigations need to be anticipated relative to average climatic conditions. Reducing irrigations from 400 mm to 240 mm, *i.e.* a reduction of 40% in the season irrigation depth leads to RYL ~ 30% for surface irrigation and RYL ~ 26% for sprinkling (Tables V and VI).

Figure 8 shows the soil water content simulated when the season irrigation depth is 240 mm for both surface and sprinkler irrigation. Reducing applied depths to half of optimal irrigation depths would produce RYL ≥ 30 %, while applying only 80 mm would lead to RYL > 42 % (Tables V and VI). Larger reductions in applied depths may be easier practiced with sprinkler irrigation. When available water is severely restricted for the season, irrigations may be easier applied by sprinkling at planting and by the mid season while the same water volume may do not be enough for applications at both timings when surface irrigation is practiced.

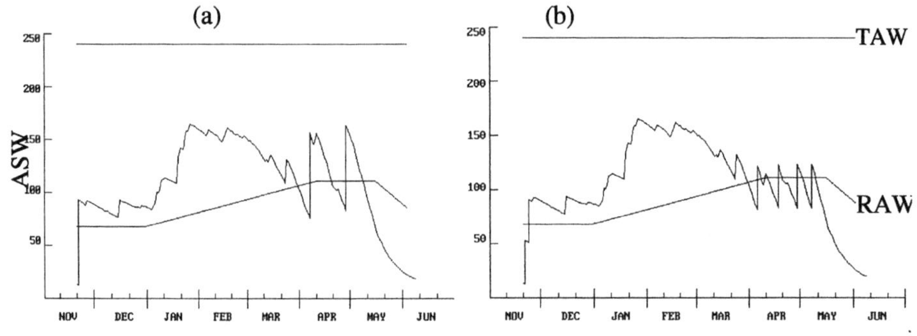

Fig. 8. Simulated available soil water (mm) when 240 mm supplemental irrigation is applied during the wheat crop of season irrigation under high climatic demand conditions using: (a) surface irrigation; and (b) set-sprinkler systems

The simulation analysis show that when high climatic demand conditions prevail an irrigation at planting is recommended and spring irrigation should be applied early April to control water deficits at the critical water stress periods, *i.e.* to ensure that the appropriate number of grains per head be formed and grain filling be successful. Under very high climatic demand, as for severe droughts, reducing the optimal season depth from 560 mm to 320 mm, would produce RYL of the same magnitude (35 %) of the non irrigated crop under average conditions. Under severe drought, not only soil water is insufficient at planting, but also rainfall is largely scarce along the crop season. Thus crop water stress is generally apparent from February. These conditions impede rainfed production and imply larger season applied depths to provide for successful crop production. The simulated soil water content for a season applied depth of 320 mm are shown in Figure 9 for surface and sprinkler irrigation. It shows that when sprinkler irrigation is used less water is applied at planting and irrigation should start earlier than for surface methods. A more frequent irrigation until the yield formation stage may lead to lower RYL than for surface irrigation.

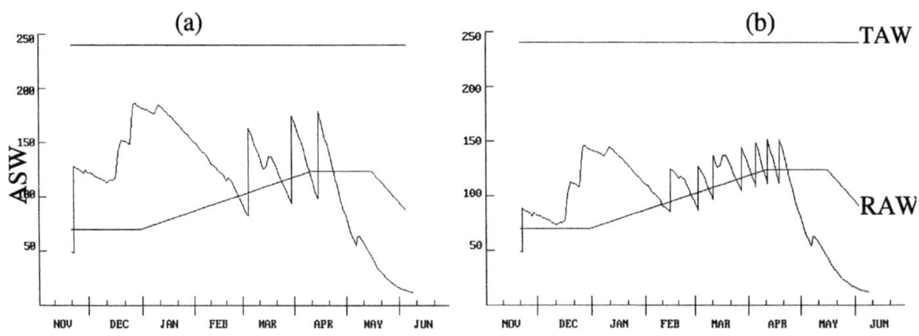

Fig. 9. Simulated available soil water (mm) when 320 mm supplemental irrigation is applied during the season irrigation under very high climatic demand conditions: (a) surface irrigation; and (b) set-sprinkler systems

Under very high climatic demand as for severe droughts, less water may be applied but generally causing RYL > 40 %. In any case, the supplemental irrigation schedules consider irrigation at planting and at starting elongation in order to ensure respectively crop establishment and the formation of an adequate number of heads and the potential seed number per head. Next, irrigation is required at heading to provide for appropriate number of seeds per head. Finally, irrigation is also desirable at yield formation to avoid incomplete grain filling. Results are similar to those obtained for Central Tunisia (Zairi *et al.*, 2002). Therefore, despite an economical analysis has not been performed, it may be concluded that deficit irrigation to increase water productivity may be economically feasible.

4. Conclusions

The analysis presented in this paper shows the usefulness of using simulation models such as the ISAREG model to establish alternative supplemental irrigation schedules for wheat in relation to the available seasonal irrigation water and to evaluate the feasibility of their adoption to cope with drought and water scarcity in relation to the relative yield losses expected. The modelling approach may also be useful to interpret the field experimental results and to extend them to wider climatic conditions. However, it is important to underline that the information obtained by these models would better be complemented with an economical analysis.

Results of this study show different responses of the wheat crop to supplemental irrigation when the climatic demand conditions are diverse. Under average climatic demand conditions, irrigation at planting is desirable and spring irrigation is required to achieve adequate grain filling. When moderate droughts occur, irrigation requirements increase: not only planting irrigation has to be considered but also the spring irrigation has to be anticipated to avoid a small number of grains per head. Under reduced severe drought, irrigation at planting is also necessary and spring irrigations have to be anticipated to the elongation stage to avoid stress at heading.

Simulation results show differences concerning sprinkler and surface irrigation. Sprinkling may be more favorable when more severe deficits are considered.

Acknowledgments

This research is funded by the Bundesministerium f. Wirtschaftliche Zusamenarbeitung through the Program on Water Management in West Asia and North Africa carried by ICARDA and the research contract ERBIC-18CT-970169 with the European Union.

References

Allen, R.G., Pereira, L.S., Raes, D. and Smith, M. (1998) *Crop Evapotranspiration. Guidelines for Computing Crop Water Requirements.* FAO Irrig. Drain. Pap. 56, FAO, Rome, 300p.

Doorenbos, J. and Kassam, A.H. (1979) *Yield Response to Water.* FAO Irrig. Drain. Pap. 33, FAO, Rome.

Kirkham, M.B. and Kanemasu, E.T. (1983) Wheat. In: I.D. Teare and M.M. Peet (Eds.) *Crop-Water Relations.* John Wiley & Son, Inc., pp. 187 - 211.

Martin, D.L., Stegman, E.C. and Fereres, E. (1990) Irrigation scheduling principles. In: G.J. Hoffman, T.A. Howell and K.H. Solomon (Eds.) *Management of Farm Irrigation Systems.* ASAE, St. Joseph, MI, pp. 155-203.

Oweis, T. (1997) *Supplemental irrigation, a highly efficient water use practice.* ICARDA, Aleppo, 16p.

Oweis, T. (2001) Coping with Increased Water Scarcity in Dry Areas: Increased Water Productivity. *Proceedings of the international workshop "New Approaches to Water Management in Central Asia",* Aleppo, Syria, 6-11 Nov 2000. UNU Desertification Series No 3, pp. 75-96.

Oweis, T., Hachum, A. and Kijne, J. (1999). *Water Harvesting and Supplemental Irrigation for Improved Water Use Efficiency in Dry Areas.* SWIM Paper 7, ICARDA and IWMI, Colombo, 41p.

Oweis, T., Zhang, H. and Pala, M. (2000). Water use efficiency of rainfed and irrigated bread wheat in a Mediterranean environment. *Agron. J.* 92(2): 231-238.

Pereira, L.S. (2002) Irrigation demand management to cope with drought and water scarcity. (This issue).

Pereira, L.S., Teodoro, P.R., Rodrigues, P.N. and Teixeira, J.L. (2002) Irrigation scheduling simulation: the model ISAREG. (This issue).

Rodrigues P.N., Pereira L. S. and Machado T. G. (2000) KCISA, a program to compute time averaged crop coefficients. Application to field grown vegetable crops. In: M.I. Ferreira and H.G. Jones (Eds.) *Proceedings of the 3rd International Symposium on Irrigation of Horticultural Crops.* ISHS, Acta Horticulturae Nr. 537, Leuven, vol. I, pp. 535-542.

Rodrigues, P.N, Machado, T., Pereira, L.S., Teixeira, J.L, El Amami, H. and Zairi, A. (2002) Feasibility of deficit irrigation with center-pivot to cope with limited water supply in Alentejo, Portugal. (This issue).

Stewart, J.L., Hagen, R.M. and Pruit, W.O. (1976) *Water production functions and predicted irrigation programs for principal crops as required for water resources planning and increased water use efficiency.* University of California, Dept. Land, Air and Water Resources, Davis, and USDI/Bureau of Reclamation, Denver, CO.

Teixeira, J.L. and Pereira L.S. (1992) ISAREG - An irrigation scheduling model. *ICID Bulletin.* 41(2): 29-48.

Zairi, A., El Amami, H., Slatni, A., Pereira, L.S., Rodrigues, P.N. and Machado, T. (2002) Coping with drought: evaluation of deficit irrigation strategies for cereals and field horticultural crops in Central Tunisia. (This issue).

Zhang, H. and Oweis, T. (1999). Water-yield relations and optimal irrigation scheduling of wheat in the Mediterranean region. *Agric. Water Manag.* 38: 195-211.

Part 4

Modelling water supply system operation under drought conditions

DERIVATION OF OPERATION RULES FOR AN IRRIGATION WATER SUPPLY SYSTEM BY MULTIPLE LINEAR REGRESSION AND NEURAL NETWORKS

A. CANCELLIERE, G. GIULIANO, A. ANCARANI and G. ROSSI
Department of Civil and Environmental Engineering, University of Catania, Italy

Abstract: An approach based on optimization-multiple regressions or neural networks is applied for deriving the operating rules of an irrigation supply reservoir. Two Dynamic Programming models are adopted to determine the optimal releases by using objective functions based on the sum of squared deficits, subject to various constraints. The two models differ for the presence of a penalty term enforcing solutions which consider system management criteria, based on social and political constraints. Then, monthly operation rules are derived by expressing the optimal releases as functions of significant variables by regression or neural networks approach. Both regression equations and neural networks parameters are derived from optimization results on a long period including severe drought events; this period is also used to select the best rules. Then, the behaviour of such operation rules is assessed on a different shorter period, by simulating reservoir operation and computing several performance indices of the reservoir and crop yield through a soil water balance model. Results show that neural networks approach appears to improve the reservoir operation and that operating rules based on optimisation with constraints resembling real system management criteria, yield good performance both in normal and in drought periods, reducing maximum deficits and water spills.

Key words: dynamic programming, neural networks, irrigation reservoir, operating rules

1. Introduction

In the last years, the occurrence of severe and frequent droughts has emphasized the need to improve operation rules to be adopted by water resources managers in the decision making process, in order to face more effectively the risk of dramatic water deficiency to users (Cancelliere *et al.*, 1998). In this perspective, the use of optimization algorithms to derive the operation rules represents an important advancement in the field of water resources management.

Optimization techniques have been often successfully applied in the past to study the operation and the design of reservoirs systems. Available methods include linear programming, dynamic programming (DP) and nonlinear programming, to be chosen according to the mathematical modelling of reservoir system and availability of data. DP technique has been widely applied because of the possibility to model nonlinear and

G. Rossi et al. (eds.), *Tools for Drought Mitigation in Mediterranean Regions*, 275-291.

stochastic features of water resources systems, and to reduce highly complex problems to a multistage decision process.

Definition of operating rules can be carried out by a dynamic programming- regression approach, as proposed by Young (1967). In particular, optimisation technique allows to compute, on the basis of a given inflow series to the reservoir, the corresponding optimal set of releases, i.e. releases which minimise or maximise a given objective function (Yeh, 1985; Simonovic, 1992; Wurbs, 1993). Then, such series of optimal releases are expressed as a function of reservoir state variables and hydrologic input (storage, inflows, etc.) by regression equations, which ultimately allow the water managers to determine the water to be released as a function of available information (Bhaskar and Whitlach, 1980; Karamouz and Houck, 1987; Rossi *et al.*, 1999).

Application of regression techniques however requires the explicit definition of the link between independent and dependent variables. To partly overcome such difficulty, a new interpolation method based on Neural Networks (NN) has been recently applied in several fields of research; it can also be considered as a candidate technique to the determination of operating rules. NN are able to approximate a wide range of multivariate linear and non-linear functions, while maintaining very good generalisation capabilities, and therefore they are ideal candidates if the relationship among the variables is unknown. They have been applied for interpretation, prediction, diagnosis, planning, monitoring, debugging, repair, instruction, and control, but the most successful applications are in categorization and pattern recognition; in fact, they have been shown to perform successfully where other methods do not, recognizing and matching complicated, vague, or incomplete patterns.

In last years NN have found more and more application in hydraulics and hydrologic field, and also in water management studies. Among the recent state-of-arts on the subject the following are worth to be mentioned. The first part of a study published in ASCE (2000a) focused on setting a conceptual framework for the application of NN to related hydrologic issues, by comparing the modelling capabilities of NN with other traditional modelling approaches. Applications of NN to many nonlinear hydrologic processes such as rainfall-runoff modelling, streamflow forecasting, ground water, precipitation forecasting, and water quality issues, can be found in the second part of the ASCE study (ASCE, 2000b), as well as in Govindaraju and Ramachandra (2000). NN have showed a great potential in prediction problems, although a good understanding of the underlying physical processes seemed to be a key factor for successful applications. The study also pointed out the need for further investigations, to be conducted when applying NN, about the choice of their architecture, the optimal training data set, etc.

In water management, NN have been applied for deriving reservoir operating policies with reference to different types of water supply systems. Raman and Chandramouli (1996) applied dynamic programming algorithm to a single irrigation reservoir, then trained neural networks by back-propagation algorithm for deriving operating rules from optimization results. Initial storage, inflow and demand during the period were selected as independent variables. The performance of this model was then compared with a stochastic dynamic programming model and the standard operating policy.

Neelakantan and Pundarikanthan (1999) applied a simulation-optimisation procedure to improve the operation of a municipal water supply reservoir under deficit conditions. The operation policies were based on a hedging rule, able to avoid extreme deficit conditions and to increase the system sustainability. In order to speed up the optimization process, NN were used to replace traditional models in the reservoirs system simulation. This allowed for a more efficient screening of the potential operation policies.

Rossi *et al.* (1999) compared the linear regression approach vs. the neural network approach, for deriving operating rules of a reservoir supplying an irrigation district. Monthly optimal releases were obtained by DP, assuming the sum of square deficits as objective function. Reservoir storage at the beginning of the current month, streamflow and release during the previous month were selected as independent variables. Results indicated that NN performed better than multiple linear regression.

Jain *et al.* (1999) applied NN for two different tasks. First, NN were used for predicting inflows to a reservoir part of a multipurpose project devoted primarily to irrigation and generation of electric power. Then reservoir operation policies were formulated relating optimal releases with storage, inflow and demand trough linear and nonlinear regressions and NN.

Cancelliere *et al.* (2000) analysed the effects of the presence of severe drought years in the definition of the operating rules of an irrigation reservoir through the use of DP and NN techniques. The selected rules were compared on the basis of reservoir performance indices and crop yield losses.

Chandramouli and Raman (2001) made an attempt to extend the DP-NN model, previously applied to a single reservoir, to a multireservoir project. A DP formulation of three reservoirs working in conjunction was developed, and operating rules were derived using regression models and NN.

The present study, developed within the EC INCO-DC project "DSS Drought", concerns the development of a procedure for improving the management of irrigation reservoir under drought conditions. The operation rules are derived by a combined use of dynamic programming, multiple regression and neural network techniques. In particular, water demands are computed by averaging monthly irrigation requirements obtained through a soil water balance model (Pereira *et al.*, 2002) . Then the optimal set of releases is computed by the application of a Dynamic Programming (DP) model, adopting the sum of squared deficits as objective function. Further, the above algorithm has been slightly modified by introducing a penalty term enforcing more realistic solutions. i.e. reservoir releases equal to irrigation demands are imposed when reservoir storage exceeds a fixed percentage of maximum storage. Operating rules are derived by tracking the input-output data of the DP model, applied on a long training period containing a severe drought period, using multiple linear regression (MLR) and NN techniques. In order to make the model easier to use, only known variables are considered as input, while optimal releases are the output of the model. Then the obtained operating rules are validated by simulating the behaviour of the reservoir on a different short period. The behaviour of the different operating rules is assessed by simulation in terms of crop yield indices, difference between actual reservoir releases and water demands and other performance indices of the reservoir.

2. Theory

2.1. DYNAMIC PROGRAMMING MODELS

Determination of optimal releases from an irrigation reservoir should take into account, either implicitly or explicitly, how such releases will affect agricultural yield losses. However, direct determination of optimal releases that minimise yield losses can be cumbersome, because of their dependence on several factors, such as the type and timing of the crop being irrigated, its stage of growth, etc.

Following a consolidated approach, optimal releases can be computed using as objective function the sum of the squared deficits with respect to the demand. Here such objective function has been adopted considering water demands patterns equal to mean monthly crop requirements series, obtained through a soil water balance model. Then an approximate estimate of agricultural yield consequent to deficits has been obtained as the ratio of actual to maximum evapotranspiration (Doorembos and Kassam, 1979).

Adopting such irrigation requirements as demand D_t, the optimisation problem can be stated as follows:

Minimise:

$$s = \sum_{t=1}^{N} (D_t - U_t)^2 \tag{1}$$

subject to the reservoir hydrological balance constraints:

$$S_t = S_{t-1} - U_t + I_t - E_t \qquad t=1,\dots,N \tag{2}$$

$$0 \leq S_t \leq K$$
$$0 \leq U_t$$

where:

- t = current time interval (month)
- D_t = irrigation demand in stage t
- U_t = release from reservoir in stage t
- S_t = volume stored in reservoir at the end of stage t
- K = maximum reservoir capacity
- I_t = reservoir inflow during stage t
- E_t = evaporation losses from reservoir during stage t
- N = total number of optimisation stages

Implementation of the above optimisation problem within a DP framework can be pursued by considering the following recursive equation for any stage t:

$$f_t^n(S_t) = min[Z_t + f_{t+1}^{n-1}(S_{t+1})] \qquad (3)$$

- where $Z_t = (D_t - U_t)^2$ when $U_t \leq D_t$
- and $Z_t = (U_t - D_t)$ when $U_t > D_t$

subject to the state equation (2) and to the constraints on stored volume and releases. Note that the two conditions on Z_t in equation (3) allow to take into account the presence of spills, which otherwise would have the same weight of deficits thus leading to unrealistic solutions.

Straight application of the above algorithm however can still lead to unrealistic sequences of releases, since a perfect knowledge of future inflows is assumed, and thus apparently unnecessary deficits might be imposed on demands because of the presence of drought periods, possibly far ahead in the future. Therefore the above algorithm has been slightly modified by introducing a penalty term enforcing more realistic solutions. In particular, reservoir releases equal to irrigation demands were imposed when reservoir storage exceeds a fixed percentage of maximum storage. In what follows this approach is called CDP (Constrained DP).

2.2. NEURAL NETWORKS MODEL: MULTILAYER PERCEPTRONES

Following a consolidated approach, the monthly operation rules were usually derived through the combined use of dynamic programming and regression methodologies, by regressing optimal releases on significant independent variables in the reservoir system. Usually linear and nonlinear functions of reservoir storage at the beginning of each period, current inflow, and inflows from previous periods were considered as representative release policies (Baskar and Withlatch, 1980).

The performance of the policies derived using the latter approach should be verified through simulation. In fact, a regression model presenting the highest coefficient of determination gives the best fit to the known data, but may exhibit lower generalisation capability than other models, i.e. lower ability to fit unknown data. This may result in a bad performance when used in real time operation.

The application of a regression model requires the verification of some statistical hypotheses on independent variables, and the adoption of appropriate functions in order to fit dynamic programming results. The nonlinearity of the functions to be fitted and the several variables to be considered constitute further limits to regression methodology. Such difficulties can be overcome by the application of a neural networks approach.

A neural network can be defined as "a massively parallel distributed processor that has a natural propensity for storing experimental knowledge and making it available for use. This technique finds inspiration from biological neural systems, so that the processing units are called neurons; they can be variously connected, composing different structures. Neural networks resemble the brain in two respects: i) knowledge is acquired by the network through a learning process; ii) interneuron connection strengths known as synaptic weights are used to store the knowledge." (Govindaraju and Ramachandra, 2000).

Neural networks can be trained directly from the data and have, under some hypotheses, the ability to generalize, i.e. to extend their decision making to novel data, not used to find the optimal weights configuration during the network's training.

The most commonly used architecture of NN is the multilayer perceptron, characterized by the presence of one or more hidden layers, whose computational nodes are called hidden neurons or hidden units. Training is usually carried out by the back-propagation algorithm which, on the basis of a comparison between targets and NN output values, determines the set of parameters, called weights, that leads to the minimum of a network error.

Back-propagation is essentially a gradient descent method that minimizes the network error function:

$$E = \sum_{P} \sum_{p} (d_i - y_i)^2 \tag{4}$$

where d_i is the calculated value, y_i is the target, p is the number of output nodes and P is the number of training patterns. The error is propagated backward through the network to each node, and the weights are adjusted based on equation:

$$\Delta w_{ij}(n) = \alpha \frac{\partial E}{\partial w_{ij}} + \eta \Delta w_{ij}(n-1) \tag{5}$$

where $\Delta w_{ij}(n)$ and $\Delta w_{ij}(n-1)$ are the weights increment between node i and j during the nth and (n-1)th epoch. In eq. (5) α and η are called learning rate and momentum.

Learning rate affects the speed of convergence, but can lead to oscillation in the weights, so the momentum factor is used to stabilize the solution.

Since back-propagation is an iterative algorithm, some index is needed to decide when it must be stopped. Usually a quadratic form index is adopted, which gives the root mean squared errors (RMSE) between targets and network outputs:

$$RMSE = \sqrt{\frac{\sum_{i=1}^{N}(d_i - y_i)^2}{N}} \quad (6)$$

The input signal propagates through the network in a forward direction, on a layer-by-layer basis, and neurons in each layer are linked only to the neurons of the preceding and following layer. The input signal to the generic neuron is constituted by a linear combination of the outputs of neurons belonging to the preceding layer; such input signal is then processed by a so called activation function, which is non decreasing and ranged between 0 and 1. The logistic function is usually used:

$$\varphi(\xi) = \frac{1}{1 + e^{-\xi}} \quad (7)$$

where ξ is the input signal to the neuron and $\varphi(\xi)$ the output (Hassoun, 1995). This function is largely used because of the convenient mathematical expression of its derivative, which allows simplification in deriving training algorithms. If the function to be fitted is linear, the following activation function, with no hidden layer, can be used:

$$\varphi(\xi) = \xi \quad (8)$$

In this case the network is equivalent to a linear regression equation.

3. Application

3.1. SYSTEM DESCRIPTION

Pozzillo reservoir on Salso river (tributary of Simeto river, the main Sicilian water course), was selected as a case study. The reservoir is devoted to the irrigation of Catania Plain irrigation district, whose water conveyance and distribution network is operated by the Catania Plain Land Reclamation Consortium. Net storage of reservoir is equal to $123 \cdot 10^6 m^3$,while the catchment area is equal to $577 Km^2$. The reservoir is part of a multipurpose system (hydroelectric, irrigation, and municipal), which includes another reservoir, three diversions, and five hydroelectric plants. Pozzillo dam was completed in 1959, while the operation of the reservoir started on 1964.

The irrigated area of the district is about 18.000 ha, and citrus is the prominent crop (more than 90%). Irrigation volumes are delivered by the Land Reclamation Consortium at fixed intervals, and microirrigation is the most common irrigation technique.

Irrigation district is particularly drought prone. For example, during 1990, at the end of a three years drought period, the annual release was only 7% of the average, due to the depletion of the water stored in previous years. Also, more recent drought events pointed out the need to redefine operating rules, in order to manage effectively water deficiency and to mitigate the worst irrigation deficits.

Available data for the observation period 1959-1998 include:

- monthly streamflow series of Salso river at Pozzillo reservoir, estimated by means of a hydrological balance of the reservoir;
- monthly evaporation rates calculated as a function of the mean monthly temperatures;
- monthly areal precipitations over the irrigation district;
- monthly reference and maximum evapotranspiration for the irrigation district, computed by the Hargreaves model.

Annual streamflow series (hydrological year October-September), shown in Figure 1, presents a high variability between the maximum value equal to $373.6 \cdot 10^6 m^3$ ($647.5mm$), which occurred on 1972 and the minimum value $3.3 \cdot 10^6 m^3$ ($5.7mm$), which occurred on 1989. The mean value is $94.0 \cdot 10^6 m^3$ ($162.9mm$). The most severe drought period was experienced during the three years 1988-90, during which total streamflow volume was below $58 \cdot 10^6 m^3$, and crop yield decreased significantly.

Since citrus is the prominent crop in the irrigation district, irrigation demands (Table I) have been computed by taking the average of the irrigation requirements obtained by a simple soil water balance model. Losses in conveyance and distribution have been assumed to be equal to 25%.

Applying the model with mean values of pedological features of irrigated soil and monthly values of precipitation and temperature related to period 1959-91, the following values were obtained:
- mean yearly demand of 357.5 mm ;
- minimum yearly demand of 181.0 mm in 1972;
- maximum yearly demand of 600.0 mm in 1981.

Table I. Average monthly irrigation demands

	May	***June***	***July***	***August***	***September***	***October***	***Total***
Demand *($10^6 m^3$)*	*3.5*	*11.6*	*24.6*	*25.5*	*17.3*	*5.5*	*88.0*

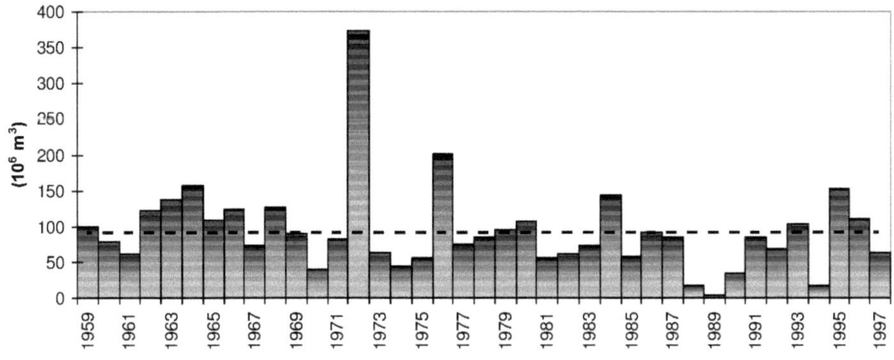

Fig. 1. Annual streamflow series (hydrologic year) of Salso river at Pozzillo dam during the years 1959/60-97/98 and mean annual streamflow

3.2 DERIVATION OF OPERATING RULES

3.2.1. Multiple linear regression equations

Dynamic Programming and Constrained Dynamic Programming were applied using historical inflow series during 1959-91 period, thus obtaining two series of optimal reservoir releases Er_t during the irrigation season. Er_t equals the total release U_t if spills do not occur. The CDP model was run using a stored volume of $100 \cdot 10^6 \, m^3$ as a threshold above which the monthly demand must be fully satisfied.

The results of the optimisation models DP and CDP were used to estimate the parameters of the regression equations. A multiple regression approach was applied to determine optimal releases as a linear function of available information at the beginning of the month. Since the determination of the operating rules was carried out by deriving six equations (one for each irrigation month), the monthly demand D_t was excluded from the set of independent variables. Furthermore, the selection of independent variables was limited to storage volume, release and inflow at previous month.

The choice of the best combinations of input variables was carried out on the basis of the C_p statistic (Helsel and Hirsh, 1995).

$$C_p = k + \frac{(n-k)(s_k^2 - \hat{\sigma}^2)}{\hat{\sigma}^2} \tag{9}$$

where:
n = number of observations;
k = number of coefficients (explanatory variables plus 1);
s_k^2 = mean square error (MSE) of the p coefficient model;

$\hat{\sigma}^2$ = minimum MSE among the 2^k possible models.

Lower values of the statistic indicate better MLR models.

The application of the *Cp* criterion, suggested considering three alternative sets of such variables:
- the storage volume at the beginning of May, the release at previous month for the remaining months (June-October);
- the storage volume at the beginning of May, the release at previous month and the storage volume for the remaining months;
- the storage volume at the beginning of May, the release at previous month and the storage volume for the remaining months except for October, for which the inflow at previous month is also included.

After determining the MLR equations, a simulation model that computes releases according to the MLR operating rules was applied and the following reservoir performance indices were computed:
- sum of squared deficits $\Sigma(D_t\text{-}Er_t)^2$;
- volumetric reliability $\Sigma Er_t/\Sigma D_t$;
- deficit frequency;
- minimum of the ratios of monthly actual to maximum evapotranspiration, $\min(ETa_t/ETm_t)$, which was assumed as a proxy of agricultural yield. Actual monthly evapotranspiration has been computed by means of a soil water balance methodology similar to the one used to derive the irrigation demands (Cancelliere *et al.*, 2002).

In Table II for each set of MLR equations, the independent variables, the sum of squared deficits, the volumetric reliability, the deficit frequency and the minimum of the ratios of monthly *ETa* to *ETm* in 1959-1991 period are reported. The results of DP and CDP optimisation runs are also shown for comparison.

It can be inferred that MLR (DP), trained using as data pattern DP results, are more conservative than MLR(CDP), thus leading to a lower sum of squared deficits and volumetric reliability. However, MLR(DP) never satisfy fully the monthly demand (deficit frequency 100%), regardless of the storage levels, which is obviously unacceptable from a practical standpoint. This is due to the features of the adopted optimisation models. In fact, DP algorithm may lead to deficits many years in advance of a drought period. CDP algorithm, on the other hand, is forced to release the demand D_t when storage is bigger than the fixed threshold, thus leading to a lower deficit frequency and a slightly greater volumetric reliability and sum of squared deficits.

Regarding the crop yield index, the performance of MLR equations are very similar because of the three years period 1988-1990, whose very low inflows produce high deficits with any operation rule, and consequently similar levels of water stress in crops.

Table II. Independent variables, sum of squared deficits, volumetric reliability, deficit frequency and minimum monthly ratio between maximum and actual evapotranspiration during 1959-91 period for analysed linear regression equations based on optimisation models DP and CDP

	Months			$\Sigma(D_t-Er_t)^2$	$\Sigma Er_t/\Sigma D_t$	**Deficit frequency**	**Min** (ETa_t/ETm_t)
	May	June-Sept.	October	$(10^6\ m^3)^2$	(%)	(% months)	(%)
MLR(DP)	S_{t-1}	Er_{t-1}	Er_{t-1}	4225	78	100	26
	S_{t-1}	S_{t-1}; Er_{t-1}	S_{t-1}; Er_{t-1}	4129	79	100	26
	$\boldsymbol{S_{t-1}}$	$\boldsymbol{S_{t-1}; Er_{t-1}}$	$\boldsymbol{S_{t-1}; I_{t-1}; Er_{t-1}}$	4120	79	100	26
MLR(CDP)	$\boldsymbol{S_{t-1}}$	$\boldsymbol{Er_{t-1}}$	$\boldsymbol{Er_{t-1}}$	4219	82	86	25
	S_{t-1}	S_{t-1}; Er_{t-1}	S_{t-1}; Er_{t-1}	4249	82	80	26
	S_{t-1}	S_{t-1}; Er_{t-1}	S_{t-1}; I_{t-1}; Er_{t-1}	4253	82	80	26
DP		-		2555	83	81	36
CDP		-		3718	84	60	27

Based on these results, the third set of MRL (DP) rules and the first set of MRL (CDP) rule, indicated in bold in Table II, have been selected for validation.

3.2.2. Neural networks based rules

Like for MLR approach, six different networks were trained (one for each irrigation month), having the same input variables as the selected linear regression equations. The optimal number of neurons in the hidden layer was determined by heuristics, following the indications of a previous study (Cancelliere *et al.*, 2000). In particular, networks with logistic activation functions and three neurons in the hidden layer were adopted. Training was carried out by back-propagation algorithm, considering a stopping criterion based on the RMSE threshold.

The same indices of the MLR approach were adopted to evaluate reservoir performance. For each set of NN considered, the input variables, the sum of squared deficits, the volumetric reliability, the deficit frequency and the minimum of the ratios of monthly *ETa* to *ETm* in 1959-1991 period are reported in Table III. The results of DP and CDP optimisation runs are also reported for comparison.

It can be inferred from the table that NN(DP), trained using as data pattern DP results, are more conservative than NN(CDP), thus leading to a lower sum of squared deficits and volumetric reliability. In analogy with the MLR approach, NN(DP) never satisfy fully the monthly demand (deficit frequency 100%). NN(CDP) on the other hand, release the demand *Dt* when storage is greater than the fixed threshold in CDP optimisation, thus leading to a lower deficit frequency and a slightly greater volumetric reliability. Same considerations as for MLR approach apply to the crop yield index.

Table III. Input variables, sum of squared deficits, volumetric reliability, deficit frequency and minimum monthly ratio between maximum and actual evapotranspiration during 1959-91 period for analysed neural networks and optimisation models DP and CDP

	Months			$\Sigma(D_t-Er_t)^2$	$\Sigma Er_t/\Sigma D_t$	**Deficit frequency**	**Min** (ETa_t/ETm_t)
	May	June-Sept.	October	$(10^6\ m^3)^2$	(%)	(% months)	(%)
NN(DP)	S_{t-1}	Er_{t-1}	Er_{t-1}	4025	80	100	26
	S_{t-1}	S_{t-1}; Er_{t-1}	S_{t-1}; Er_{t-1}	4044	81	100	27
	S_{t-1}	S_{t-1}; Er_{t-1}	S_{t-1}; I_{t-1}; Er_{t-1}	3830	80	100	27
NN(CDP)	S_{t-1}	Er_{t-1}	Er_{t-1}	4319	83	89	25
	S_{t-1}	S_{t-1}; Er_{t-1}	S_{t-1}; Er_{t-1}	4285	82	77	27
	S_{t-1}	S_{t-1}; Er_{t-1}	S_{t-1}; I_{t-1}; Er_{t-1}	4297	82	94	27
DP	-			2555	83	81	36
CDP	-			3718	84	60	27

The third set of NN(DP) rules and the second set of NN(CDP) rules, indicated in bold in the Table III, have been selected for validation.

The difference of behaviour between the DP and CDP based rules can be inferred from Figure 2, where the release vs. initial storage in the month of May are plotted for the two optimisation runs and for the selected NN and MLR rules.

The plot shows that DP releases present a fairly large spread even for full reservoir levels (from no release to full demand), and consequently the trained rules NN(DP) and MLR(DP) never release fully the demand, since they are not able to capture such variability. CDP releases on the other hand, are always equal to the demand for storage volumes above $100 \cdot 10^6\ m^3$ and NN(CDP) reflects such behaviour, thus leading to more realistic operating rules.

4. Validation of operating rules

Validation of the selected operating rules (MLR and NN) was performed by simulating reservoir operation during 1992-1998 period, not included in the training period. For comparison purposes, a simulation applying the so called SOP (Standard Operating Policy) rule, i.e. a policy that releases as much water as available without any hedging, in order to meet the current demand, was also carried out. In some extent, such policy can be considered similar to the management practices currently in use for Pozzillo reservoir, since the analysis of historical releases showed that the multi-annual regulation of the reservoir was almost never put into practice.

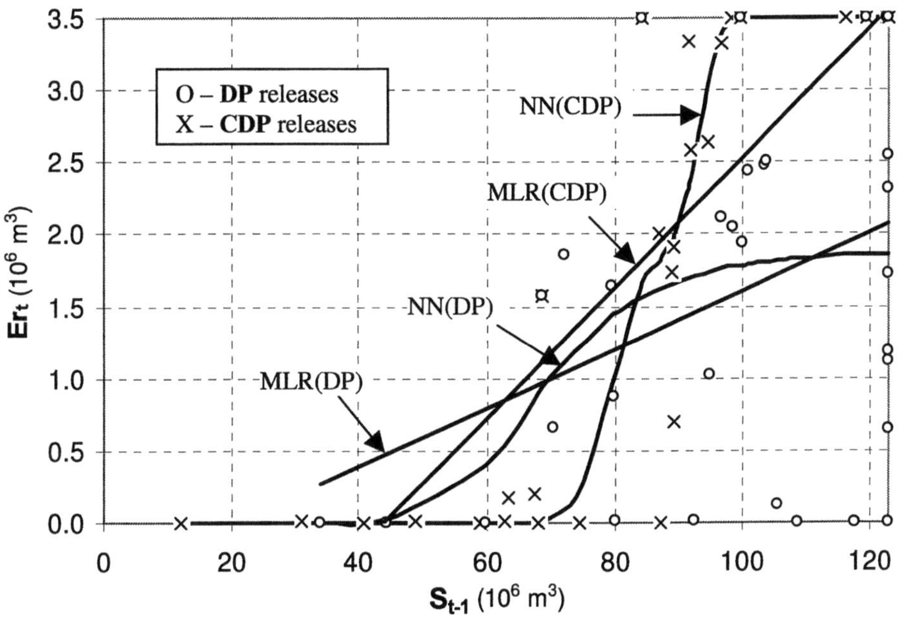

Fig. 2. DP and CDP optimal release (May), MLR and NN operating rules as functions of the storage volume

Furthermore, optimisation of releases during the same period using DP and CDP was carried out. Such optimisation runs represent a theoretical optimal management, under the perfect knowledge of the future and as such, they are useful only for comparison purposes.

Results of simulation and optimisation runs were compared on the basis of the following reservoir performance and crop yield indices:

- volumetric reliability $\Sigma Er_t/\Sigma D_t$;
- sum of squared deficits $\Sigma(D_t-Er_t)^2$;
- deficit frequency;
- deficit mean length;
- maximum seasonal deficit;

– total spills;
– minimum (ETa_t/ETm_t)

The above indices are reported in Table IV. It can be inferred that NN(DP) and MLR(DP) rules lead to very similar results, although the NN(DP) rules obtain a lower sum of squared deficits, which is indicative of a more uniform distribution of the deficits during the irrigation season. MLR(CDP) rules are more conservative than the analogous NN(CDP), thus leading to lower deficits and released volumes, greater spills and higher values of crop yield index. Both NN and MLR rules outperform the SOP rule in terms of smaller sum of squared deficits, smaller maximum seasonal deficit and greater minimum ratio between actual and maximum evapotranspiration. On the other hand, SOP leads to a greater volumetric reliability, lower average length of deficit periods and lower spills. The above findings are consistent with the fact that the developed rules are conservative, while the SOP does not impose deficits, unless there is not enough water to meet the demand.

Table IV shows also that unconstrained NN(DP) leads to overall best results in terms of sum of squared deficits, as expected, and crop yield index. However, NN(CDP) approach performs better in terms of greater volumetric reliability, and lower deficit frequency and total spills, although the sum of squared deficits, the maximum seasonal deficit range and the crop yield index are worse. These results are consistent with those obtained by direct application of dynamic programming runs. Indeed DP, which gives the minimum sum of squared deficits, lower maximum seasonal deficit and better crop yield index, leads to more conservative rules while CDP ensures higher volumetric reliability, lower deficit frequency and equal values of spills.

In Figure 3 the time series of the sum of squared deficits and seasonal releases are shown for the selected rules, the optimisation models and the SOP rule. The presence in the validation period of a particularly severe drought year (1995), emphasizes the opportunity to use operating rules providing some degree of hedging in place of SOP rule, whose simulation run leads to maximum sum of releases, but also maximum seasonal deficit and minimum annual crop yield index.

5. Conclusions

Monthly operation rules for an irrigation reservoir have been derived by interpolating the results of the application of two Dynamic Programming models (DP and CDP), by means of linear regression and neural networks techniques. The two models differ for the presence in CDP of a penalty term enforcing real system management criteria, taking explicitly into account the necessity to release water demand when reservoir storage volume exceeds a fixed threshold.

Table IV. Performance indices of the reservoir operation and crop yield index during 1992-98 period by using rules obtained by the neural networks NN(DP) and NN(CDP), multiple regression MLR(DP) and MLR(CDP), and comparison with values related to SOP rule and optimization models.

	$\Sigma Er_t/\Sigma D$ (%)	$\Sigma(D_t-Er_t)^2$ $(10^6\ m^3)^2$	*Deficit frequency (% months)*	*Deficit mean length (months)*	*Maximum seasonal deficit* $(10^6\ m^3$ *-% dem.)*	*Total spills* $(10^6\ m^3)$	*Minimum* ETa_t/ETm_t *(%)*
NN(DP)	*78*	*639*	*100*	*6*	*45 – 51%*	*43*	*46*
NN(CDP)	*82*	*784*	*76*	*3*	*55 – 63%*	*35*	*41*
MLR(DP)	*78*	*725*	*100*	*6*	*45 – 51%*	*43*	*51*
MLR(CDP)	*81*	*774*	*86*	*5*	*52 – 59%*	*36*	*50*
SOP	*85*	*1600*	*21*	*3*	*65 – 74%*	*35*	*36*
DP	*83*	*465*	*57*	*6*	*30 – 34%*	*35*	*50*
CDP	*84*	*712*	*55*	*6*	*52 – 60%*	*35*	*42*

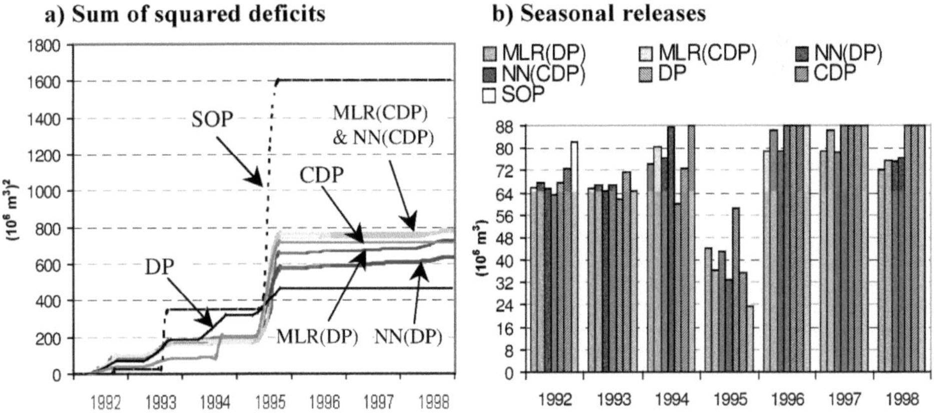

Fig. 3. Comparison between operating rules obtained by neural networks NN(DP) and NN(CDP), multiple linear regression MLR(DP) and MLR(CDP), SOP, DP and CDP optimization during 1992-98 period: a) Sum of squared deficits b) Seasonal releases

First, the best operating rules have been selected by analysing their performance during the optimisation period. Then, the selected operating rules have been validated by simulation of the reservoir operation over a shorter period.

Validation of selected rules over a seven years period showed the superiority of rules based on CDP pattern. In fact, rules based on DP pattern resulted more conservative than others, and never met the total demand, regardless from reservoir storage conditions. On the contrary, rules based on CDP pattern showed the ability to reproduce the CDP operation criteria, thus obtaining best results in terms of reservoir operation indices, and performing only a little bit worst in terms of sum of squared deficits and crop yield index.

NN were shown to be able to improve the reservoir performances during drought conditions, although the results were similar to the ones obtained by the MLR based rules. This confirmed the general agreement on the fact that NN may sometimes enhance the solution of many engineering problems, and yet, that they are not a panacea for every problem.

Ackowledgement

The financial support from European Community grant INCO-DC Project Contract n.IC18-CT97-0169 DG12-CDPE and from Italian Research Council, Group for Prevention of Hydrogeological Disasters is gratefully acknowledged.

References

ASCE Task Committee on Application of Artificial Neural Networks in Hydrology (2000a) Artificial neural networks in hydrology. I: preliminary concepts. *J. Hydrologic Engng.* 5(2): 115-123.

ASCE Task Committee on Application of Artificial Neural Networks in Hydrology (2000b) Artificial neural networks in hydrology. II: hydrologic applications. *J. Hydrologic Engng.* 5(2): 124-137.

Bhaskar, N. and Whitlatch, E. (1980) Derivation of monthly reservoir release policies. *Water Resour. Res.* 16 (6): 987-993.

Cancelliere, A., Ancarani, A. and Rossi, G. (1998) Susceptibility of water supply reservoirs to drought conditions. *J. Hydrologic Engng.* 102 (4): 140-148.

Cancelliere, A., Ancarani, A., Giuliano, G. and Rossi, G. (2000) Determination of reservoir operating rules by neural networks (in Italian). *Proceedings of the XXVII Convegno di Idraulica e Costruzioni Idrauliche*, September 12-15, Genoa, Vol. 3, pp. 141-150.

Cancelliere, A., Ancarani, A., Giuliano, G. and Rossi, G. (2002) A neural networks approach for deriving irrigation reservoir operating rules. *Water Resour. Manage.* 16: 71-88.

Chandramouli, V. and Raman, H. (2001) Multireservoir modelling with dynamic programming and neural networks. *J. Water Resour. Plann. Managt.* 127(2): 89-98.

Doorenbos, J. and Kassam, A.H. (1979) Yield response to water. *Irrigation and Drainage Paper n°33,* FAO, Rome.

Govindaraju, R.S. and Ramachandra Rao, A. (2000) *Artificial Neural Networks in Hydrology.* Kluwer Academy Publishers, the Netherlands.

Hassoun, M.H. (1995) *Fundamentals of Artificial Neural Networks.* The MIT Press, Cambridge.

Helsel, D.R. and Hirsch, R.M. (1995) *Statistical Methods in Water Resources.* New York: Elsevier, Amsterdam.

Jain, S.K., Das, A. and Srivastava, D.K. (1999) Application of ANN for reservoir inflow prediction and operation. *J. Water Resour. Plann. Managt.* 125(5): 263-271.

Karamouz, M. and Houck, H. (1987) Comparison of stochastic and deterministic dynamic programming for reservoir operating rule generation. *Water Resour. Bull.* 23(1): 1-9.

Neelakantan, T.R. and Pundarikanthan, N.V. (1999) Hedging rule optimisation for water supply reservoirs system. *Water Resour. Manage.* 13(6): 409-426.

Pereira, L.S., Teodoro, P.R., Rodrigues, P.N. and Teixeira, J.L. (2002) Irrigation scheduling simulation: the model ISAREG (This issue).

Raman, H. and Chandramouli, V. (1996) Deriving a general operating policy for reservoirs using neural network. *J. Water Resour. Plann. Managt.* 122(5): 342-347.

Rossi, G., Ancarani, A. and Cancelliere, A. (1999) A comparison between neural network approach and multiple regression procedures for deriving reservoir operation policies. In: V.P. Singh, Il Won Seo and J.W. Sonu (Eds.) *Water, Environment, Ecology, Socio-Economics, and Health Engineering* Water Resources Publication, Colorado.

Simonovic, S.P. (1992) Reservoir systems analysis: closing gap between theory and practice. *J. Water Resour. Plann. Managt.* 118(3): 262-280.

Wurbs, R. A. (1993) Reservoir-system simulation and optimization models. *J. Water Resour. Plann. Managt.* 119(4): 445-472.

Yeh, W. (1985) Reservoir management and optimization models: a state of the art review. *Water Resour. Res.* 21(12): 1797-1818.

Young Jr., G. K. (1967) Finding reservoir operating rules. *J. Hydraulic Div.* 93(6): 297-321.

OPTIMISATION OF RESERVOIR MANAGEMENT IN SEMI-ARID COUNTRIES: CASE OF LAKHMESS, TUNISIA

F. LEBDI[(1)], M. BERGAOUI[(2)] and M.A. BOUSLIMI[(1)]
[(1)]*Institut National Agronomique de Tunisie, Tunisia*
[(2)] *École Supérieur des Ingénieurs de l'Équipement Rural, Tunis*

Abstract: The paper presents a model for the management of Lakhmess reservoir on Siliana River in Tunisia. Two objectives were identified namely the satisfaction of the agricultural demand and the guarantee of minimum storage volume in the reservoir at the end of the optimisation horizon. The management rule requires the specification of weights to be given to each objective. Stochastic dynamic programming (SDP) has been applied as optimisation technique. Results indicate the feasibility of the proposed approach for reservoir management in semi-arid countries.

Key words: Reservoir management rules, optimisation, stochastic dynamic programming, shortage risk.

1. Introduction

In Tunisia, water resources management is becoming more and more of primary concern for the decision-makers in order to achieve a better satisfaction of ever increasing water needs. The following example shows the importance of water uses for irrigation in terms of mobilised volumes:

- Irrigated area = 1,000 Ha
- Average flow = 0.8 l/s/ha
- Approximated daily volume = 69,000 m^3

If the objective of management would be water supply of a city and if the daily consumption would be of 200 l/d/inh., a daily volume of 69,000 m^3 would correspond to the water supply of a 345,000 inhabitants town.

An important issue related to irrigation in Tunisia is that the management rules of an irrigation reservoir must attempt to find a compromise between the water quantity to be released during summer and volumes of resources to be stored essential to the growth of future autumnal sowings. These rules of inter-annual management are needed because of the scarcity of the resource as well as because of its variability and lag between the period when water is available and the one when water requirements is needed mostly. The problem of the water resources management is strongly stochastic in the semi-arid zones due to the difficulty of modelling the physical processes (rainfall-runoff process) and to the generally incomplete information, often affected by errors.

G. Rossi et al. (eds.), Tools for Drought Mitigation in Mediterranean Regions, 293-304.

A number of mathematical methods have been developed in the field of water resources management. Yakovitz (1982) drew up a complete overview of the methods developed as well as of the methods based on dynamic programming. Yeh (1985) reviewed the state of art on the matter. Yazicil et al. (1983) set up a multi-objective and multi-reservoir management (Green River Basin) that minimised through linear programming the deviations from fixed drawdown curves of reservoirs. However, regardless of the used technique, modelling runs up against many obstacles, namely:

- The difficulty of quantifying the management objectives;
- The multi-objective aspects of management, where the uses are multiple and objectives cannot be described by only one criterion;
- The stochastic feature of the hydrological series and the behaviour of the resource-users system that must be integrated in the management modelling.

A simplifying choice is to use dynamic programming on possible scenario evolutions (Goussebaille and Roche, 1986) or to test by simulations or adjustment a restricted class of management rules (Bhaskar and Whitlach, 1980). Besides, the concepts of risk, reliability and failure of management can also be taken into account although this involves more complex algorithms (Richard and Turgeon, 1998).

In the present study, a model for the management of the Lakhmess reservoir, located on the basin of Siliana river in Tunisia (Area of Siliana in the north west of Tunisia) is presented. Two objectives of management were selected namely the satisfaction of agricultural demands and the guarantee of a minimum storage volume in the reservoir at the end of irrigation season. The optimisation of the management rules has been carried out by dynamic programming through the specification of the weights of each partial objective.

2. Description of Lakhmess reservoir system

The Lakhmess reservoir is located on the catchment area of Siliana River and its main objective is to supply a nearby irrigated area (Figure 1).

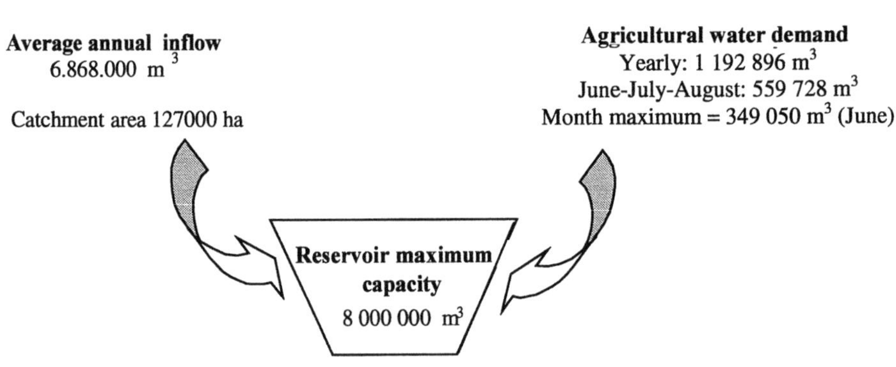

Fig. 1. Lakhmess reservoir main characteristics

On average, the inflows are sufficient to satisfy the agricultural needs, but the inter-annual variability of the flows is very strong. Although the agricultural demand is lower than the average inflows there is a risk of shortage in the summer period, and therefore the reservoir must on one hand to satisfy the agricultural demand from March to August while keeping enough water for the next season.

3. Reservoir water balance

On a given interval of time, the general equation of the water balance is the following :

$$\Delta V = (V_r + V_{ecs} + V_p) - (V_{ev} + V_d + V_{inf} + V_u) \quad (1)$$

- ΔV: variation in the storage volume.
- V_p: contribution due to precipitation on the reservoir surface, estimated from the rainfall observations through the height-surface curve of the reservoir.
- V_{ev}: evaporation volume estimated on the basis of the daily measurements of vats and average surface the same day. This estimate remains subject to error because of the choice of the selected conversion coefficient of vat evaporation to evaporation from the reservoir.
- V_d: spill volume.
- V_u : bottom-draining.
- V_r: inflows, which is a first unknown factor of the balance because there is no gauging stations upstream the reservoir.
- V_{inf}: infiltrated volume, which is a function of the hydraulic load in reservoir and thus of stored volume. It is the second unknown factor of the balance.
- V_{ecs}: a contribution by underground flow into the reservoir and it is the third unknown factor of the balance.

A hydro-meteorological station was installed on the dam in March 1996, allowing to collect daily data of rainfall, evaporation, and water levels in the reservoir.

Inflows series were extended making use of the rainfall-runoff model GR3 of CEMAGREF (software EMILE2, IRD). The model was calibrated on the basis of three years of streamflows estimated from the reservoir water balance. A generation of inflows was then carried out on the basis of rainfall observations in nearby stations, leading to a long series of 24 years of streamflows.

Figure 2 shows the comparison between observed and estimated inflows during the 3 years with available observations. The plot indicates a good agreement between observed and estimated inflows at decadal time step.

Only one bathymetry was surveyed on the reservoir during its construction. The water surface is computed from the height (H) by linear interpolation between two points of reference. The relation volume (V in m3)-height (H in m) is expressed by fourth degree polynomial smoothing.

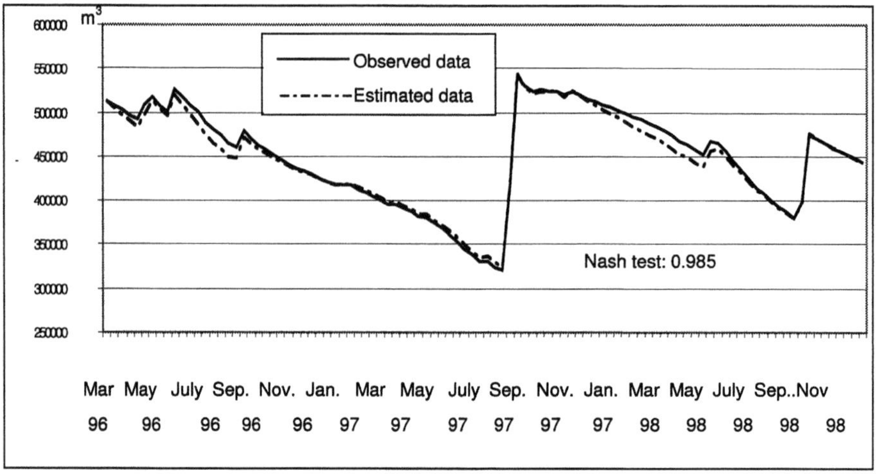

Fig. 2. Comparison between observed and estimated inflows by the rainfall-runoff model GR3

$$V = 151.56\,H^4 - 3757.6\,H^3 + 57586H^2 + 161036H + 1E{+}06 \quad (2)$$

In semi-arid climate, the change in levels of reservoirs are violent and short in time. At a daily time step, we can consider that in the balance of the reservoir infiltration and evaporation are negligible compared to the inflows. The balance equation is then simplified. Recorded volumes are smoothed because of the error in the recordings of the liquid level (limningraphe) recorder (day of high wind and absence of rising) which poses problem during the calculation of the terms of the daily balance.

The evaporation from the reservoir surface is estimated from a Colorado vat located on the study site, by taking a conversion coefficient equal to 0.8 (windy zone). Evaporation values are deduced from a series observed over three years. These values are correlated with the volume stored in reservoir, through a linear expression of the type $V_{ev} = a\,V + b$.

Infiltration can be estimated during no inflow periods as:

$$V_{inf} = \Delta V - V_{ev} \quad (3)$$

The balance conditions are an absence of rain during the last fifteen days and a negative variation of stored water. This infiltrated volume is correlated with the storage level in the reservoir, according to a power expression $V_{inf} = c + V^b$.

4. Optimisation of the reservoir management

4.1. BASIC ELEMENTS OF THE MODEL

The development of the optimization model was based on the identification of the state variables, the decision variable, on the formulation of the reservoir water balance constraints and of the objective function. The formulation of the optimisation model is as follows:

- The state variables are reservoir storages at the beginning of the generic interval and are related to the terms of the water balance (inflows, evaporation, etc.);
- The decision variable is the release at the interval *t*;
- The state equation expresses the change of state variable at a given interval consequent to the decision taken and reflects the dynamic character of the management;
- The constraints express the minimum and maximum values of water storage in the reservoir and flow capacities of the releases:

$$V_{min} \leq V(t) \leq V_{max} \tag{4}$$

$$L_{min} \leq L(t) \leq L_{max} \tag{5}$$

Optimization of releases was carried out making use of an objective function with two weighted objectives, namely the squared deviation of storage from target storage levels and the squared deficit of releases with respect to demand. The first objective attempts to conserve water in the reservoir in order to face future deficits while the second attempts to satisfy the requests of water from the reservoir. The mathematical formulation is as follows:

$$MIN \sum_{t=1}^{11} \left[\alpha \left(\frac{S_{t+1} - S_{consig}}{S_{max}} \right)^2 + (1-\alpha) \left(\frac{u_t - B_t}{B_{max}} \right)^2 \right] \tag{6}$$

S_{t+1}: Final storage at the end of each interval.
S_{max}: Maximum exploitable volume (net capacity of the reservoir).
S_{consig}: Target storage at the end of the interval.
α: Weighting coefficient ranging between 0 and 1.
u_t: Optimised release between time i and $(i+1)$.
B_t: Water demand at the beginning of the management-period.
$t = 0$: corresponds at the beginning of September.

The above objective function does not lead to direct optimisation of an economic benefit, which would provide directly the management solution. Rather it achieves a compromise between the two objectives based on the distribution of the relative weight of each objective (to satisfy requests from the reservoir and to conserve a strategic storage). By selecting different values of the weighting coefficient α different operating rules can be found, among which the one that best fit the desire of the water manager can be identified. Such a choice is carried out by plotting the indices of performance of the reservoir as a function of α, thus facilitating the choice of the value corresponding to the best compromise between the two management objectives.

4.2. OPTIMISATION BY STOCHASTIC DYNAMIC PROGRAMMING (SDP)

Reservoir management is by nature stochastic since the inflows are not known for each step of time. Thus, for each storage level at the interval (t+1) there corresponds an inflow value and a probability of occurrence of such value. For an initial storage level at the interval t, the system can evolve up to n possible final states, being n the number of possible discretized inflow values. The process of optimisation by SDP is based on a recursive equation that makes it possible to optimise the objective function by taking into account the probabilities of occurrence of the inflows on a given time horizon. The time horizon chosen for calculation is one year while the adopted time discretization interval is the month, since it is generally assumed as the minimum duration to simulate the managing of the reservoir correctly. Further, monthly inflows are uncorrelated from month to month, as shown by the test of auto-correlation of Wald - Wolfowitz. Inflows and storage volumes are discretised in classes of 35,000 m^3, which correspond to the minimal agricultural volume request per month and which is taken as basic unit in calculation.

At every iteration of the optimisation procedure, the optimal objective function value is calculated for each state of the system. When this value becomes constant from a period to another, the criterion of convergence is fulfilled. Such procedure of optimisation is known as decision/hazard, where the variable inflows is the origin of hazard. Figure 3 depicts a scheme of such optimization where C_{ir} is the cost of transition between S_i et S_j, P_{rj} is the probability of the transition from S_i to S_j, C_{rj} is the value of this transition, u_{it} is the decision release at time t from the storage and $V_{F(t+1)}$ is the value of the objective function at time t+1.

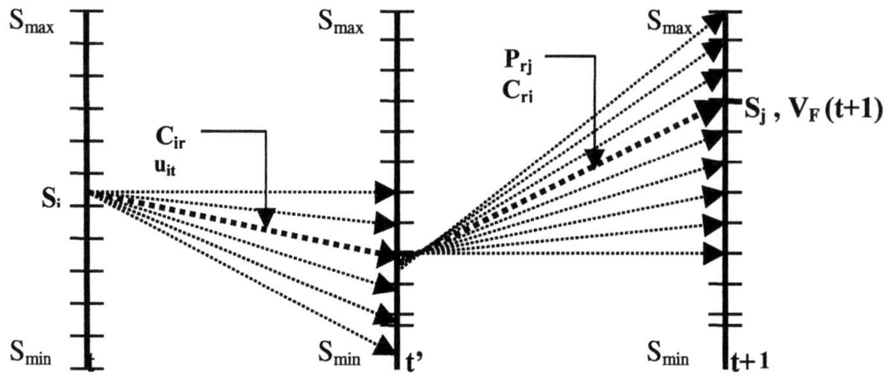

Fig. 3. Scheme of the Decision/hazard optimisation

5. Definition and choice of the system performance indices

Different performance indices can be adopted in order to describe the system behaviour. The interest here lies on the failure during reservoir management.

In reservoir management one could be interested to different failures according to the objective of minimizing water shortages with reference to the demand and to the objective of reducing the deviation between the real volume and the target volume at the end of management period. In more general terms, failure could indicate any event like reaching an alarming storage level, which occurs out of a range of pre-specified limits or levels of targets. Thus the following failures have been considered here.

a. ***Failure of storage***: occurs when the reservoir is completely empty before the end of the period of management. This situation depends on the variability of hydrological inputs and on the policy of management followed until such as date (Parent et al., 1992).
b. ***Non-satisfied demand***: occurs when at a given interval, the storage in the reservoir is not enough to satisfy the demand.
c. ***Spill***: this occurs when the reservoir storage exceeds the capacity. Though is not strictly a failure still is a useful parameter, because it is the logical complement of the first type of failure.

Various indices of performance can be associated with each of the above failures:

1. The risk: it is the probability that at a given interval, the given failure occurs.
2. Impact strength: it is an estimate of average time required by the system to return to a satisfactory state after a failure occurred.
3. Vulnerability: it measures the average severity of failure for the users of the resource.
4. Duration of the permanence of the system in a state of failure.
5. Stationary probability that the system is in a state of failure .
6. The time of the first passage towards a state of failure.
7. Average time between two states: total satisfaction and failure.

Figure 4 shows a plot of reservoir storage levels vs. time, with the indication of failure periods.

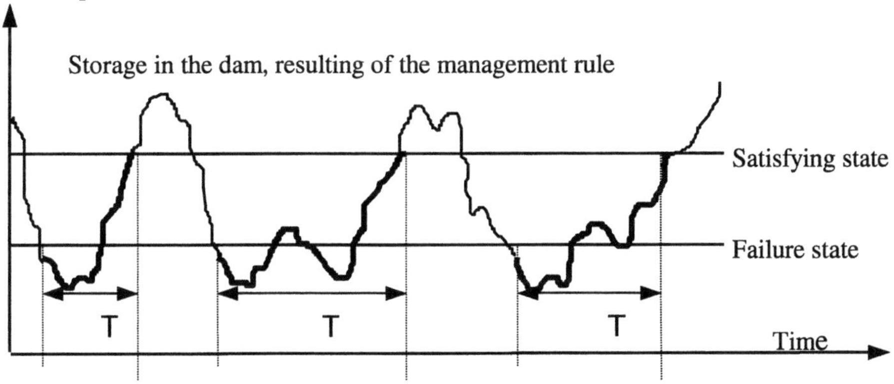

Fig. 4. Representation of the system impact strength

With reference to Figure 5, which represents the evolution of the system in time, let's indicate:

$$Vdfs = \frac{1}{N}\sum_{t=1}^{n}\left(S(NC1) - S_t\right) \qquad for\ S_t \leq S(NC1) \tag{7}$$

$$Vdfs = \frac{1}{N}\sum_{t=1}^{n}\left(S(NC2) - S_t\right) \qquad for\ S_t \leq S(NC2) \tag{8}$$

$$Vdev = \frac{1}{N}\sum_{t=1}^{n}\left(S_t - S\ (NC3)\right) \qquad for\ S_t \geq S(NC3) \tag{9}$$

where N: number of years of simulation and $S(NC1)$; $S(NC2)$; $S(NC3)$ are the storage thresholds corresponding to the different types of incidents.

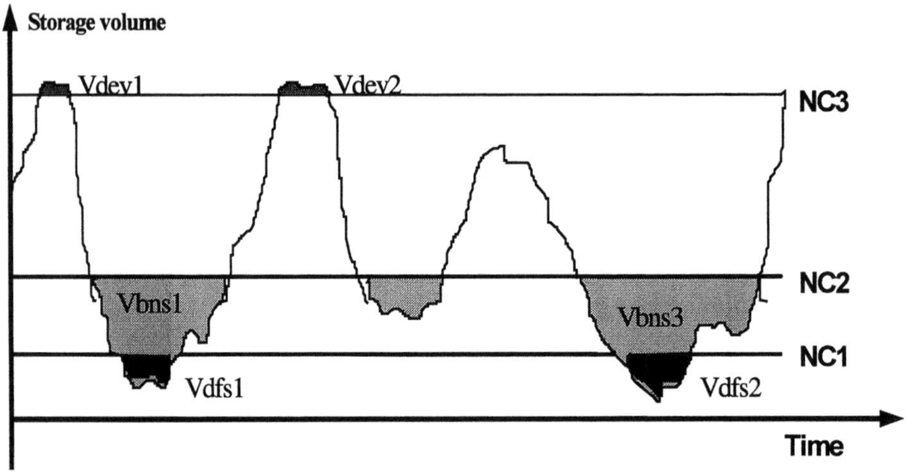

Fig. 5. Representation of the vulnerability or of the deficit

The performance of one management rule versus another is estimated by comparing values of the same index of performance calculated for each rule, and for two levels of failure events.

6. Simulation

The performance of the operating rules have been assessed through simulation. In particular, the following performance indices were used:

- Risk of failure of storage (RDF), of spill (RDV) and of non-satisfied demand (RBN).
- Vulnerability of failure (VDF), of discharge (VDV) and non-satisfied demand of the request (VBN).
- Impact strength of failure (RESDF) and of spill (RESDV).
- Release of water in addition to the demand (water not used and of surplus): VSUP

These indices all are standardised between 0 (very good performance) and 1 (very bad performance) by a threshold of acceptability by means of the following formula:

$$\text{Standardised Indices} = \frac{\text{indices - acceptable threshold}}{\text{unacceptable threshold} + \text{acceptable threshold}}$$

Figures 6, 7 and 8 depict the output windows of the simulation program specifically developed, corresponding to values of α=1.0, 0.5 and 0.0 respectively. The windows report the plot of the storage volumes vs. time, as well as the values of the corresponding performance indices, not standardized.

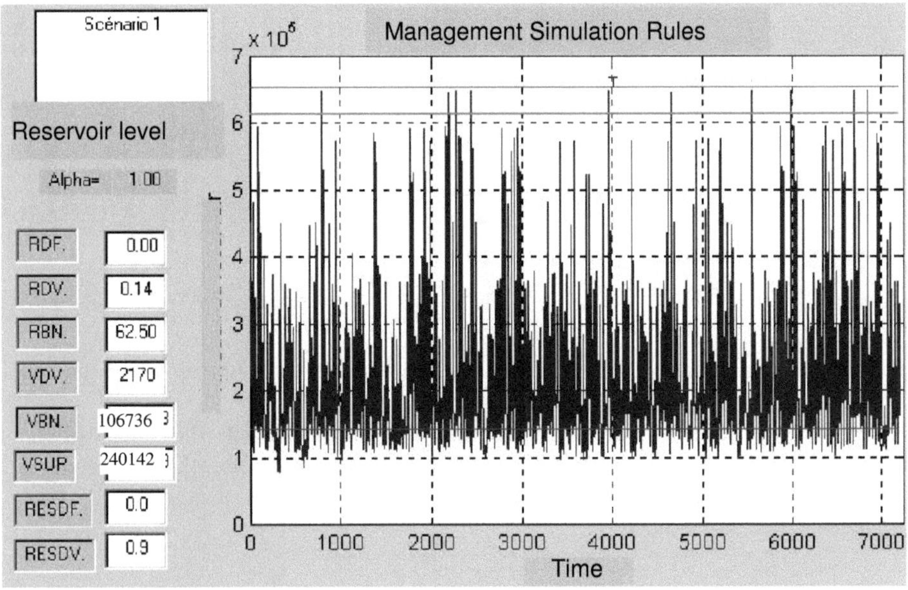

Fig. 6. Series of storage levels and values of performance indices (α = 1.0)

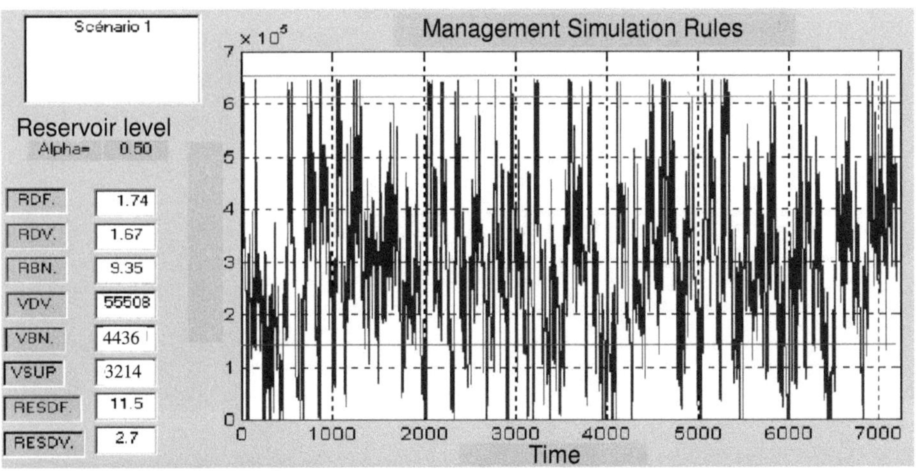

Fig. 7. Series of storage levels and values of performance indices (α = 0.50)

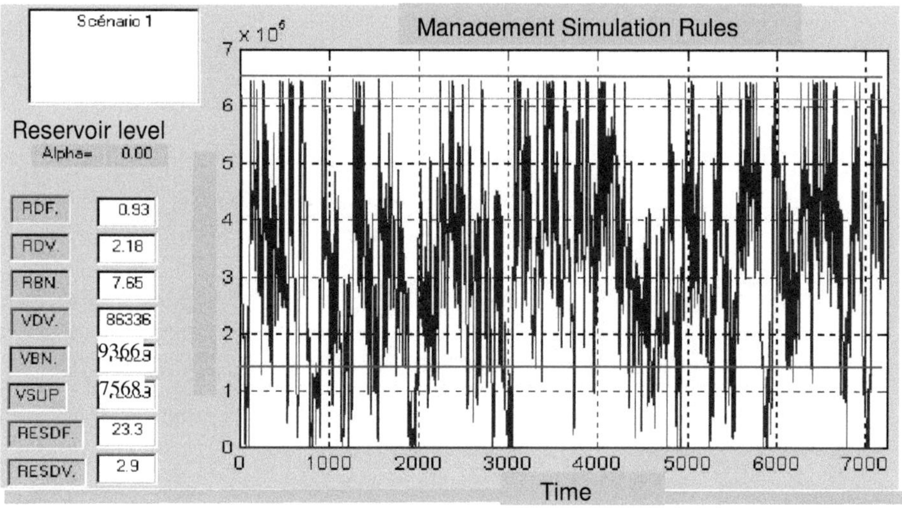

Fig. 8. Series of storage levels and values of performance indices ($\alpha = 0$)

To simplify the interpretation of the results, Figure 9 depicts the values of the different standardized indices vs. α. When the rule has a standardised index beyond 1, it is rejected. Two optimised rules are thus acceptable compared to the thresholds of the fixed indices ($\alpha = 0.2$ and 0.3).

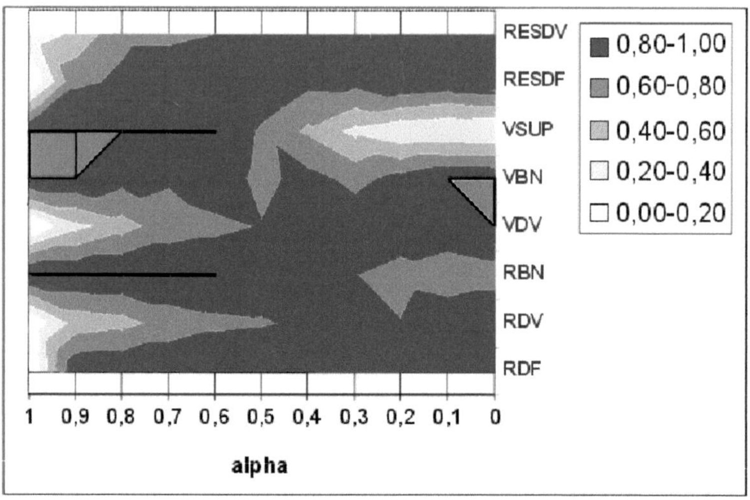

Fig. 9. Performance indices vs. α

A comparison of the two rules, corresponding to $\alpha = 0.2$ and 0.3 is reported in Figure 10. In particular, the upper figure reports the deficits while the lower figure reports the deficits expressed as a percentage of the demand. In both cases, the failures of supply are weak and thus it appear beneficial to give a preferential weight to the agricultural requests (α <.5).

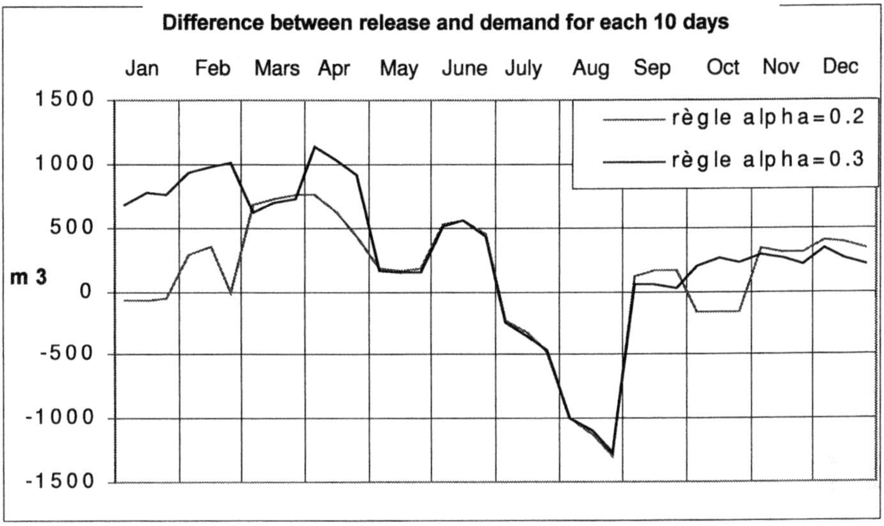

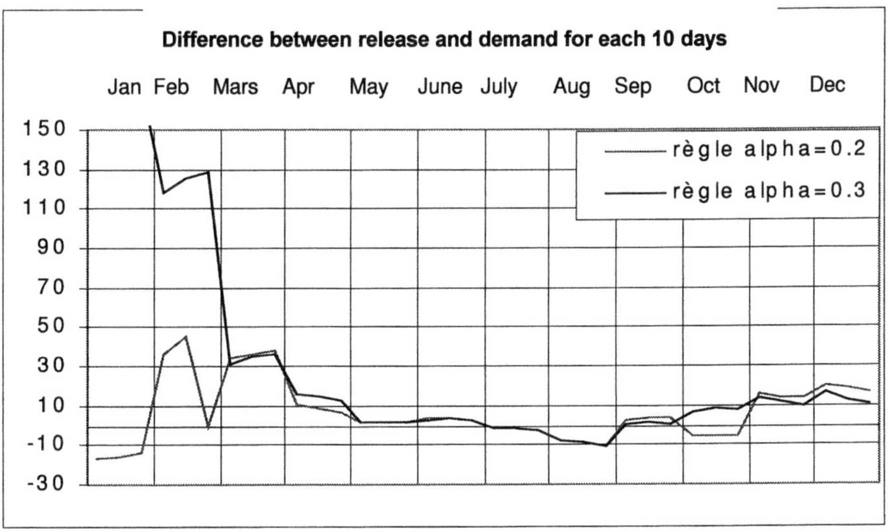

Fig. 10. Results of the optimised management

The results also show that from May to August, the two rules of management satisfy the request in the same way. On the other hand, they differ from January to April. The rule (α=0.3) always gives a surplus of water, except during the dry months (July-August). The rule (α=0.2) is closer to an average deficit equal to 0 (less surplus of water) and presents a slight deficit in October (sowing). However this deficit has a low value absolute (300 m^3 per month), which is within the limit of precision of the values of the water demand. Although the choice between the two rules is not obvious, the rule α=0.3 appears preferable since it is globally less penalising (less deficit per month, minimal sum of the standardised indices).

7. Conclusions

The derived management rule is based on an objective function with a weight of 0.3 for the term related to the guarantee of a target storage in the reservoir and a weight of 0.7 for the term related to the satisfaction of the demand. The derived rule leads to a maximum risk of failure of 1.4%, a risk of spill of 2.1% and a risk of not satisfying at least 60% of the demand of 11.1%. These results are affected by the errors and uncertainties of the data which are at their origin namely the estimated inflows, water demand fixed at the time of optimisation and of the simulation of the management rules. Further they are based on the assumption that despite the variability of the supply resource, the demands are regarded as data determined by fixed pre-established scenarios always satisfied.

Despite the limitations of the results due to the uncertanty of the data, the application shows the feasibility of the proposed methodology for reservoir management in semi-arid countries.

Acknowledgments

The research has been carried out within the European Community INCO-DC project “A DSS for drought mitigation in Mediterranean regions” (DSS-DROUGHT, IC18-CT97-0169).

References

Bhaskar, N. and Whitlach, E. (1980) Derivation of monthly release policies. *Water Resour. Res.* 16: 987-993.

Goussebaille, J. and Roche, P. A. (1986) Gestion d'un réservoir de soutien d'étiage et d'écretage de crues. Applications des techniques de programmation dynamique. *Rapport EDF.*

Parent, E., Lebdi, F. and Hurand, P. (1992) Strategic managing of a water resource system : application to the NESTE water resource system. *Revue des Sciences de l'Eau.* 4: 543-568.

Richard, D. and Turgeon, A. (1988) Analyse de contraintes probabilistes dans la gestion d'un système hydroélectrique. *Revue des Sciences de l'Eau.* 3: 171-178.

Yakovitz, S. (1982) Dynamic programming applications in water resources. *Water Resour. Res.* 18(4): 673-696.

Yacizil, H., Houcks, M. and Toebes, G. H. (1983) Daily managing of a multipurpose dam system. *Water Resour. Res.* 19(1): 1-13.

Yeh, W.W.G. (1985) Reservoir management and operations models: a state of the art review. *Water Resour. Res.* 21(12): 1797-1818.

DECISION SUPPORT SYSTEM FOR THE EVALUATION OF DROUGHTS AND DROUGHT MITIGATION MEASURES

A. CANCELLIERE[(1)], G. GIULIANO[(1)] and A. LONGHEU[(2)]
[(1)] *Department of Civil and Environmental Engineering - University of Catania, Italy*
[(2)] *Department of Informatics and Telecommunications Engineering - University of Catania, Italy*

Abstract: Successful management of water supply system under drought conditions requires the effective use of advanced tools able to support water managers in the decision process. This paper presents the Decision Support System (DSS) developed within the EC INCO-DC project "A DSS for drought mitigation in Mediterranean regions". The objective of the DSS is to aid decision makers in managing water systems devoted to irrigation use under drought conditions. This is accomplished through six main computational modules, namely a drought identification model, an irrigation scheduling model, reservoir optimisation and simulation models, a model that allows to derive reservoir operating rules by multiple regressions and/or neural network approach and a multicriteria analysis model, with all modules integrated in an user friendly environment in order to effectively assist in making decision in water supply systems management.

Key words: Decision Support Systems, water supply systems, drought mitigation.

1. Introduction

The term Decision Support System (DSS) is generally adopted to define a computer-based information system, oriented to help a decision maker by providing him with all relevant data in a form that is easy to be understood (Mittra, 1986). More specifically a DSS consists of a set of computer graphics-based applications, generally several interconnected mathematical optimization and/or simulation models, sometimes together with a qualitative rule-based algorithms, all referring to the same database, within an user friendly environment. The above definition, although general, identifies the features that differentiate a DSS from a simple computer based mathematical tool, i.e. the presence of three subsystems, the modelling subsystem, the database subsystem and the user interface subsystem, interconnected with each other.

Similar structural definitions, focusing on the architectural structure or components present in most DSS rather than on the objectives or activities supported by DSS has been adopted by several authors. According to Sprague (1983), a DSS should attempt to integrate models or analytical techniques with data access and retrieval functions, focusing on features that make them easy to use to non computer-experienced people, and emphasising flexibility and adaptability to accommodate changes in the environment and in the decision making approach of the user. Integration principle in building DSS is also adopted by Reitsma *et al.* (1996), who define a Decision Support System as: "a computer-based system which integrates state information, dynamic or process information, and plan evaluation tools into a single software implementation".

G. Rossi et al. (eds.), Tools for Drought Mitigation in Mediterranean Regions, 305-318.

Databases and/or Geographic Information Systems implement the state information task, process information includes the mathematical representation of the system (simulation and/or optimisation models), and evaluation tools refer to modules able to produce relevant information for decision making process.

The approach to develop a DSS has been traditionally to identify the mathematical models to solve a particular problem and then to add a pre- and post-processor, usually a graphical interface, for the input and editing of data and for visualisation of the models results. A new kind of development approach that is emerging mainly focuses on the decision making process and how a decision support system can aid that process (Loucks and Da Costa, 1991). This approach tends to emphasize the capabilities of the DSS to help understanding a given problem, explore the consequences of different alternatives and facilitate sensitivity analyses.

The adoption of a DSS can reveal very useful in many areas where decision-making problem is crucial, either for strategic planning, or management control, or operational control; such questions are all present in particular in water resources management (Labadie, 1993), e.g. in managing an irrigation reservoir system under extreme events, such as droughts (Raman *et al.*, 1992).

In the present paper, the DSS developed within the EC DSS-DROUGHT project "A DSS for drought mitigation in Mediterranean regions" is presented. The objective of the DSS is to aid decision makers in managing water systems devoted to irrigation use under drought conditions (Rossi, 2002).

In particular, the DSS allows to perform the following main tasks:

- to analyse and assess the severity of past droughts at a site and over a region;
- to determine irrigation scheduling based on the simulation of the soil water balance;
- to determine the optimal series of releases from a reservoir by means of dynamic programming;
- to derive and validate reservoir operating rules to be used by system managers in order to improve the decision process of water supply over time;
- to simulate the performance of a water supply system during a given historical period;
- to evaluate alternative drought mitigation measures by means of multicriteria analysis.

In the following, first a description of the DSS as a whole is given. Then, details on each DSS component are reported, giving an overview of each single task module features, describing the main interface through which users access all DSS modules, and finally illustrating the database main features.

2. Structure of the DSS

The DSS has been developed with the aim to provide potential users with an easy to use software, able to assist the different tasks of a comprehensive approach to cope with drought, from hydrological drought identification to alternative mitigation measures assessment. A secondary aim for its development has been to foster exchange of experiences among the different partners of the DSS-DROUGHT project through the use of a common tool to perform the research activities.

From a conceptual point of view, the DSS can be considered articulated into three main components:

- computational modules specific to the various DSS tasks;
- user interface to manage the input of the necessary information and the output of the results in tabular and graphical format;
- system database.

A functional scheme of the DSS, depicting the data connections between the different components is reported in Figure 1.

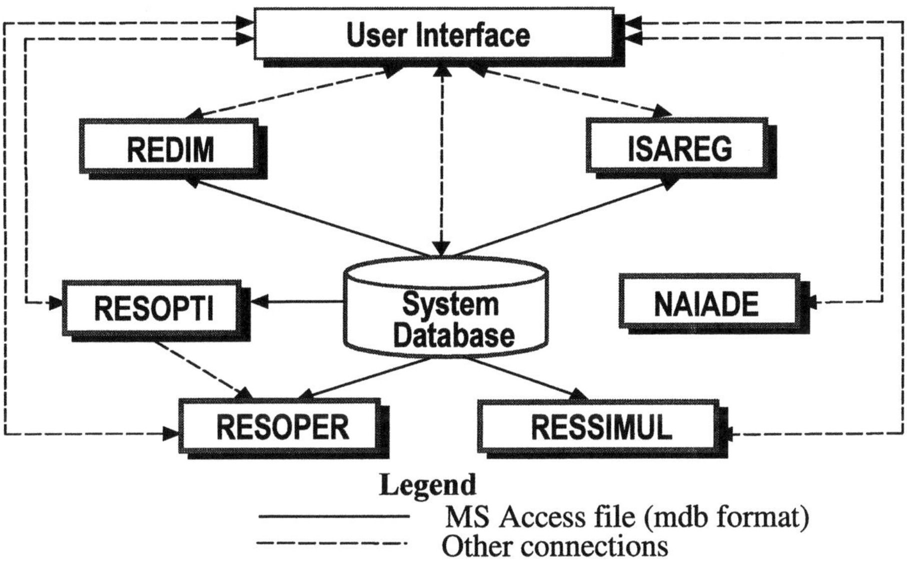

Fig. 1. Structure of the DSS

The DSS makes use of six different computational modules to accomplish its tasks, namely:

- REDIM: drought analysis at a site and over a region;
- ISAREG: irrigation scheduling based on the simulation of the soil water balance;
- RESOPTI: optimal releases determination by means of dynamic programming;
- RESOPER: reservoir operating rules determination by means of multiple linear regression and/or neural networks;
- RESSIMUL: simulation of water supply reservoir operation;
- NAIADE: multicriteria analysis.

Some of the above modules can be used either in stand-alone mode, as well as to provide input for other modules. For example, REDIM module allows to perform the drought analysis either to assess the drought risk over the region under investigation (stand-alone mode), or to identify the most severe droughts in the past to be used as input to determine the optimal series of releases from a reservoir. In this case, REDIM output are given as input to RESOPTI module.

In the following, each module is briefly described; for further details about the methodological aspects of REDIM, ISAREG, RESOPTI-RESOPER-RESSIMUL the reader is referred to Rossi and Cancelliere, (2002), Pereira *et al*, (2002) and Cancelliere *et al.*, (2002) respectively, whereas details about the software can be found in the Decision Support System Reference Guide (DSS, 2001) .

3. Computational modules

3.1. THE REDIM MODULE.

The REDIM module allows to perform drought analysis on hydrological series both at a site and over a region using the method of run. In addition, it allows to test statistically for the existence of trends and non-randomness in a time series, whose presence could lead to misleading drought analyses. The software, written in Visual Basic, is structured as a succession of dialog boxes which guide the user throughout the analysis of stationarity and drought identification and characterisation processes. The main features of REDIM can be listed as follows:

- Data files are in MS Access format, to guarantee the exchange of data with the other components of the DSS;
- Different aggregation time scales can be used (monthly, three monthly and yearly), with the possibility to select the initial month of aggregation in order to take into account water years and/or different three months periods;
- Testing for stationarity and/or randomness in hydrological series is carried out by means of three statistical tests: Student's t-test for trend, Kendall's • test (rank correlation test) and turning point test;
- Identification and characterisation of drought at a site is performed by means of run method, and three different threshold levels can be selected, namely the mean, the median, the mean minus the standard deviation or the quantile corresponding to a given frequency level.
- Regional drought analysis as performed by using hydrological series at several stations and by considering regional drought a period where the sum of the areas corresponding to stations presenting deficits is greater than a critical area.
- Numerical output of results include the identification and characterization of local and regional droughts which can be used as input for RESOPTI.
- Graphical output of results allows to easily identify droughts on a given time series, and/or region.

3.2. THE ISAREG MODULE

This model allows to simulate several irrigation scheduling options. It is based on the soil water balance computation methodology proposed by Doorenbos and Pruit (1977), as developed by Texeira and Pereira (1992). The data are organised in:

- Meteorological files including data on precipitation, reference evapotranspiration and other meteorological variables;
- Crop files indicating the development stages, dates, the crop coefficients, root depth evolution and the depletion factor;
- Soil files with the values of potential depth for root extraction, field capacity and wilting point, defined for each soil layer.
- Groundwater file indicating the values of the parameters necessary to compute the groundwater contribution to soil water balance;
- Irrigation option and water restrictions files, to help managing the different simulations.

ISAREG model is combined with two other programs, EVAP56, to compute reference evapotranspiration, and KCISA to compute crop parameters. The irrigation scheduling options are performed through two thresholds: one defining when an irrigation is required, the other defining the upper limit that quantifies the irrigation depth. The irrigation options can be subject to restrictions like, for example, a minimum interval between irrigations or water supply restrictions. In particular the program allows to:

- Schedule irrigations, calculating the day and depth of each irrigation, the yield reduction if the crop is subject to water stress, and the peak continuous design discharge;
- Determine the net irrigation requirements;
- Evaluate a given irrigation schedule;
- To search for an optimal irrigation scheduling under conditions of limited water supply;
- To compute the water balance without irrigation.

3.3. THE RESOPTI MODULE

The RESOPTI module allows to determine optimal releases from a reservoir by means of dynamic programming technique. Referring to the scheme shown in Figure 2, the input is given as:

- reservoir characteristics, i.e. maximum and minimum storage capacity and the storage-area relationship coefficients;
- hydrological series including inflows to the reservoir, evaporation depths and/or temperatures (to estimate evaporation);
- irrigation demands, which can be either provided by the user (directly through dialog boxes, or using project text files), or imported from ISAREG module output;

- constraints and optimization parameters, as the optimization years range and storage bounds.

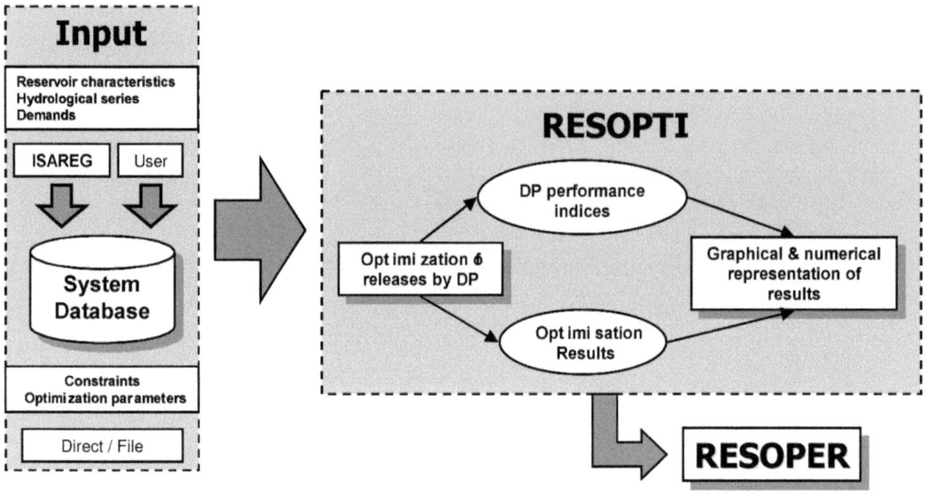

Fig. 2. RESOPTI scheme

Given all these data, RESOPTI computes the series of optimal releases that minimize a given objective function, using a deterministic discrete dynamic programming technique (for details see Cancelliere *et al.*, 2002).

The output of the module consists of various plots (inflows, demands, releases, reservoir storage volumes etc.), as well as tabular information; results also include a graphical and numerical representation of reservoir performance measures, namely reliability and vulnerability indices. Optimal releases and demands can be plotted (Figure 3 shows an example), as well as storage volumes, evaporations, inflows and spills.

Finally, note that results can also be stored in order to be used to derive reservoir operating rules by means of the module RESOPER.

Figure 4 shows the internal scheme of RESOPTI. The core of RESOPTI is the computational engine, which is actually strictly related to the CSUDP module implementing the dynamic programming algorithm (Labadie, 1999).

The engine communicates on one hand with the CSUDP module, in order to compute the series of optimal releases, and it also interacts with:

- the Graphical User Interface (GUI) to manage user input data as well as results;
- the File Manager module, to store and retrieve project text files;

- the Database Manager module, which allows access to database for querying hydrological data series.

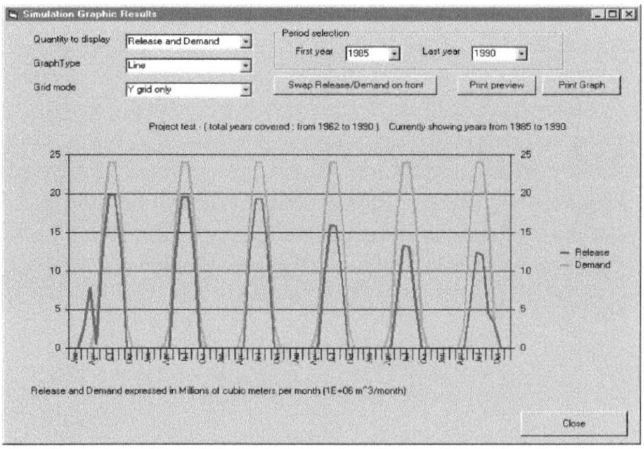

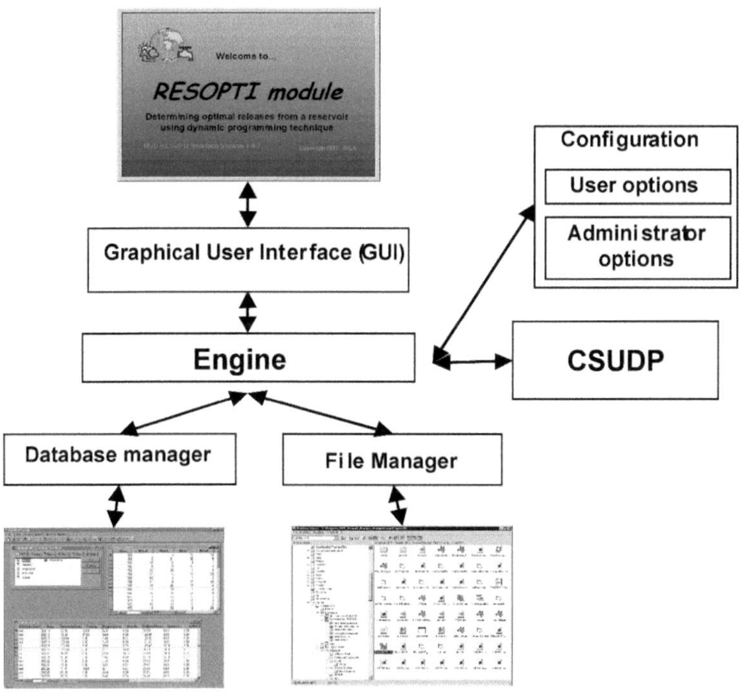

Fig. 3. RESOPTI graphical results example

Fig. 4. RESOPTI internal scheme

Finally, note that RESOPTI can be configured through the configuration module: some parameters can be set by users (as for instance database table names or default values for forms), whereas internal parameters (as those used to communicate with CSUDP) can be set only by an administrator user.

3.4. THE RESOPER MODULE

The RESOPER module allows to derive reservoir operating rules on the basis of the results of the dynamic programming (module RESOPTI). More specifically, it allows to express the optimal releases computed by dynamic programming as a function of state variables of the reservoir such as stored volume and of inflows and releases in the previous time intervals by means of either multiple regressions (MLR) and/or neural networks (NN). Figure 5 shows the functional scheme of RESOPER, together with the interaction with RESOPTI; note that input data are the same as for RESOPTI.

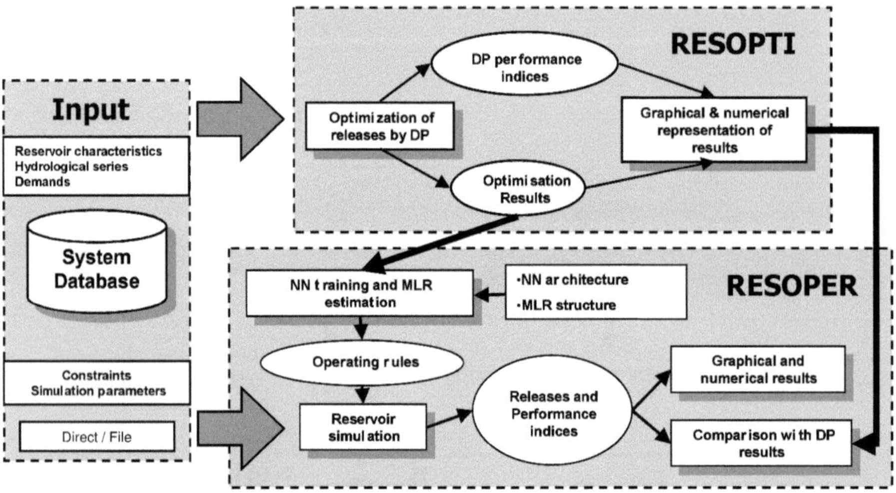

Fig. 5. RESOPER scheme

RESOPER can perform three standard tasks namely fitting, validation and real time decision making. Fitting consists in determining the parameters of the multiple linear regression or of the neural networks such that the model fits with a minimum error the DP results; according to this, the solid line in Figure 5 shows the use of optimal releases (RESOPTI results) for the NN training / MLR estimation inside RESOPER. As an example, in Figure 6 a neural network with 3 hidden neurons is used to express the optimal releases in the month of May using the storage volume as input variable. The network weights are adjusted through the back-propagation algorithm, whose parameters can be set by the user, and an overall information about the goodness of fitting is provided by the RMSE (root mean squared errors).

Validation allows to simulate the fitted multiple regression or the trained networks (i.e. in both cases, the resulting operating rules) over an historical period to test their performance. Such a validation is carried out by simulating the reservoir using the derived operating rules, and computing a set of indices which express the reservoir performance (e.g. reliability and vulnerability). Such indices can be displayed both in numerical and graphical form. Figure 7 shows numerical representation as an example.

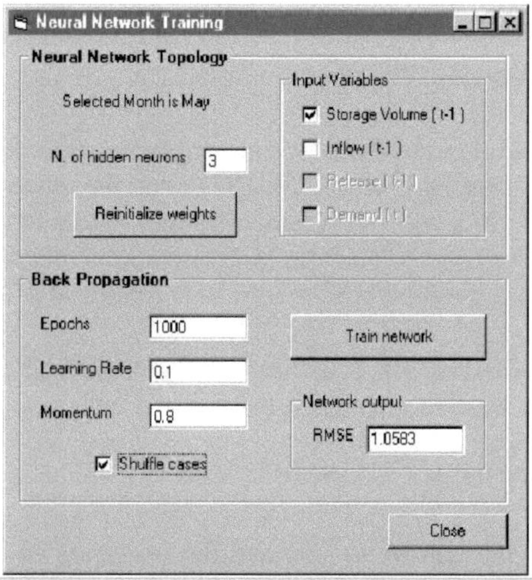

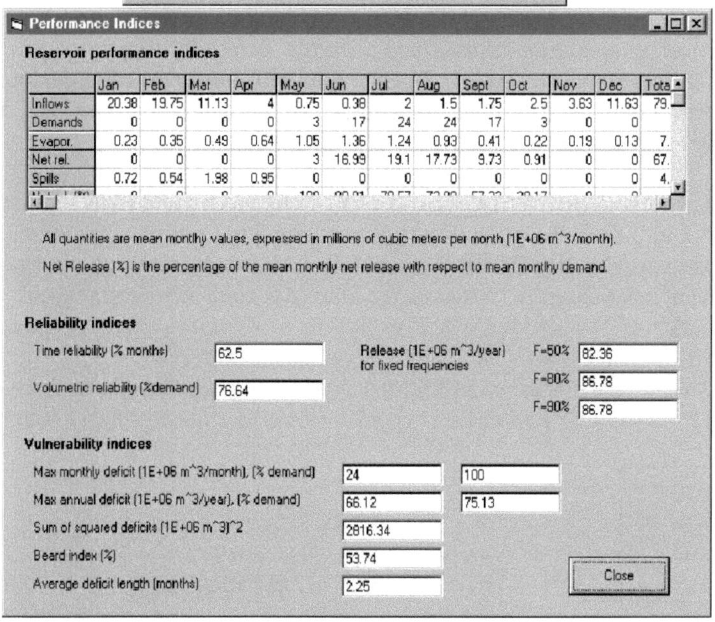

Fig. 6. RESOPER Neural Network Training

Fig. 7. RESOPER Numerical Results (performance indices of reservoir operation)

During real time decision making (the third task which is possible to perform with RESOPER), the user can enter the state variables, input the current month and estimate how much water to be released in the next month, using the previously defined operating rules. The structure of the multiple regression (choice of input variables), as well as the architecture of the network (e.g. input variables, number of hidden neurons) can be completely specified by the user in order to compare alternative MLR structures or NN architectures during the validation stage. The historical period used for validation can also be defined by the user, for example on the basis of the most severe droughts identified by the module REDIM.

Finally note that it is also possible to compare simulation results of RESOPER with optimization results coming from RESOPTI, over the same input data. This is provided through a simple function (represented by the "Comparison with DP results" module in Figure 5) which allows to perform both simulation and optimization over the same input data, then showing a form with graphical comparison of both results.

The internal scheme of RESOPER is the same as that of RESOPTI shown in Figure 4, except for CSUDP module, which is replaced with MLR and NN computation modules.

3.5. THE RESSIMUL MODULE

The RESSIMUL module allows to simulate the operation of a reservoir during a predefined time horizon. More specifically, given a series of inflows to the reservoir, temperatures or evaporation depths, and monthly demands, the program computes reservoir volumes, actual releases, spills, etc. Different operating rules can be adopted to assess alternative water rationing policies under drought conditions, and Standard Operating Policy can also be implemented for reference.

Several reservoir performance indices are computed on the basis of the actual releases, and in particular reliability and vulnerability indices. The output of the module consists in various plots (inflows, releases, demands, reservoir storage etc.), as well as tabular information in order to facilitate the comparison among the different operating rules.

Figure 8 shows the scheme of RESSIMUL, where input data (similar to those provided for RESOPTI and RESOPER) are given to perform simulation and to obtain results.

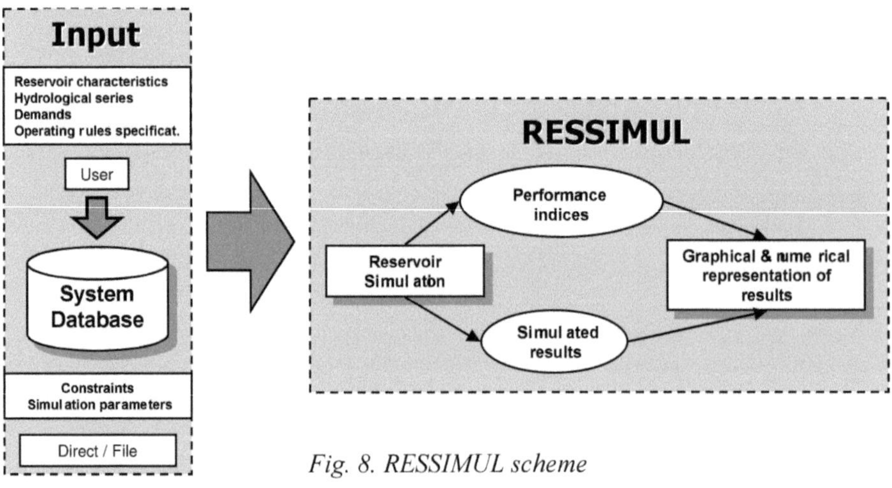

Fig. 8. RESSIMUL scheme

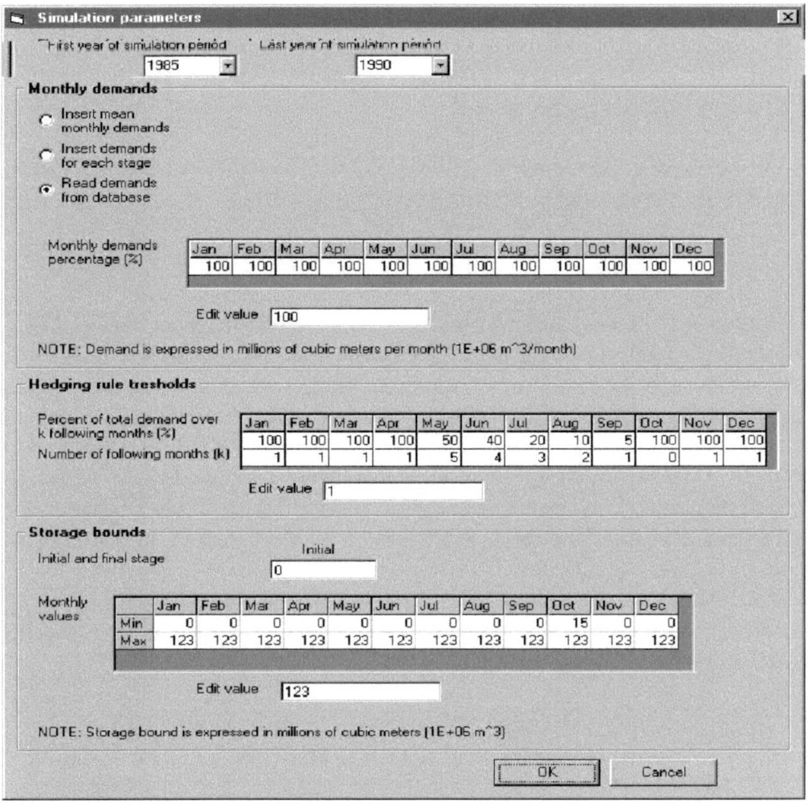

Fig. 9. RESSIMUL Simulation Parameters

3.6. THE NAIADE MODULE

NAIADE (Novel Approach to Imprecise Assessment and Decision Environments) (Munda, 1995) is a discrete multicriteria method oriented to evaluate alternatives for resources management and/or environmental protection. The evaluation matrix may include either crisp, stochastic or fuzzy measurements of the performance of an alternative a_n with respect to a judgement criterion g_m, thus it is very flexible for real-world applications.

The main features of the NAIDE method can be synthesised as follows:

- The method is based on some aspects of the partial comparability axiom, in particular, a pairwise comparison between alternatives is carried out which takes into account intensity of preference;
- Use of a conflict analysis procedure allows to seek for justifiable decisions which express a certain degree of consensus from different interest groups;
- Input includes the list of alternatives, the criteria parameters and the list of interested groups;

- Output consists of the ranking of alternatives and of the dendrogram of the coalition formation process.

4. User interface and system database

The user interface represents the connection between the hydrological database, the computational modules and the output of the results.

Note that some module can be used either as a single, independent tool, as well as in conjunction with other modules, having in this case the already mentioned I/O data transfer between different modules. This leads to two different kinds of user interface, i.e. on one hand user interfaces for all independent modules, named as *local* interfaces, that allow to use such modules in a stand-alone mode, whereas on the other a global interface has also been developed, in order to provide a uniform environment both to call each module, as well as to manage the shared database. When the global interface starts, the main window shown in Figure 10 is displayed. This window contains a set of buttons, each named with the name of an application included in the DSS (plus a data manager explained in the following). Each button is used to start the corresponding application, by simply clicking onto it.

Local interfaces are simple Microsoft Windows-based applications and Microsoft Visual Basic language has been used to develop most of the interfaces.

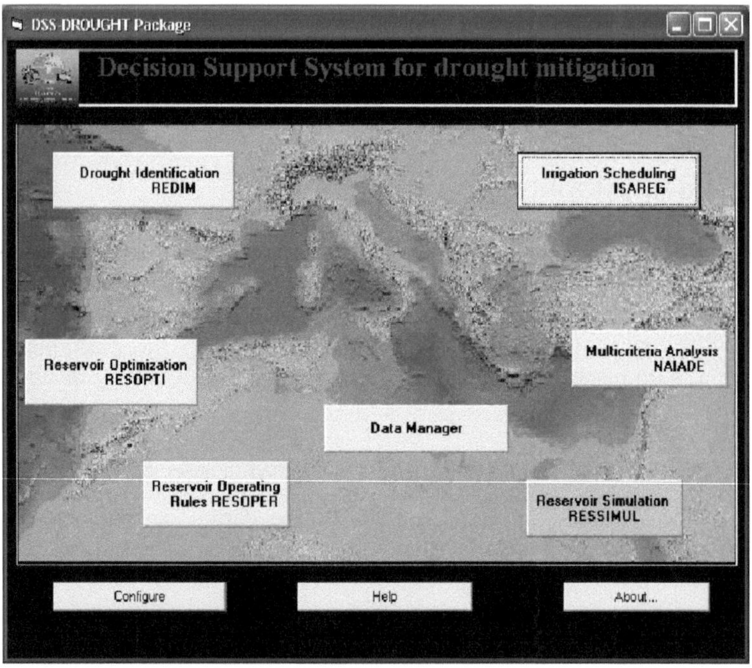

Fig. 10. Global User Interface (DSS main window)

Note that each application can be installed separately, and therefore the DSS will run even if some application may be not currently present in the system. This allows flexibility, since each module can be easily updated when new versions are released, without reinstalling the whole package.

The system database allows to input, manage and retrieve all the available hydrological data and other data necessary to describe water system characteristics for the case study under investigation. It stores besides all the input data, part of the output data of the computational modules, therefore allowing for a full interconnection between the modules.

The system database is based on the Microsoft Access database engine, in order to avoid the need of expensive specific database management system or database architecture (e.g. SQL Server, Oracle). MS Access has been chosen for its widespread use and because it allows full compatibility with the Jet database engine under Visual Basic, which is the programming language that is used to develop most of the user interfaces, as mentioned in the previous section.

The data is organised in a site-oriented fashion, i.e. for each site a number of tables can be created to store information such as precipitation, streamflows, temperature, demands, etc. This particular form of storing the data is motivated by the fact that for certain components of a water supply systems, more than one type of information is available at a given site. For example, in the case of a reservoir, one may want to store under the same site streamflows (inflows to the reservoir), precipitation (areal rainfall over the watershed) and temperature. Note that in the case of other types of sites (such as precipitation gages), the tables corresponding to non existing data (e.g. streamflows) will simply be empty.

Data Import software modules are also provided to allow to create automatically MS Access database files from existing ASCIII text files, thus avoiding tedious database creation processes, also assuring compatibility with other databases storing existing data (all databases indeed provide export functionalities of data as simple ASCIII text).

Finally, note that actually each module can use any number of databases, while on the other hand, a given database can be shared by several module. This approach ensures that each DSS module achieves the maximum flexibility and independence from others.

5. Conclusions

Management of water supply systems can effectively improve through the application of advanced tools, especially when facing drought conditions. Such advanced tools however must fulfil some requirements in order to be used by water managers and not remain academic exercises, namely ease of use and flexibility to accommodate changes in the environment and in the decision making approach. The integration of several mathematical models within an user friendly Decision Support System environment can effectively satisfy these requirements, thus contributing to the implementation of a more efficient use of water resources able to cope with recurrent drought conditions.

Acknowledgements

The research has been carried out within the European Community INCO-DC project "A DSS for drought mitigation in Mediterranean regions" (DSS-DROUGHT, IC18-CT97-0169).

References

Cancelliere, A., Giuliano, G., Ancarani, A. and Rossi, G. (2002) Derivation of operation rules for an irrigation water supply system by multiple linear regressions and neural networks. (This issue).

Doorembos, J. and Pruitt, W. O. (1977) Crop water requirements. *Irrigation and Drainage paper no. 24,* FAO, Rome.

DSS (2001) *Decision Support System Reference Guide.* INCO-DC project DSS-DROUGHT Final Report. Department of Civil and Environmental Engineering, Catania University and Department of Rural Engineering, Lisbon Technical University.

Labadie, J. W. (1993) Decision support systems in water resources. In: J.B. Marco, R. Harboe and J.D. Salas (Eds.) *Stochastic hydrology and its use in water resources systems simulation and optimisation,* Kluwer Academic Publishers, Dordrecht, The Netherlands, pp. 373-396.

Labadie, J. W. (1999) Generalized Dynamic Programming: CSUDP. Documentation and User Manual. Department of Civil Engineering, Colorado State University, Colorado.

Loucks, D. P. and Da Costa, J. R. (1991) *Decision Support System, Water Resources Planning.* NATO ASI Series.

Mittra, S. S. (1986) *Decision Support Systems. Tools and Techniques,* John Wiley, New York.

Munda, G. (1995) *Multicriteria evaluation in a fuzzy environment. Theory and applications in ecological economics,* Physica-Verlag, Heidelberg.

Pereira, L.S., Teodoro, P.R., Rodrigues, P.N. and Teixeira, J.L. (2002) Irrigation scheduling simulation: the model ISAREG. (This issue).

Raman, H., Mohan, S. and Rangacharya, N.C.V. (1992) *Decision support for crop planning during droughts. Water Resour. J.* 75-82.

Reitsma, R.F., Zagona, E.A., Chapra, S.C. and Strzepek, K.M. (1996) Decision Support Systems for Water Resources Management. In: L.W. Mays (Ed.) *Water Resources Handbook,* McGraw-Hill, New York, pp. 33: 1-35.

Rossi, G. (2002) An integrated approach to drought mitigation in Mediterranean regions (This issue).

Rossi, G. and Cancelliere, A. (2002) At-site and regional drought identification with REDIM model. (This issue).

Sprague, R. (1983) A framework for the development of decision support systems. In: W. House (Ed.) *Decision support systems.* Petrocelli Books, New York.

Teixeira, J. L. and Pereira, L. S. (1992) ISAREG, an irrigation scheduling model. *ICID Bulletin* 41(2): 29-48.

LONG-TERM AND SHORT-TERM MEASURES FOR COPING WITH DROUGHT

B. DZIEGIELEWSKI
Department of Geography, Southern Illinois University, Carbondale, USA

Abstract: Drought-induced shortages of water supply have adverse impacts on human activities and ecological systems. A number of measures can be undertaken to avoid or lessen such impacts. Both long-term and short-term measures, which are aimed toward increasing supply or reducing demand, should be considered. This chapter provides a review of alternative approaches for protecting the established uses of water against future droughts and summarizes the practice and status of drought protection and preparedness in the United States.

Key words: Drought, drought mitigation, USA, water supply system.

1. Introduction

There is a large array of measures for avoiding or lessening the adverse impacts of drought-induced shortages of water supply. A range of adjustments can be undertaken during three distinct drought planning and management activities: water supply planning, drought contingency planning, and drought management. Effective plans for coping with drought should seek an optimal combination of both long-term and short-term measures, which are aimed both toward increasing supply and reducing demand. This paper reviews the alternative approaches for drought management and planning and summarizes the practice and status of drought protection and preparedness among the U.S. water supply systems.

The natural variability of seasonal and annual precipitation necessarily leads to situations when the cumulative volume of precipitation falls below its average levels. These accumulated precipitation deficits result in general declines in soil moisture, streamflow, reservoir storage, and groundwater levels. When the declines from normal levels exceed certain values, the established uses of water cannot be satisfied. When the magnitude of these water shortages reaches a point where adverse impacts on human activities and ecological systems become apparent the situation is generally defined as "drought". In most regions of the world the presence of periodic droughts is a major concern because human settlements, industry and agriculture depend on the availability of water supply that is associated with long-term average patterns of precipitation.

The social and economic consequences of drought depend on the severity and duration of the drought episode, as well as the regional dependence on rain-fed agriculture and the proportion of the annually renewable resources that is utilized in supporting the established

G. Rossi et al. (eds.), Tools for Drought Mitigation in Mediterranean Regions, 319-339.

uses of water. Adverse impacts may propagate upwardly from local areas to the regional, national and even global levels of the social system during severe droughts that affect large regions of a country, or regions encompassing two or more countries. As the geographical extent and severity of drought increases, so does the likelihood that impacts will be felt at higher levels of the social system (Warrick *et al.*, 1975). These impacts can range from the loss of income, bankruptcy and dislocation, to international conflict, starvation and famine in large regions of the world.

It is useful to distinguish between the impacts of drought on rain-fed agriculture and its impacts on water supply. There is normally very little that can be done to mitigate the effects of low soil moisture on the survival of crops or crop yields in rain-fed agriculture. Crop production is dependent on precipitation, so even short-duration droughts during the growing season may destroy crops or significantly reduce yields. Because few farmers can install irrigation systems and acquire access to water just in time to save their crops, the full impacts of soil moisture deficit will occur. Therefore, the impacts of drought on rain-fed agriculture cannot be managed through additional water supply and are managed primarily with drought relief funds or other financial assistance, which at least partially replaces the lost income of farmers whose crops are damaged.

Unlike the agricultural impacts of soil moisture deficits, the impacts of drought on water supply can be mitigated through appropriate water supply planning and management of water resources in general. In the United States, the recently established National Drought Policy Commission noted the dual nature of the problem of drought by distinguishing between "natural water" droughts and "stored water" droughts (NDPC, 2000).

Natural water droughts mostly affect farmers and other businesses that depend on direct precipitation and natural streamflow and their adverse impacts may arise after just a few weeks or months of below-normal rainfall. Stored water droughts affect users of water supply systems that depend on water stored behind dams, in natural lakes and groundwater aquifers. Because such systems, in principle, are designed to withstand the lowest precipitation and streamflow periods in the hydrologic record, they are only affected by drought events that are more severe than the design drought. However, because the severity and duration of an ongoing drought cannot be known in advance, water managers may decide to curtail the withdrawals from storage in anticipation of a drought that exceeds the design drought. The risk-averse behavior of water managers can be justified by the increasing vulnerability to shortages caused by declining storage capacities due to reservoir siltation and increasing withdrawals to support greater populations and economic activities. Also, global climate change, which is expected to result in a more oscillating climate, may increase the probability of a severe drought significantly above the probability that can be derived from the recent hydrologic record.

The array of available measures for coping with droughts consists of both structural measures, such as water storage reservoirs, groundwater recharge basins or aqueducts, and nonstructural measures, such as crop insurance, restrictions on water use, pricing systems and education of water users. In order to evaluate the feasibility, cost and efficacy of individual measures, it is important to distinguish between the long-term (or strategic) and short-term (tactical and emergency) adjustments to drought. Also, some measures apply primarily to the mitigation of impacts caused by natural droughts while others apply primarily to stored water droughts that affect established systems of water supply. Finally, the kinds of coping measures available to managers of individual water supply system are different from those that are available for managing the impacts of drought at the regional or national levels. The relationship between the long-term and short-term measures for coping with droughts is the main focus of this paper.

2. Coping with drought: a framework

While drought itself is a natural phenomenon that cannot be substantially modified by human actions, its consequences can, and often are, mitigated through appropriate human adjustments to drought. In the case of water supply, water managers can deal with the problem of drought through three interrelated activities, namely: water supply planning, drought contingency planning, and drought management. In the literature on natural hazards, these three activities are known, respectively, as mitigation, preparedness and response. The relationship between these activities is depicted in Figure 1.

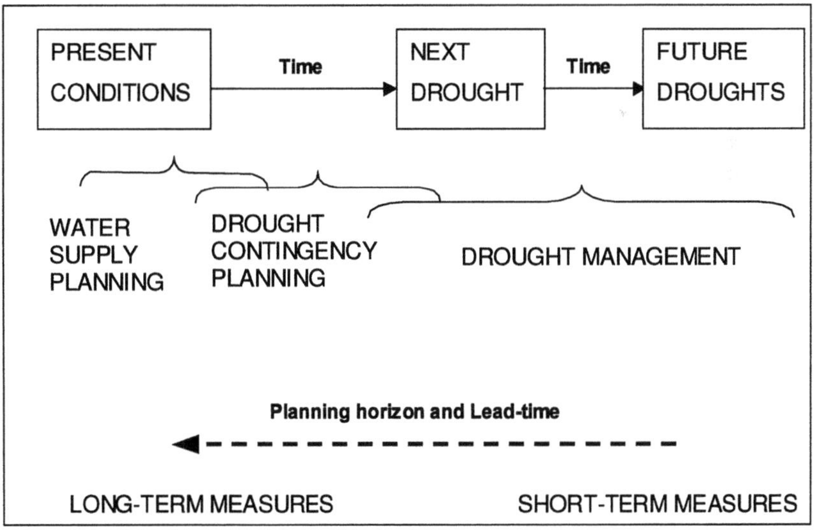

Fig. 1. A sequence of drought coping activities

The first activity is water supply planning, which employs long-term adjustments to drought, such as selecting water supply sources that are not affected by drought, or providing sufficient storage capacity to compensate for deficient streamflow or groundwater recharge during droughts. Because long-term water supply plans can rarely eliminate the risk of water shortages during severe and prolonged droughts, they are often supplemented with drought contingency plans, which specify conditions and actions that would be taken in case of a severe drought. The third activity, referred to as "drought management", begins when a drought has been recognized and water supplies can no longer be managed as if the hydrologic conditions were normal. A more precise term for this activity is "water management during drought", since the purpose is not to influence the drought itself but to make sure that available water (diminished in quantity) will be used judiciously. These three activities are interrelated and, to some degree, overlapping. The interrelationships are important since all three activities are necessary in order to achieve success in dealing with drought.

Figure 1 also displays the relationship between the planning and management activities and the type of measures for coping with drought. Implementation of water supply planning measures often requires long lead times and employs longer planning horizons. Such measures are effective only when implemented well in advance of a drought. Drought contingency planning and drought management rely primarily on short-term measures that can be implemented quickly. Since the purpose of drought contingency planning is to prepare water supply systems for the next drought, the preparedness measures are undertaken prior to drought.

Figure 2 shows examples of measures for coping with drought within each of the three activities.

Given the large array of measures for coping with drought, it is important to know which measure, or which combination of measures, should be relied upon in drought planning and management. Current U.S. approaches vary by state, climatic region, and level of governmental planning. In semi-arid western states, there is a noticeable shift from long-term structural measures toward better preparedness and drought management (Werick, 1991; U.S. Army Corps, 1995). With respect to the level of government, generally, local water supply systems seem to favor drought mitigation as part of their water supply planning.

Governmental agencies at the regional, state and federal levels, whose primary mandate is the stewardship of state water resources, seem to emphasize drought contingency planning and drought management.

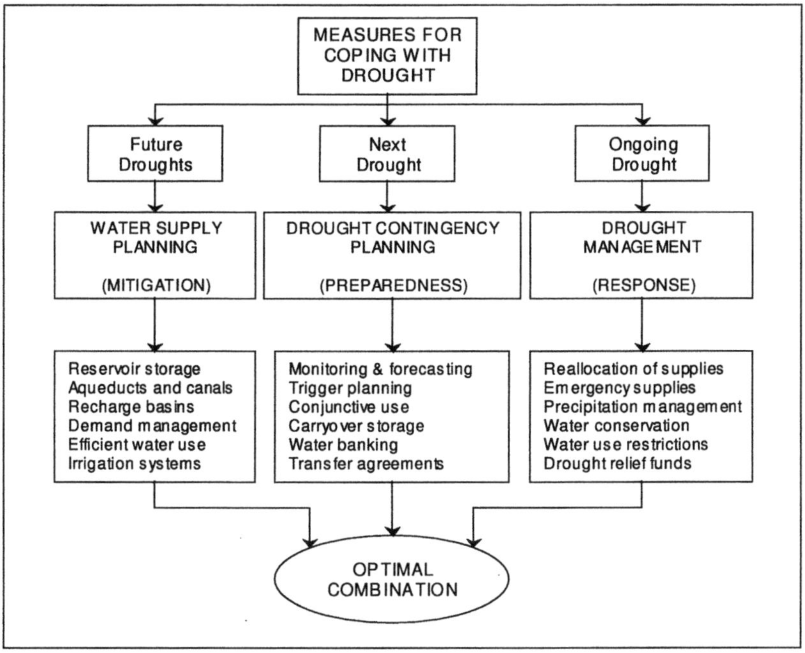

Fig. 2. Classification of measures for coping with drought.

3. Drought mitigation through water supply planning

3.1. LONG-TERM MITIGATION MEASURES

In a broad sense, water supply is the provision of water for municipal, industrial, and agricultural uses and the provision of minimum streamflow to support aquatic resources and protect water quality. During a prolonged drought, declines in streamflow and water storage limit the amount of water that can be withdrawn to supply cities, industries and irrigated agriculture or released for in-stream uses. The impacts of drought on water supply can be mitigated through long-term planning for water supply. The type of long-term measures for drought protection depends on whether the planning is conducted at the local or regional level. The specific long-term mitigation measures, which can be considered as alternatives in long-term water supply planning, fall into five broad categories:

- *Enlarging water storage capacity:* With added storage capacity more water can be captured and stored during droughts. Traditional ways of storing water include building new surface water impoundments, adding storage to existing reservoirs by installing spillway gates or dredging silt deposits, or constructing "terminal" storage for water that is imported from distant sources. More innovative types of storage capacity include storage of surface water or reclaimed wastewater in depleted aquifers, and off-stream storage of water by high-flow skimming.

- *Regional collaborative extra-supply options:* Local water supply systems can develop cooperative regional water storage and transfer schemes, which utilize joint facilities for storing and transporting water. Examples of structural measures include construction of water conveyance grids and adding water supply storage in multi-purpose reservoirs. Nonstructural options include regional water banking programs, and regional conjunctive use of surface water and groundwater.
- *Improvements in efficiency of supply systems:* In some situations, storage capacity can be enhanced by measures, which are aimed at reducing the unproductive losses of water in storage and transmission. Such measures include lining of water transmission canals, reservoir evaporation suppression or reduction of seepage and dam leaks.
- *Improvements in efficiency of water use:* Improvements in technical efficiency of water use can free up significant quantities of water by meeting the existing needs of individual users and uses with less water. Examples of technologies and efficient water use practices that can be employed to reduce water use for the major sectors of users are listed below:
    - Domestic use: low-flow showerheads, shower-flow restrictors, toilet-tank inserts, faucet aerators, low- and ultra-low flush toilets, dual flush toilets, insulation of hot-water pipes, horizontal axis washing machines, low-pressure supply connections, pressure-reducing valves, water-efficient landscape designs, efficient landscape irrigation practices and other devices.
    - Industrial use: counter-flow washing and rinse systems, reuse of process water, recirculation of cooling water, ozone treatment for cooling towers, treatment and reuse of blowdown, water recycling.
    - Agricultural irrigation: micro-spray and drip irrigation systems, soil-moisture sensors, laser-assisted field leveling, and evapotranspiration-driven irrigation schedules.
    - All uses: metering of water use, rehabilitation of water delivery systems, leak detection and repair, pressure reduction in distribution systems.
- *Water demand management:* In addition to improvements in water-use efficiency, there are several measures that cam modify or reduce the future growth in water demand, which would increase the level of long-term protection against drought. Such measures include efficient water pricing strategies, or adoption of zoning ordinances and growth policies to control the number of water users in areas with limited supplies.

Although some of these measures can be used for balancing water demand and supply under normal weather conditions, they can also be implemented for the purpose of long-term drought protection. However, despite this wide range of available long-term measures, traditional water supply planning tends to rely primarily on the provision of extra storage capacity.

3.2. LOCAL PLANS FOR WATER SUPPLY

In highly engineered local water supply systems, the long-term protection against water shortages caused by drought is accomplished through "safe yield" planning. This long-term drought protection method involves "drought-proofing" a supply system with extra storage capacity to withstand all but the most severe droughts.

Safe yield is a characteristic of a source, or a combination of sources, which is defined as the amount of water that could be extracted on a regular and permanent basis without the depletion of the reserve storage. Another concept that complements safe yield is the reliability, or the probability of a drought that would make the system incapable of providing the safe yield. A reservoir of a given storage volume can support a range of withdrawals that can be sustained during a period of drought depending on the severity of the event. Higher withdrawal rates (or drafts) will have lower reliability and additional storage will be required in order to increase the rate of withdrawal. As the severity of a drought increases, the level of reliable withdrawals will decrease. The levels of protection of a water supply system against future drought can be very high, especially when the reliability level of safe yield is chosen at 98 or 99 percent, certain optimistic assumptions are used in projecting future demands, and the actual rate of withdrawals is substantially below the safe yield.

The determination of safe yield for a given level of storage is usually based on long-term records of streamflow. These records are only 40 to 100 years long for many streams in the United States. The reliability of safe yield is not commonly expressed in terms of probability but simply stated as the reservoir yield during the "drought of record" for a given location. For groundwater supplies, safe yield is the amount of naturally occurring groundwater that can be withdrawn from an aquifer without impairing the quality of groundwater or causing environmental damage. Safe yield of a groundwater body cannot be measured as precisely as the safe yield of a reservoir system, but the degree to which groundwater level declines during severe droughts is usually known and can be managed.

Figure 3 shows an example of reliability analysis for a water supply system in one of the Mid-Atlantic States in the U.S. The system can currently store 15 billion gallons (56.8 million m^3) of water in several interconnected reservoirs. Based on the assumption that the lower one third of reservoir volume is unusable and cannot be withdrawn during drought, the safe yield (as an average daily draft) is estimated at 56 million gallons per day (mgd) or 212,000 m^3/day. The results of flow and withdrawal simulation conducted at a daily time step for a 162-year historical record indicate that as the demand, and therefore average withdrawal, from the system increases, the reliability of the supply decreases.

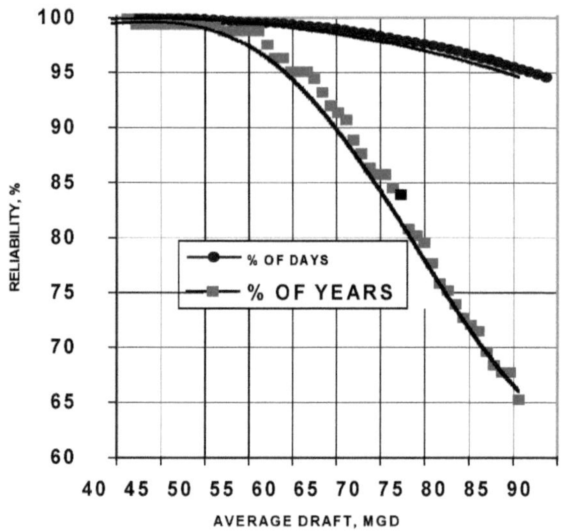

Fig. 3. Percent of days and years without supply shortfalls as a function of average draft during a 162-year record

The two lines on Figure 3 are used here to illustrate one problem of the working definition of reliability of safe yield. When reliability is defined as the percent of time during which the demand is met, large differences in reliability can result when different measures of time are used. For example the system used in the example could meet the average withdrawal of 90 mgd for 95 percent of days but only for 65 percent of years during the 162-year record used in the simulation.

The traditional way of maintaining high reliability of supply as the demand on the system grows is to increase the volume of storage. Because many smaller reservoirs are already designed to capture as much of the natural streamflow as possible, new storage reservoirs have to be constructed on nearby streams.

In practice, few public water supply systems in the United States use high levels of "drought proofing" through safe yield planning. A survey of 547 water supply systems serving some 31 million urban residents in the states of Alabama, California, Florida, Oklahoma, Tennessee, and Wyoming showed that only one-third of the respondents in the three eastern states, and one-fourth in the three western states, reported a built-in protection against the drought of record (Dziegielewski *et al.*, 1992b). Table I shows the distribution of drought protection levels among the survey respondents.

Table I. Safe yield protection in a sample of 547 water supply systems

Recurrence Interval of Design Drought	Number of Systems Affected	Percent of Systems, %
5-year	142	29.2
10-year	64	13.1
20-year	49	10.1
50-year	39	8.0
Drought of record	143	29.4
100-year	36	7.4
1000-year	14	2.9
All respondents	547	100.0

Source: Dziegielewski *et al.*, 1992b

The results in Table I indicate that more than half of the surveyed systems are likely to respond to droughts, which are equal or less severe than a 20-year event. In the same survey, nearly half of all systems have experienced problems during the droughts of the 1980s and 9 out of 10 systems affected by drought implemented measures either to reduce demand or increase supply. The type of measures implemented by these systems is shown in Table II.

Table II. Actually implemented drought response measures

Drought Management Measure	Number of Systems	Percent of Systems, %
Mandatory restrictions on selected uses	133	60.0
Mandatory rationing plan	51	23.0
Restrictions on wholesale deliveries	11	5.0
Water price adjustments	34	15.3
Acquisition of emergency supplies	23	10.4
Voluntary conservation and other	70	31.5
Totals	222	100.0

Source: Dziegielewski *et al.*, 1992b

Surveys of drought experiences suggest that the majority of water supply systems does not have sufficient storage, or do not trust their estimates of the probability of safe yield and as a result, many systems implement drought management measures during mild or moderate droughts (Dziegielewski, *et al.*, 1992b). This situation highlights the need for developing drought contingency plans.

The long-term protection of local systems can be severely constrained in regions with intensive utilization of the existing freshwater resources. The competition for available sources of supply can make it difficult, if not impossible, to achieve high levels of drought protection without affecting other suppliers or other uses of water. In such regions, regional water authorities and state or national water resources agencies must coordinate the planning for water supply with built-in drought protection.

3.3. REGIONAL WATER SUPPLY PLANNING

Planning for the protection of water supply at the regional or state level, while rarely performed, has great potential for mitigating the consequences of drought-induced shortages of water supply. The need for a higher level of planning is becoming more critical as the local entities seek to appropriate ever-increasing proportions of the existing resources to enhance the reliability of their individual supplies. State and regional water resources agencies can assume the responsibility for developing water storage and transmission facilities and for finding ways to satisfy the most critical uses of water when the supplies are diminished because of drought. The need for governmental planning is important in regions or countries with limited general availability of freshwater resources.

While the development of water storage and extensive transmission systems is possible and practiced in some regions, it is not done for the sole purpose of drought protection. Obviously, construction of a multibillion-dollar aqueduct that would be used to transfer water between regions only during a severe drought would be a highly uneconomical way of long-term drought protection. Such systems are built prior to the development of irrigated agriculture or urban areas and their cost is recovered by their constant use.

For example, in California, the State Water Project (SWP) and federal Central Valley Project (CVP) provide "dependable supplies" to the major population centers and agricultural areas of the state by exporting water from the Sacramento-San Joaquin Delta. The SWP facilities consist of 22 reservoirs (including 2 reservoirs shared with CVP) with a total storage of 7.86 km^3. These reservoirs are connected to 445 miles (716 km) of aqueducts and 21 pumping plants with design flows ranging from 2.3 m^3/s to 374 m^3/s. The CVP facilities include 20 reservoirs with a combined storage capacity of 16.8 km^3 from which the stored water is distributed by eight aqueducts with a combined length of 613 miles (987 km). In addition to the SWP and the CVP, there are four large regional projects with several hundred of miles of aqueducts. Many local systems in the state are protected against water shortages during drought by the availability of water imports from the regional and state storage reserves. However, the ability of the regional suppliers to increase drought protection through additions to the existing storage is severely constrained by a general opposition to water development within the state (Dziegielewski *et al.*, 1993).

Regardless of the level of available storage capacity, when the supply is less than demand, the biggest challenge for regional and state-level planning is to reallocate the available supplies in a manner that would minimize the social impacts of drought. Legal and institutional arrangements for water management during drought are the primary tool for minimizing these impacts. In the United States, each state is free to develop its own laws concerning the use of its water resources. Federal involvement is limited to the control of water quality, federal and Indian reserved water rights, regulation of navigable waters for navigation, commerce and public access, and interpretation of water resources activities in accordance with the federal Constitution.

4. Drought preparedness through contingency planning

4.1. FEATURES OF DROUGHT CONTINGENCY PLAN

Drought contingency planning is an anticipatory approach to drought, which involves the preparation of a detailed plan for prompt and efficient response to drought conditions. Drought contingency planning can be undertaken as part of water supply planning where the drought response actions are considered in conjunction with long-term water supply options and with respect to the actual risk of shortage. It can also be developed for any existing system regardless of the risk of actually experiencing drought-induced shortages of water supply. Many researchers strongly recommend such a proactive drought management approach in contrast to reactive activities that are more characteristic of crisis management (Wilhite and Wood, 1985; Dziegielewski, 1986; Easterling and Riebsame; 1987, Riebsame *et al.* 1990; Grigg and Vlachos, 1989; Wilhite, 2000a).

The preparedness stage of coping with drought is not limited to the "plan for action." It can also include implementation of water management measures in anticipation of the next drought. The essence of a contingency plan is a determination of which drought response actions will be implemented under what condition of water shortage. The development of drought contingency plans can reduce the uncertainty in drought response decisions. Such plans can provide guidelines for decision makers as to the proper times to employ drought response measures, and which measures to employ, depending on the status of demand and supply. This aspect of contingency planning is sometimes referred to as "trigger planning" as it makes an advanced determination of what levels of drought indices or reservoir storage will trigger a response action. The proper use of triggers requires that weather conditions be closely monitored so that drought conditions can be recognized as early as possible. Drought forecasting can play an important role by facilitating early drought management decisions. Some drought contingency measures represent actual changes in water management practices that are implemented during periods of normal weather. The purpose of such preparedness measures is to maximize the reserves of stored water at all times and prevent the depletion of these drought reserves during normal weather conditions. Another form of such preparedness measures is to make arrangements for the purchase or exchange of water during drought periods. Specific examples of the preparedness measures that can be implemented before a drought occurs include:

- *Maintenance of storage reserves:* An effective practice is to maintain a reserve of stored water at the end of each water year, which is known as "carryover" water.

Another measure is to maximize the use of surface water sources during normal supply conditions thus allowing the local and regional aquifers to be replenished through natural infiltration or artificial recharge.

- *Construction and maintenance of standby supplies:* Some sources of water supply have high cost of operation and are too expensive to be used on a continual basis. However, their use only during periods of shortage can be justified. Examples of standby supplies include water desalination plans, abandoned wells that require expensive wellhead treatment or other sources of water with impaired quality, which require blending or expensive treatment before use.
- *Water transfer and exchange agreements:* Opportunities for making agreements for temporary water transfers and exchanges are more likely to exist during normal weather than during a drought. Urban systems can secure contracts with irrigation districts or individual farmers for the purchase of agreed-upon quantities of water during drought.

These investments and arrangements could be effective means of mitigating water supply shortages when a drought occurs. There are many examples of their successful use during recent droughts in the U.S. An example of conjunctive use of water is shown in Figure 4.

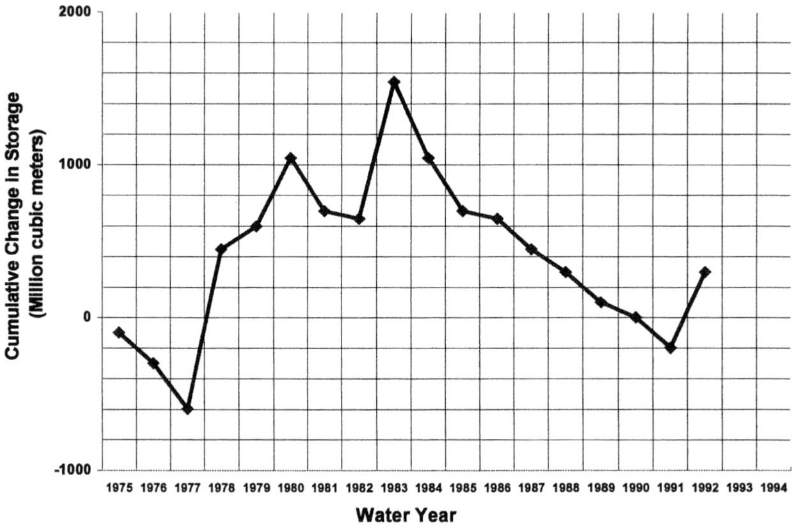

Fig. 4. Long-term changes in groundwater storage under a conjunctive use program in Southern California.

The plot of the cumulative change in groundwater storage in Figure 4 indicates that after the initial depletion in 1975, its conjunctive use and recharge over six consecutive years resulted in an additional storage reserve of 2,138 million m^3 in 1983. This was followed by an 8-year period of storage depletion which near the end of the six-year drought in California 1991 reached 1,542 million before showing some recovery in 1992. Many other measures that are listed above, were actually employed during the 1987-1992 drought, including a construction of a 7.5 mgd (28.4 thousand m^3/day) desalination plant in the coastal city of Santa Barbara.

4.2. LOCAL DROUGHT CONTINGENCY PLANS

A majority of water supply systems in the U.S. have a formal drought contingency plan. Past surveys of water supply systems conducted by Moreau and Little (1989), DalMonte (1991), and Dziegielewski *et al.*, (1992b) found that approximately one-half of water supply agencies had developed drought contingency plans. Also, several states have passed laws, which require that water supply agencies develop drought contingency plans.

A critical deficiency in drought preparedness often arises due to an inadequate understanding of the legal and administrative aspects of local drought response decisions. Many water supply agencies do not have an accurate knowledge of their legal entitlements to water supply sources or about the formula by which their entitlement would be reduced during a drought. Another deficiency concerns the legal authority to implement temporary restrictions on water use or other drought management measures. Many water supply agencies are subject to outside involvement in the decision-making process due to financial, legal or political considerations. Although some water suppliers might be faced with legal and administrative requirements imposed by multiple levels of government, generally most drought response activities remain well within the scope of local jurisdictions. Systems are usually required to obtain local government approval and provide procedural due process (usually public notice and hearing) to citizens.

4.3. REGIONAL AND STATE-LEVEL CONTINGENCY PLANNING

Historically, state governments devoted minimal resources to statewide drought planning. Wilhite and Wood (1985) reviewed state response to the 1988 drought and concluded that in most states the responses took the form of short-term emergency measures to deal with the crisis at hand. Walker *et al.*, (1991) found that many state governments placed drought management within a state agency, which only has the authority to respond to disasters after they have occurred, rather than within an agency that is responsible for water resources and can undertake drought contingency planning activities to reduce the cost of future droughts. Since 1990, this situation has changed, and as of 1997, 27 states had some form of a drought contingency plan (Wilhite, 2000b).

Since state governments have the authority to allocate water among competing uses, including water needed for the environment, they are in a position to affect the ability of local water supply systems to cope with droughts. The most important measures that can be included in state-level drought contingency plans and preparedness activities include:

- Establishment of monitoring programs and data collection networks to provide early warning system and appropriate triggers for drought severity.
- Assessment of vulnerability of existing water supply systems and economic sectors to drought.
- Review of existing state water laws and statutes pertaining to water management and drafting of emergency legislation to facilitate drought management.
- Development of legislation to require the development of local and regional drought contingency plans.
- Establishment of technical assistance programs for management of existing supply systems and water conservation methods, such as model conservation ordinances for local governments and water authorities.
- Establishment of a state-sponsored water banking program and streamlining of the approval process for water transfers among user sectors and between regions.
- Establishment of economic incentives (and abolishment of existing subsidies and disincentives) to improve long-term efficiency in water use.

The major focus of existing state drought contingency plans is on enhancing drought monitoring and early warning systems, impact assessment, and coordination of drought preparedness and response efforts within state agencies as well as between local, state and federal governments.

5. Drought management

The term "drought management" is often applied to all activities that are undertaken during and after the drought with a purpose of alleviating the adverse impacts of drought. Such activities can be oriented toward reducing water demand and thus conserving the supplies that are still available, or toward increasing the available supply from emergency sources. Table III shows a roster of short-term drought management measures that can be used in dealing with either actual or impending shortages of water.

These actions, whether aimed at increasing supply or reducing demand, may result in a temporary increase in the cost of water or may cause water users to suffer significant economic losses when their normal supplies are curtailed. The phrase "water management during drought" should be understood to represent both the management of water resources and management of water demand. It implies that the sources of water supply should be managed differently during drought than under normal hydrological conditions and also demands for water should be modified to reflect the condition of water supplies. In essence, the primary water management activity during drought is to conserve the existing supply by limiting water withdrawals.

Table III. Examples of short-term measures for increasing supply and reducing demand

Measures for Increasing Water Supply	*Measures for Reducing Water Demand*
Water system operations: Providing pumped reservoir storage Transfer of water to non-spilling reservoirs Reservoir evaporation suppression Recirculation of filter wash water Meter repair and replacement programs Leak detection and repair Discontinuing hydrant and main flushing Reduction of water pressures Emergency and auxiliary supply sources: Emergency interconnections Importation of water by trucks Importation by railroad cars Construction of temporary pipelines Utilization of small ponds and quarries Utilization of dead reservoir storage Temporary in-channel storage Reactivation of abandoned wells Drilling of new wells Emergency water transfers and diversions: Temporary transfer of agricultural supplies Emergency water banks Water exchange agreements Cross-purpose diversions: Reduction of releases for hydropower Postponing releases for flood control Withdrawals from recreational lakes Relaxation of minimum streamflow Other options: Precipitation management Technological innovations	Public information and education campaigns: Mass media campaigns on conservation School education programs Community outreach: exhibits, tours Water service restrictions: Moratorium on new customer connections Suspension of wholesale water deliveries Discontinuation of sales to water haulers Restrictions on non-essential uses of water: Filling of swimming pools Lawn watering and car washing Water hosing of pavements Irrigation of parks and street medians Irrigation of golf courses Operation of public fountains Closing non-recycling car washes Emergency conservation programs: Residential indoor plumbing retrofits Home water audits for high-quantity users Plumbing retrofits for public buildings Commercial/industrial/institutional audits Drought emergency pricing of water: Drought surcharge or excess-use levy Steeply increasing block rate Customer conservation credits Water rationing programs: Odd/even-day watering Per capita or per household allocations Customer water budgets Rationing through inconvenience

Source: Adapted from Dziegielewski (2000)

For systems with well-developed drought contingency plans, drought management could be limited to the monitoring of drought conditions and implementation of appropriate short-term measures when the pre-established trigger levels are reached. However, drought contingency plans, while invaluable in supporting drought management decisions, are not likely to foresee all possible circumstances or remove the inherent uncertainties in dealing with droughts whose duration and severity cannot be predicted.

In 1990, a National Study of Water Management During Drought was undertaken in order to examine existing water management systems and to develop a strategy to improve water management during drought (Werick, 1991). The results of this study provided the impetus for the National Drought Policy Act of 1998. This Act established the National Drought Policy Commission with a mandate "to provide advice and recommendations on the creation of an integrated, coordinated Federal policy designed to prepare for and respond to serious drought emergencies." The main intention of the Act is to formulate a national drought policy based on preparedness, mitigation, and risk management rather than on crisis management.

6. Finding the best drought plan

Formulating the best plan for coping with drought is an enormous endeavor given the reality of drought and the existing legal, institutional and social environments within which the drought-management decisions must be made. While a single conceptual model of drought management may not be feasible, the goal of reducing the social and environmental costs of droughts can be addressed through economic analysis and evaluation of the potential measures for coping with drought. Figure 5 illustrates the general relationship between the long-term cost of water shortages (or damages) and the cost of water supply infrastructure to reduce this cost.

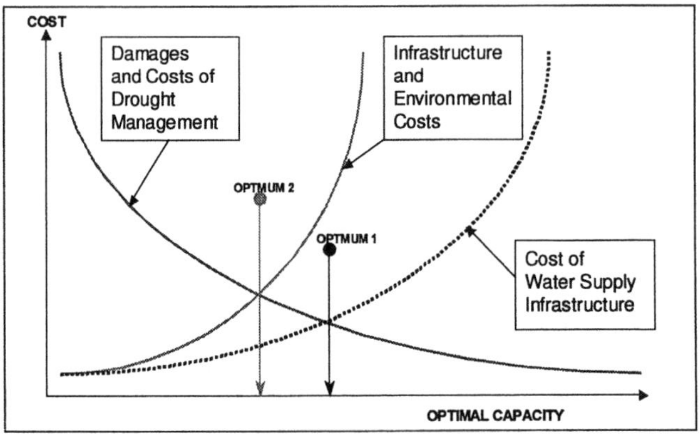

Fig. 5. Derivation of the optimal level of long-term protection from droughts (After Russell et al. 1970)

According to the hypothetical relationships in Figure 5, the optimal level of the long-term protection is found by summing the cost of damages and the cost of long-term measures (e.g., storage capacity) and finding the minimum value (Optimum 1). However, as the opportunities for water development are exhausted, additional development may have significant external cost in the form of environmental impacts. Under such conditions the optimal point is moved to the left (Optimum 2) where the optimal level of long-term protection is lower and damages are higher. The following section describes a practical approach for judging the economic effectiveness of long-term and short-term measures for dealing with drought.

6.1. EXPECTED COST OF COPING WITH DROUGHTS

Selection of the safe yield standard of water supply reliability based on the "design drought" or the most severe recorded drought does not require an analysis of the cost of the last increments of investments in water storage or transmission facilities, or the cost of damages that may occur during droughts that are more severe than the design drought. Also, it relies only on a single long-term measure for coping with droughts to the exclusion of other long-term and short-term measures. Without considering other coping alternatives for dealing with future droughts, it is possible that the level of drought protection, which is based solely on the design drought standard, is sub-optimal. The cost of water shortages that would be caused by droughts more severe than the design drought can be much higher, or much lower, than the incremental cost of additional storage capacity.

Theoretically, the optimal level of drought protection can be determined by minimizing the sum of: 1) the cost of long-term measures, such as additional water storage; 2) expected damages during droughts more severe than the design drought; 3) the cost of attaining a certain level of success in coping with actual droughts (Dziegielewski *et al.* 1991). A simplified approach to this optimization problem is to examine the tradeoffs between the investment in long-term drought protection and the cost of coping with periodic droughts of various severities during the planning period. A detailed procedure for calculating the drought protection investment tradeoff can be found in Dziegielewski *et al.* (1992a) and Dziegielewski (2000).

The coping cost can be determined by summing the costs of drought management for each year of the planning period, which includes the expenditures for drought response measures and the residual economic losses that cannot be mitigated. The expected value of this cost can be calculated as:

$$E(C_t) = \sum_{i=1}^{k} p_{it} C_{it} \tag{1}$$

where:

$E(C_t)$ = expected value of total coping cost in any year t of the planning period T

p_{it} = probability of drought with severity level i
C_{it} = cost of coping with water shortages caused by drought of probability level i
k = number of discrete levels of drought severity analyzed

The expected value of coping cost gives recognition to the fact that while the cost of dealing with a 100-year drought may be, for example, $3 million, the risk-adjusted cost is only $30,000. By summing the risk-adjusted costs of dealing with droughts with different levels of severity (e.g., 10-, 25-, 50-, 100-, 200-year events), we will obtain the expected value of total cost of coping with "statistical droughts" in any given year. The present worth of the coping cost during the entire planning period can be calculated as:

$$PW(C) = \sum_{t=1}^{T} E(C_t) \cdot (1+r)^{-t} \qquad (2)$$

where:
$PW(C)$ = present worth of the expected cost of coping with shortages during planning period T
r = discount rate
T = number of years in the planning horizon

In order to arrive at the value of total coping cost for a given water supply systems it is necessary to project future demands and calculate the magnitude of water shortages, which would be associated with droughts of different return intervals. Next, for each level of shortage one must estimate the cost of dealing with the shortage. It is reasonable to assume that because the magnitude and duration of shortage is known, a least-cost plan consisting of short-term drought management measures can be found for each level of drought severity. Some costs of coping with shortages are easily estimated. These include the actual expenditures by water supply agencies on emergency water supplies, or water conservation campaigns. Other costs, especially damages caused by lost supplies, are estimated using special data and assumptions. For example, recent studies found the cost of water use restrictions to residential water users, during the fifth year of a six-year drought in California, were less than $5 per household per week in the San Francisco Bay area and less than $2.50 per week in Los Angeles (U.S. Army Corps, 1994).

6.2. TRADEOFFS BETWEEN LONG-TERM AND SHORT-TERM COSTS

In principle, each dollar spent on long-term drought protection should produce a return in the form of reduced future expenditures on drought management measures. For some systems, the additional drought protection obtained from investing in extra reservoir storage could be worth less than the cost of obtaining that protection. However, there may be other long-term drought protection measures, which would produce higher returns than the investment in extra storage capacity. Finally, in systems with a high level of long-term drought protection, the cost of coping with all probabilistic droughts during the planning period may be relatively small (for example, the cost to consumers of a 10 percent cutback from normal demands for a period of 2 months every 25 year) and there may not be any long-term investments that would be cost-effective in eliminating this risk.

An application of drought protection tradeoff analysis in assessing long-term protection measures for the City of Phoenix in Arizona indicated that some drought protection investments have high returns when measured by the reduction in coping costs with periodic future droughts (Dziegielewski *et al.*, 1992a). Table IV shows the estimated coping costs and returns on investment in long-term drought protection measures for two water-planning areas within the system, which are required by law to secure independent water supplies.

The two water supply subsystems, although similar in size have different present values of the expected coping cost because the demand in subsystem A is projected to grow four times faster than in Subsystem B. All three long-term measures in subsystem A would substantially reduce the unmitigated coping cost of $112.6 million. A long-term water conservation program is the most cost-effective alternative for long-term drought protection with a return of $17.3 in reduced cost of drought management per $1 of the investment.

Table IV. Drought-protection investment tradeoffs for two water subsystems in Phoenix, Arizona

Long-term Measure	Investment Cost 1000$	Coping Cost, 1000$	Reduction in Coping Cost 1000$	Trade-off Ratio
Subsystem A				
No investment	0	112,600	0	--
Drilling new wells	20,500	25,200	87,400	4.3
Renovation and drilling of wells	22,000	33,600	79,000	3.6
Long-term conservation program	5,600	15,800	96,800	17.3
Subsystem B				
No investment	0	25,100	0	--
New wells without nitrate treatment	8,000	11,000	14,100	1.76
New wells with nitrate treatment	17,700	11,000	14,100	0.80
Long-term conservation and new wells	6,100	7,500	17,600	2.89
Agricultural land-fallowing program	100	11,500	13,600	136.0

Source: Adapted from Dziegielewski *et al.*, (1992a)

In subsystem B, the most cost-effective alternative is a land-fallowing program, under which the city would enter into interruptible water supply contracts with owners of irrigated agricultural land who share the same source of water supply. The city would pay the farmers compensation equal to average profits of about $500 per acre, or $250 per acre-foot of water. At a moderate up-front cost of $100,000 to establish the administrative mechanism for the program, the city could obtain a very inexpensive source of supplemental supplies during drought.

The present worth of the expected cost of coping with future droughts provides a useful guideline for assessing the cost effectiveness of long-term measures for drought protection. In the case of subsystem A in Phoenix, the coping cost of $112.6 is associated with an average supply deficit of 81,000 acre feet during a severe drought during the planning period. At this level of deficit, an approximate unit cost of drought management is $1,390/acre-foot. Therefore, those long-term measures, which can provide additional supplies or reduce long-term water demand at a unit capacity cost below $1,390, would be cost-effective. For subsystem B, the respective ratio of the coping cost to the maximum level of deficit is $326/acre-foot, and only additions to supply capacity costing less than $326/acre-foot would be cost-effective.

7. Conclusions

Droughts have significant impacts on rain-fed-agriculture, which can affect not only the livelihood of the farmers but also the national economies of countries where agricultural production makes a major contribution to the national economy. Even greater economic impacts can be caused by drought-induced shortages of water supply, which affect irrigated agriculture, industry and production of hydropower. The adverse economic and environmental impacts of drought are likely to increase in regions where large proportions of available water resources are extracted for off-stream uses creating strong competition between economic and environmental uses.

Better planning and effective management of water resources during drought can mitigate the impacts of drought on water supply. The traditional long-term measure for drought mitigation is the provision of extra storage of water in reservoirs and underground aquifers. However, this option is not available in all localities and other options such as water conservation and importation can be used to improve the supply and demand balance during drought.

The integrated management of available resources over large regions or entire countries can provide options for transferring the available supply from areas with water surplus to areas with shortages. However, the investment in conveyance grids to facilitate the movement of water over long distances for the sole purpose of protection against severe droughts is not economical and other cost-effective options are usually explored. Voluntary water transfers between users, where physically possible, with appropriate institutional arrangements that permit such transfers within a controlled water market, are one such option. A temporary reduction in urban or agricultural demands is another viable alternative for mitigating potentially serious consequences of drought-induced shortages of water supply.

References

DalMonte, J.A. (1991) *Water Utility Managers' Response to Drought in the Midwest.* Masters thesis. Department of Goography. Southern Illinois University. Carbondale, Illinois.

Dziegielewski, B. (1986) Drought Management Options. *Drought Management and its Impacts on Public Water Systems.* Water Science and Technology Board, National Research Council. Washington, D.C., National Academy Press.

Dziegielewski, B. (2000) Drought Preparedness and Mitigation for Public Water Supplies. In: D.A. Wilhite (Ed.) *Drought: A Global Assessment. Volume II.* Routledge, London and New York, pp. 194-205.

Dziegielewski, B., Lynne, G.D., Wilhite, D.A. and Sheer, D.P. (1991) *National Study of Water Management During Drought: A Research Assessment.* U.S. Army Corps of Engineers, Institute for Water Resources, IWR Report 91-NDS-3. Alexandria, Virginia.

Dziegielewski, B., Mee Jr., W.R. and Larson, K.R. (1992a) Developing a Long-Term Drought Plan for Phoenix. *Journal of the American Water Works Association* 84(10): 46-51.

Dziegielewski, B., Ferrell-Dillard, R. and Beck, R.E. (1992b) Coping with Urban Water Shortages During Drought: The Effects of Legal and Administrative Factors. *Research Report USGS Grant 14-08-001-G1897.* Southern Illinois University, Carbondale, Illinois.

Dziegielewski, B., Garbharran, H.P. and Langowski, J.F. (1993) The Great California Drought of 1987-1992: Lessons for Water Management. *Report of the National Study of Water Management During Drought.* U.S. Army Corps of Engineers, Institute for Water Resources Report IWR Report 93-NDS-5.

Easterling, W.E. and Riebsame, W.E. (1987) Assessing Adjustments in Agriculture and Water Resources Systems. In: D.A. Wilhite, W.E. Easterling and D.A. Wood (Eds.) *Planning for Drought: Toward a Reduction of Social Vulnerability.* Boulder, Colorado: Westview Press.

Grigg, N.S. and Vlachos, E.C. (1989) *Drought Water Management: Preparing and Responding to Drought.* Ft. Collins, CO: International School for Water Resources, Colorado State University.

Moreau, D.H. and Little, K.W. (1989) *Managing Public Water Supplies During Droughts: Experience in the United States in 1986 and 1988.* University of North Carolina: Water Resources Research Institute.

NDPC (2000) Preparing for Drought in the 21st Century. *Report of the National Drought Policy Commission.* U.S. Department of Agriculture. Washington, D.C.

Riebsame, W., Changnon, S. and Karl. T. (1990) *Drought and Natural Resource Management in the United States: Impacts and Implications of the 1987-1990 Drought.* Natural Hazards Research and Applications Information Center, University of Colorado, Boulder.

Russell, C.S., Arey, D.G. and Kates, R.W. (1970) *Drought and Water Supply.* Johns Hopkins University, Baltimore, Maryland.

U.S. Army Corps of Engineers (1994) *Human and Environmental Impacts: California Drought. Institute for Water Resources.* IWR Report 94-NDS-10. Alexandria, Virginia.

U.S. Army Corps of Engineers (1995) *National Study of Water Management During Drought: The Report to the U.S. Congress.* Institute for Water Resources, IWR Report 94-NDS-12. Alexandria, Virginia.

Walker, W.R., Hrezo, M.S. and Haley, C.J. (1991) Management of Water Resources for Drought Conditions. In: R.W. Paulson, E.B. Chase, R.S. Roberts and D.W. Moody (Eds) *National Water Summary 1988-89: Hydrologic Events and Floods and Droughts.* U.S. Geological Survey Water Supply Paper 2375, pp. 147-156.

Warrick, R.A., Trainer, P.B., Baker, E.J. and Brinkman, W. (1975) *Drought Hazard in the United States: A Research Assessment.* Washington, D.C. US Department of Commerce, National Technical Information Service.

Werick, W.J. (1991) *The National Study of Water Management During Drought: Report of the First Year of Study.* IWRA Report 91-NDS-1. U.S. Army Corps of Engineers. Fort Belvoir, Virginia.

Wilhite, D.A. and Wood, D.A. (1985) Planning for Drought: The Role of State Government. *Water Resources Bulletin.* 21(1): 31-38.

Wilhite, D.A. (2000a) Preparing for Drought: A Methodology. In: D.A. Wilhite (Ed.) *Drought: A Global Assessment. Volume II.* Routledge, London and New York, pp. 89-104.

Wilhite, D.A. (2000b) State Actions to Mitigate Drought: Lessons Learned. In: D.A. Wilhite (Ed.) *Drought: A Global Assessment. Volume II.* Routledge, London and New York, pp. 149-157.

Appendix

Report and recommendations of the Expert Consultation and Workshop on Drought Mitigation for the Near East and the Mediterranean (Aleppo, May 27-31, 2001)

REPORT AND RECOMMENDATIONS OF THE EXPERT CONSULTATION AND WORKSHOP ON DROUGHT MITIGATION FOR THE NEAR EAST AND THE MEDITERRANEAN (ALEPPO, MAY 27-31, 2001)

In the last few decades, the occurrence of severe droughts in the Near East and in the Mediterranean region has pointed out the need for appropriate mitigation measures, able to reduce the negative effects of such extreme events. In order to identify appropriate approaches to the drought mitigation problem, a "FAO-EC Expert Consultation and Workshop on Drought Mitigation in the Near East and the Mediterranean" was organised by ICARDA and the Department of Civil and Environmental Engineering, University of Catania, from May 27 through May 31, 2001 in Aleppo, Syria.

The Meeting provided a forum for discussing drought related issues among country representatives, officials from national, regional,and international institutions as well as invited experts on water management under drought conditions. The Meeting included two parts: the presentation of the results of the INCO-DC Project "A DSS for mitigation of drought impacts in the Mediterranean regions", funded by European Commission and coordinated by the Department of Civil and Environmental Engineering (DICA), University of Catania, Italy and the discussion of several country papers from the regional and national authorities of the North Africa and Near East countries convened by FAO Regional Office for the Near East, Cairo, Egypt. A few key-note speakers were invited for enlarging the basis of the discussion about drought related problems.

Before the beginning of the Workshop, participants enjoyed an interesting field trip with a visit to water harvesting and other water management practices in Jabal El Bishri site of ACSAD, and a visit to water supply and irrigation scheme and farmers fields within the Middle Euphrates Irrigation Project, supplied by Assad reservoir at Tabka dam.

The workshop was officially opened by H. El. Ahmad, Deputy Minister of Agriculture of the Syrian Government, G. Hammadallah Representative of FAO RNE, J.-L. Araus Ortega, Representative of European Commission, G. Rossi, coordinator of INCO-DC Project, and W. Erskine, Assistant Director General for Research at ICARDA.

Invited key-note speeches were given by: B. Dziegielewski from Southern Illinois University, USA who introduced the concept of the best mix of long-term and short-term measures for coping with drought, T. Estrela from Centro de Estudios Hydrographicos, CEDEX, Spain, who presented the methods used for regionalisation

G. Rossi et al. (eds.), Tools for Drought Mitigation in Mediterranean Regions, 343-347.

and identification of droughts within an other EC research project recently completed (ARIDE: Assessment of the Regional Impact of Droughts in Europe). A comprehensive analysis of the research programs in water saving and in drought management developed by ICARDA was presented by T. Oweis. Also a speech about the European programs on water management and drought was given by J.L. Araus-Ortega. The role of the World Health Organisation for reducing the drought impacts on health was presented by A. Musam, and the activities developed by the Euromediterranean Water Information System in the last years were outlined by M. Mazzola, on behalf of EMWIS.

The sessions devoted to the presentation of the results of the INCO-DC Project were structured according to the three main tasks of the project, namely drought identification and characterisation, modelling irrigation management, and water supply system operation and Decision Support System. The project has been specifically oriented to the development of adequate tools to be provided to decision makers at different level for water management of irrigation systems under drought conditions in arid and semiarid regions. The application of the developed models and software packages was carried out through several case-studies selected in the participants countries

After an introductory lecture, devoted to a general outline of the methodology and of the main results of the project ("An integrated approach to drought mitigation in Mediterranean regions" by G. Rossi, Italy) a presentation of appropriate approaches to the water use for irrigation in water scarcity regions and of emerging technologies concerning models and decision support tools was given by L. Pereira, Portugal.

Then the methodology developed and applied to identify and analyse local and regional drought, based on the run concept and implemented in the software package REDIM (Regional Drought Identification Method) were discussed (G. Rossi and A. Cancelliere). The results of the applications of such a tool were presented through a set of papers prepared by each Project partner for case-studies in Portugal (A.A. Paulo, L. Pereira, P. Matias), in Tunisia (M. Bergaoui), in Jordan (A. Salihi), in Italy (A. Cancelliere), The preliminary analysis of stationarity on precipitation series indicated a generally negative trend affecting precipitation in all investigated regions, which partly explains the increasing frequency, duration and severity of droughts recently observed in all areas, altough contemporaneity of drought periods was not observed especially among the farthest regions (Portugal and Jordan).

The main outcomes of the second task of the Project were presented through a lecture on the role of the irrigation scheduling simulation for selecting irrigation strategies under limited water availability and on the features of the ISAREG model (L. Pereira, P. Teodoro, P. Rodrigues, J. L. Teixeira, Portugal). The comparison of the results obtained for different crops in different areas included the analysis of irrigation strategies in the Tunisian case-study for cereals and horticultural crops with a supplementary economic evaluation (A. Zairi, El Amami, A. Slatmi from Tunisia and L. Pereira, P. Rodrigues, T. Machado and J. Teixeira from Portugal), in Portuguese case-study for cereals and summer crops (P, Rodrigues, T. Machado. L. Pereira and El Amami, A. Zairi), in Italian case-study for citrus orchards (I. Alba) and in Jordan case study for cereals and horticultural crops (M. Shatanawi, Y. Al-Zu'bi, O. Al Jayyonsi).

The research results allowed to validate ISAREG model in different experimental conditions and to evaluate through the use of the model the best irrigation strategies. In particular for some annual crops the comparison of the yield reductions following different irrigation scheduling showed that deficit irrigation (either watering depth or frequency) is feasible for moderate drought conditions, whereas under severe drought events it is better to reduce the irrigated area and to apply near optimal parameters. Although such reduction of irrigated area cannot be applied to multi-year crops, as the citrus orchards, results indicated that an appropriate irrigation scheduling can limit yield losses during drought periods.

Within the framework of the third task of the project, two different approaches to derive operation rules of an irrigation water supply reservoir were discussed, namely the combined use of an optimisation algorithm (Dynamic Programming) and a Neural Network technique through a set of software modules (RESOPTI, RESOPER and RESSIMUL) developed by Italian partner (A. Cancelliere, G. Giuliano, A. Ancarani, G. Rossi) and a stochastic dynamic programming developed by Tunisian partner (L. Fethi, M. Bergaoui, M. Bouslimi). The use of multicriteria analysis in comparing different deficit irrigation strategies was presented by Jordan group (M. Shatanawi, O. Al Jayyousi, Y. Al-Zu'bi). Finally the features of the whole Decision Support System developed for the evaluation of drought and drought mitigation measures were discussed with the presentation of the user's guide (A. Cancelliere, G. Giuliano).

A substantial part of the Meeting was devoted to the expert consultation convened by FAO, RNE. It included an overview of the impacts of recent drought events in the countries of the Near-East and North Africa regions and a detailed proposal of the elements of a drought mitigation plan including the role of national government and international organisations (G. Hammadallah). A key-note address on "Inferences of a drought mitigation action plan" by M. Bazza focused on the major steps of a pro-active approach for coping with drought, analysing the actions to be taken to incorporate drought mitigation measures in regular planning process and to establish contingency plans for drought periods with particular attention to agricultural, range and livestock sectors.

Present situation of drought problems and successful experiences in mitigation measures were discussed in a set of country reports related to Egypt (A. El-Tantawy), Iran (H. Siadat and M. Sharinti), Lybia (M. Razzaghi), Morocco (M. Karrou), Pakistan (M. Cheema), Algeria (M. Kelkouli), Sudan (M. E. AbdelRahman), Yemen (A. Haidar), Syria (I. Al Aghawani), Oman (A. Al-Ajmi).

The Meeting was concluded with the approval of some recommendations (reported below), including the proposal of establishing a Drought Information Network for the Region, to improve drought preparedness and mitigation through national action plans and to develop a Drought Watch and Early Warning System for a timely alert.

Recommendations approved by the Workshop

The participant at "FAO-ICARDA-EU Expert Consultation and Workshop on Drought Mitigation in the Near East and the Mediterranean" concluded that, due to the recurrence of drought events with their detrimental socio-economic and environmental consequences to communities, serious and concerted efforts are to be exerted from all stakeholders in the Region to adopt and implement long-term preparedness and mitigation plans. The main recommendations made by the Meeting are the following:

1. *Drought Information Network*
   1.1 Call for the establishment of a Network on Drought for the WANA Region (which includes FAO Near East countries) and interested member countries of the European Union (EU), for facilitating the exchange of knowledge and experience relevant to the aspects of drought, including: early warning, drought characterization, agro-meteorological and hydrological data and analysis, methodologies, impact assessment, evaluation and monitoring.
   1.2 ICARDA should coordinate this effort and convene a Task Force Group meeting ,with FAO and EU participation, within two-three months to discuss initiation of the Network.

2. *Technology Transfer and Capacity Building*
   2.1 Call upon member countries, as well as international organizations and donors to lend adequate support to research and technology transfer related to Drought Mitigation activities, including: drought tolerant germplasm, drought adapted animal breeds, biotechnology, capacity building and human resources development.
   2.2 Call for the initiation of a "Regional Project", starting in 3 - 4 countries, to serve as a pilot project for testing, validating, adapting and adoption of appropriate technologies for water saving and other drought coping mechanisms, with joint responsibility of FAO, ICARDA and EU, based on the comparative advantage of each party.

3 *Water Resources Development and Management*
   3.1 Elaborate medium- and long-term water resources development and management plans, or amend existing ones as required, to adopt techniques and methodologies targeting the reduction of vulnerability to drought.
   3.2 Develop adequate contingency plans to address the immediate communities' needs for water during drought periods.
   3.3 Promote research work on soil and water conservation and irrigation management with focus on supplementary irrigation, deficit irrigation, water harvesting, reduced tillage, cultural practices, etc.
   3.4 Promote research and development activities related to the safe and efficient use of low quality water resources (brackish, wastewater and drainage or blended water).

4 *Drought Preparedness and Mitigation*

4.1 Urge member countries to take necessary actions for formulating National Action Plans for Drought Mitigation, within the overall development schemes for communities in drought-prone areas.

4.2 Call upon member countries to commit support at the highest political level possible, for the timely adoption and implementation of drought mitigation measures.

4.3 Call for providing the necessary resources to establish a "Drought Watch and Early Warning System" for a timely alert the country.

4.4 Involve and mobilize all community sectors, including NGOs, women and youth, for launching an awareness and education campaign to sensitize all stakeholders to drought impacts ; as well as identifying each sector's role in the relief and rehabilitation activities.

4.5 Urge member countries to provide facilities and enact required legislation and policies to allow creating the enabling mechanisms that would be entrusted with drought mitigation measures to be implemented at district, provincial and national levels.

4.6 Call on member countries to establish adequate liaison with regional and international organizations, such as UNCCD, UNSO and others, and interest groups for the follow-up of drought mitigation activities at regional and global levels.

4.7 Call member countries to accord due attention and consideration to the social and economic dimension of drought within the framework of the "Preparedness and Mitigation Plans", which should be comprehensive and integrated to address rehabilitation and development of other related drought affected sectors like: rangeland, livestock, forestry and crop production, within an environmental protection context.

List of Contributors

Odeh AL-JAYOUSSI
General Corporation for Environmental Protection, Amman, Jordan.
water1@ju.edu.jo

Ignazio ALBA
Dipartimento di Ingegneria Civile e Ambientale, Università di Catania, Viale A. Doria 6, 95125 Catania, Italy.
ignazio.alba@tiscalinet.it

Javier ÁLVAREZ
Centro de Estudios Hidrograficos del CEDEX, Paseo Bajo Virgen del Puerto 3, 28005 Madrid, Spain.
javier.alvarez@cedex.es

Yasin AL-ZU'BI
Applied Science University, Amman, Jordan.
water1@ju.edu.jo

Alessandro ANCARANI
Dipartimento di Ingegneria Civile e Ambientale, Università di Catania, Viale A. Doria 6, 95125 Catania, Italy.
aancaran@dica.unict.it

Muhammed BERGAOUI
École Supérieur des Ingénieurs de l'Équipement Rural, Route du Kef, 9070 ESIER, Medjez, Tunisia.
bergaoui.med@iresa.agrinet.tn

Muhammed A. BOUSLIMI
Institut National de Recherche en Génie Rural Eaux et Forets, Ministere de L'Agriculture, Rue Hédi Karray, 1004 Tunis , B.P. 10, 2080 Ariana, Tunisia.
bouslimi.med@iresa.agrinet.tn

Antonino CANCELLIERE
Dipartimento di Ingegneria Civile e Ambientale, Università di Catania, Viale A. Doria 6, 95125 Catania, Italy.
acance@dica.unict.it

Ben DZIEGIELEWSKI
Department of Geography, 4535 Faner Hall, Southern Illinois University, Carbondale, Illinois 62901-4516, USA.
bendzieg@siu.edu

Hacib EL AMAMI
Institut National de Recherche en Génie Rural Eaux et Forets, Ministere de L'Agriculture, Rue Hédi Karray, 1004 Tunis , B.P. 10, 2080 Ariana, Tunisia.
amami.hacib@iresa.agrinet.tn

Teodoro ESTRELA
Confederación Hidrográfica del Júcar, Avda. De Blasco Ibañez, 48, 46010 Valencia, Spain.
teodoro.estrela@chj.mma.es

Giuseppe GIULIANO
Dipartimento di Ingegneria Civile e Ambientale, Università di Catania, Viale A. Doria 6, 95125 Catania, Italy.
ggiuliano@dica.unict.it

Fethi LEBDI
Institut de la Recherche et de l'Einseignament Supérieur Agricole, 30 rue Alain Savary 1002 Tunis, Tunisia.
lebdi.fethi@iresa.agrinet.tn

Alessandro LONGHEU
Dipartimento di Ingegneria Informatica e Telecomunicazioni, Università di Catania, Viale A. Doria 6, 95125 Catania, Italy.
alongheu@diit.unict.it

Telma MACHADO
Departamento de Engenharia Rural, Instituto Superior de Agronomia, Universidade Teécnica de Lisboa, Tapada da Ajuda, 1349-017, Lisboa, Portugal.
aa12902@isa.utl.pt

Paulo G. MATIAS
Departamento de Engenharia Rural, Instituto Superior de Agronomia, Universidade Teécnica de Lisboa, Tapada da Ajuda, 1349-017, Lisboa, Portugal.
paumatias@isa.utl.pt

Theib OWEIS
International Center for Agricultural Research in the Dry Areas (ICARDA), P.O. Box 5466, Aleppo, Syrian Arab Republic.
t.oweis@cgiar.org

Ana A. PAULO
Departamento de Engenharia Rural, Instituto Superior de Agronomia, Universidade Técnica de Lisboa, Tapada da Ajuda, 1349-017, Lisboa, Portugal.
apaulo@isa.utl.pt

Luis S. PEREIRA
Departamento de Engenharia Rural, Instituto Superior de Agronomia, Universidade Técnica de Lisboa, Tapada da Ajuda, 1349-017, Lisboa, Portugal.
lspereira@isa.utl.pt

Paula A. N. RODRIGUES
Departamento de Engenharia Rural, Instituto Superior de Agronomia, Universidade Técnica de Lisboa, Tapada da Ajuda, 1349-017, Lisboa, Portugal.
prodrigues@isa.utl.pt

Giuseppe ROSSI
Dipartimento di Ingegneria Civile e Ambientale, Università di Catania, Viale A. Doria 6, 95125 Catania, Italy.
grossi@dica.unict.it

Adnan AL-SALIHI
Water & Environment Research & Study Center, University of Jordan, Salt Road, Amman, Jordan
mazin@go.com.jo

Muhammad SHATANAWI
Faculty of Agriculture, University of Jordan, Salt Road, Amman, Jordan
shatanaw@ju.edu.jo

Adel SLATNI
Institut National de Recherche en Génie Rural Eaux et Forets, Ministere de L'Agriculture, Rue Hédi Karray, 1004 Tunis , B.P. 10, 2080 Ariana, Tunisia.
slatni.adel@iresa.agrinet.tn

José Luis TEIXEIRA
Instituto de Hidraúlica Engenharia Rural e Ambiente (IHERA) Av. Afonso Costa 3, 1900 Lisboa, Instituto Portugal.
jlt@ihera.min-agricultura.pt

Pedro R. Teodoro
Departamento de Engenharia Rural, Instituto Superior de Agronomia, Universidade Técnica de Lisboa, Tapada da Ajuda, 1349-017, Lisboa, Portugal.
erpteodoro@yahoo.com

Abdelaziz ZAIRI
Institut National de Recherche en Génie Rural Eaux et Forets, Ministere de L'Agriculture, Rue Hédi Karray, 1004 Tunis , B.P. 10, 2080 Ariana, Tunisia.
zairi.abdelaziz@iresa.agrinet.tn

Subject index

16/3: Water-quality hydrology ISBN 0-7923-3652-6
16/4: Water resources planning and management ISBN 0-7923-3653-4
Set 16/1–16/4 ISBN 0-7923-3654-2

17. V.P. Singh: *Dam Breach Modeling Technology*. 1996 ISBN 0-7923-3925-8
18. Z. Kaczmarek, K.M. Strzepek, L. Somlyódy and V. Priazhinskaya (eds.): *Water Resources Management in the Face of Climatic/Hydrologic Uncertainties*. 1996 ISBN 0-7923-3927-4
19. V.P. Singh and W.H. Hager (eds.): *Environmental Hydraulics*. 1996 ISBN 0-7923-3983-5
20. G.B. Engelen and F.H. Kloosterman: *Hydrological Systems Analysis*. Methods and Applications. 1996 ISBN 0-7923-3986-X
21. A.S. Issar and S.D. Resnick (eds.): *Runoff, Infiltration and Subsurface Flow of Water in Arid and Semi-Arid Regions*. 1996 ISBN 0-7923-4034-5
22. M.B. Abbott and J.C. Refsgaard (eds.): *Distributed Hydrological Modelling*. 1996 ISBN 0-7923-4042-6
23. J. Gottlieb and P. DuChateau (eds.): *Parameter Identification and Inverse Problems in Hydrology, Geology and Ecology*. 1996 ISBN 0-7923-4089-2
24. V.P. Singh (ed.): *Hydrology of Disasters*. 1996 ISBN 0-7923-4092-2
25. A. Gianguzza, E. Pelizzetti and S. Sammartano (eds.): *Marine Chemistry*. An Environmental Analytical Chemistry Approach. 1997 ISBN 0-7923-4622-X
26. V.P. Singh and M. Fiorentino (eds.): *Geographical Information Systems in Hydrology*. 1996 ISBN 0-7923-4226-7
27. N.B. Harmancioglu, V.P. Singh and M.N. Alpaslan (eds.): *Environmental Data Management*. 1998 ISBN 0-7923-4857-5
28. G. Gambolati (ed.): *CENAS. Coastline Evolution of the Upper Adriatic Sea Due to Sea Level Rise and Natural and Anthropogenic Land Subsidence*. 1998 ISBN 0-7923-5119-3
29. D. Stephenson: *Water Supply Management*. 1998 ISBN 0-7923-5136-3
30. V.P. Singh: *Entropy-Based Parameter Estimation in Hydrology*. 1998 ISBN 0-7923-5224-6
31. A.S. Issar and N. Brown (eds.): *Water, Environment and Society in Times of Climatic Change*. 1998 ISBN 0-7923-5282-3
32. E. Cabrera and J. García-Serra (eds.): *Drought Management Planning in Water Supply Systems*. 1999 ISBN 0-7923-5294-7
33. N.B. Harmancioglu, O. Fistikoglu, S.D. Ozkul, V.P. Singh and M.N. Alpaslan: *Water Quality Monitoring Network Design*. 1999 ISBN 0-7923-5506-7
34. I. Stober and K. Bucher (eds): *Hydrogeology of Crystalline Rocks*. 2000 ISBN 0-7923-6082-6
35. J.S. Whitmore: *Drought Management on Farmland*. 2000 ISBN 0-7923-5998-4
36. R.S. Govindaraju and A. Ramachandra Rao (eds.): *Artificial Neural Networks in Hydrology*. 2000 ISBN 0-7923-6226-8
37. P. Singh and V.P. Singh: *Snow and Glacier Hydrology*. 2001 ISBN 0-7923-6767-7
38. B.E. Vieux: *Distributed Hydrologic Modeling Using GIS*. 2001 ISBN 0-7923-7002-3

39. I.V. Nagy, K. Asante-Duah and I. Zsuffa: *Hydrological Dimensioning and Operation of Reservoirs*. Practical Design Concepts and Principles. 2002 ISBN 1-4020-0438-9
40. I. Stober and K. Bucher (eds.): *Water-Rock Interaction*. 2002 ISBN 1-4020-0497-4
41. M. Shahin: *Hydrology and Water Resources of Africa*. 2002 ISBN 1-4020-0866-X
42. S.K. Mishra and V.P. Singh: *Soil Conservation Service Curve Number (SCS-CN) Methodology*. 2003 ISBN 1-4020-1132-6
43. C. Ray, G. Melin and R.B. Linsky (eds.): *Riverbank Filtration*. Improving Source-Water Quality. 2003 ISBN 1-4020-1133-4
44. G. Rossi, A. Cancelliere, L.S. Pereira, T. Oweis, M. Shatanawi and A. Zairi (eds.): *Tools for Drought Mitigation in Mediterranean Regions*. 2003 ISBN 1-4020-1140-7

Kluwer Academic Publishers – Dordrecht / Boston / London

MIX
Papier aus verantwortungsvollen Quellen
Paper from responsible sources
FSC® C105338

FSC
www.fsc.org

If you have any concerns about our products,
you can contact us on
ProductSafety@springernature.com

In case Publisher is established outside the EU,
the EU authorized representative is:
Springer Nature Customer Service Center GmbH
Europaplatz 3, 69115 Heidelberg, Germany

Printed by Libri Plureos GmbH
in Hamburg, Germany